INTRODUCTION TO ENGINEERING DESIGN

BOOK 2
WEIGHING MACHINES

JAMES W. DALLY

Professor, Mechanical Engineering
University of Maryland @ College Park

CONSULTING WITH:

THOMAS M. REGAN

Professor and Associate Dean
University of Maryland @ College Park

 College House Enterprises

THANKS FOR PERMISSIONS

The author, consultant and publisher thank the copyright and trademark owners providing permission for use of copyrighted materials and trademarks. We apologize for any errors or omissions in obtaining permissions. Where appropriate, we referenced similar treatments and gave credit for prior work. Errors or omissions in obtaining permissions or in giving proper references are not intentional. We will correct them at the earliest opportunity after the error or omission is brought to our attention. Please contact the publisher at the address given below.

The manuscript was prepared by the author in Word 7 using 12 point Times New Roman font. The book was printed from camera ready copy by Publishing and Printing Inc., Knoxville, TN.

College House Enterprises, LLC.
5713 Glen Cove Drive
Knoxville, TN 37919, U. S. A.
Phone (423) 558 6111
FAX (423) 584 1766

ISBN 0-9655911-1-5

To my wife Anne

and my children Lisa, Bill and Michelle

who have tolerated my many absences and inattentions while working on either textbooks or research. Their tolerance, love, patience and good spirits have supported me for many many years.

James W. Dally obtained a bachelor of science and a master of science degree, both in Mechanical Engineering from the Carnegie Institute of Technology. He obtained a doctoral degree in Mechanics from the Illinois Institute of Technology. He has taught at Cornell University, Illinois Institute of Technology, the U. S. Air Force Academy, and served as Dean of Engineering at the University of Rhode Island. He is currently a professor in the Department of Mechanical Engineering at the University of Maryland @ College Park.

Professor Dally has also worked at the Mesta Machine Co., IIT Research Institute, NIST, NRL, and IBM's Federal Systems Division. He is a fellow of the American Society for Mechanical Engineers, Society for Experimental Mechanics, and the American Academy of Mechanics. He was appointed as an honorary member of the Society for Experimental Mechanics in 1983 and elected to the National Academy of Engineering in 1984. In 1992 he was selected by his peers to receive the Senior Faculty Outstanding Teaching Award. He was a recipient of the Distinguished Scholar Teacher Award at the University of Maryland in 1994. He was also a member of the team receiving the 1996 Outstanding Educator Award sponsored by The Boeing Co.

Professor Dally has co-authored several other books including: *Experimental Stress Analysis, Photoelastic Coatings, Instrumentation for Engineering Measurements, Packaging of Electronic Systems,* and *Introduction to Engineering Design, Book 1*. He has written about 200 scientific papers and holds five patents.

Thomas M. Regan is Associate Dean of the A. J. Clark School of Engineering at the University of Maryland @ College Park and a Professor in the Chemical Engineering Department. He received his bachelor of science and doctoral degree in Chemical Engineering from Tulane University. Since 1990, he has been the principal investigator, at the University of Maryland, of the NSF sponsored ECSEL Coalition. He currently serves as the coordinator for the course ENES 100 "Introduction to Engineering Design" which incorporates active learning, hands-on experiences with students working in teams to design, build and test a prototype of a product.

Professor Regan received the Chester F. Carlson Award, for Innovation in Engineering Education from the American Association for Engineering Education. He was a lead member of the team that received the 1996 Outstanding Engineering Educator Award sponsored by The Boeing Co. He has been cited twice by the Dean of Undergraduate Studies at the University of Maryland with a certificate of Teaching Excellence. Within the College of Engineering, he was selected by his peers for the Senior Faculty Outstanding Teaching Award. He has also received the Allied Foundation Faculty Award in recognition of his outstanding contributions to undergraduate Chemical Engineering education.

CODE:

INSTRUCTOR AND STUDENT EXPECTATIONS

Learning and teaching require trust and mutual understanding between the student and the instructor. Trust and understanding lead to an enhanced learning environment. If we all recall a listing of expectations, a better relation can be developed between the instructor and his or her class that significantly promotes the learning process. While many of these expectations may be self evident, we believe the list will remind everyone of their obligations for learning, increased understanding, communication, and respect. The list, shown below, was developed by the students and instructors in the Introduction to Engineering Design course over the past six years.

Students expect the instructor to:

- Respect all students.
- Be fair in grading.
- Provide leadership in class.
- Be committed to teaching and advising.
- Provide encouragement rather than discouragement.
- Clearly define course requirements and grading algorithms.
- Balance course workload with credit hours.
- Schedule office hours, and be available accordingly.
- Provided candid and timely feedback on assignments.
- Arrive before the scheduled class time, and prepare the classroom.

Instructors expect the student to:

- Show respect to everyone involved in the program.
- Be responsible for your own progress and learning.
- Be dedicated to understanding and learning.
- Stay current with materials and issues covered in class.
- Be a good team member.
- Be inquisitive and compete within the framework of a team.
- Be interested in engineering and product design.
- Attended class or notify the instructor in advance if you intend to be absent.
- Arrive on time for class with a positive attitude.

PREFACE

This book is the second in a series dealing with Introduction to Design. We are preparing a new text each academic year for the first year engineering students at College Park. We are encouraged that the book is also being adopted by other colleges and universities.

The procedure that is followed in offering a design experience to first year students is to:

- Teach the class in relatively small sections (about 36 students in each section).
- Divide the class into product development teams with five or six members per team.
- Assign a major project entailing the development of a prototype that will require the entire semester.
- All teams develop the same product.
- In the product realization process the students:
    - Design
    - Manufacture or procure parts for the prototype.
    - Assemble the prototype.
    - Test and evaluate the prototype.
- In developing the prototype we have the opportunity to emphasize:
    - Communication skills.
    - Team building skills.
    - Graphics
    - Software applications including CAD.
    - Design methods and procedures.

The textbook is used to support the students during the semester long project. Some of the material may be covered in class or in the computer laboratory. Other material is covered by reading assignments. In other instances, the student uses the text as a reference document in independent study. Exercises are provided at the end of each chapter that maybe used for assignments when the demands of the project on the student's time are not excessive.

The course, Introduction to Engineering Design, has evolved over a period of seven years and reflects the ideas, opinions, and experiences of many faculty members who have participated in teaching the course at the University of Maryland @ College Park. The book is based on experiences that have been successful in teaching design to first year students. Through presentations at meetings sponsored by the American Society for Engineering Education and informal discussions with colleagues from other colleges of engineering, the course was focused beyond our personal experiences. Also, we were fortunate to be part of the ECSEL coalition because it gave us excellent opportunities to experiment with a curriculum that had not changed much over the past twenty five years. The Engineering Education Coalitions funded by The National Science Foundation gave us credibility as we tried to make significant changes to the content in the first course in engineering. It was, and still is in many colleges of engineering, essential to develop a course which provides students a much better educational experience in their first encounter with the engineering curriculum.

The book is organized into seven parts to present the various topics that the students need as they proceed through a significant portion of product realization process. We introduce the product realization process by considering the design of a weighing machine with an analog and a digital readout. By assigning a demanding, overarching project, we employ a holistic approach that avoids compartmentalization of knowledge. Design of a product is treated by the instructor as an opportunity to integrate a spectrum of knowledge about many topics. We have found that developing a product motivates the students. They learn much more on their own initiative that we could ever teach them in a course. The motivational benefit of hands-on participation in the design of their own product and actual building and testing the prototype significantly speeds the learning process.

Part I of the textbook covers material on weighing machines. We briefly introduce the product development process and the students role in a development team in producing a scale with analog and digital readouts. The product is described in Chapter 2 together with theory on mass and weight. Some history about measuring weight is included prior to introducing the product specification. A few concepts for the design of scales are introduced briefly. Much more detail on the design of weighing machines is given in Chapters 3 and 4. We cover the mechanical aspects in Chapter 3 by giving some of the theory for scales based on equilibrium, springs, hydraulics and electronics. More detail on the electronics aspects of the project are given in Chapter 4. We recognize that some of the analysis presented in Chapters 3 and 4 may be beyond the scope of the ordinary first year student; however, we believe that some may understand and lead others on their team to learn the material.

Part II presents three chapters on engineering graphics. We begin in Chapter 5 with a relatively complete treatment of three-view drawings. Pictorial drawings including isometric, oblique and perspective are covered in Chapter 6. A treatment of Tables and Graphs is given in Chapter 7. The emphasis in Part II is on manual preparation of the graphics. CAD preparation of drawings and spreadsheet generation of graphs is covered later.

Part III describes three different software programs useful to the first year engineering student. A computer aided drawing program, CAD KEY Complete, is described in considerable detail with several example drawings. A spreadsheet program, Microsoft Excel, is described in Chapter 9. The coverage is intended to provide the students with entry level skills. With spreadsheets we expect the student to be able to perform calculations and to plot curves and prepare several types of charts. In Chapter 10, we provide a brief description of a computer graphics presentation program, namely Microsoft PowerPoint. Our experience is that the student can learn to master this program in one to two hours if they make use of the chart wizard incorporated in the program. The coverage of the capabilities of PowerPoint is tied to preparing a design briefing on a weighing machine.

Part IV is devoted to the product development process. We first describe development teams which are new to most students who have been educated in the secondary school system to act as individuals and to avoid cooperation in learning. We cover a number of useful topics such as team member traits, positive and negative team behavior, and effective team meeting. Experience with several thousand students have shown that many of them initially have trouble adapting to the team concept. (Males appear to have more trouble than females.) However, over the semester they slowly learn how to work as effective team members and they appreciate the opportunity to do so. The social bonding that takes place on the team with the first semester Freshman students is interesting to observe. Chapter 12 deals with the product development process. We have tried to incorporate a very wide range of material in this chapter. For this reason the coverage often is brief and we refer the student to more thorough higher level books on the topic. However, we cover that part of the product development process which is important to the assigned project. We start with the product specification and cover material on design concept generation and then concept selection. Methods are introduced that the student can effectively employ to generate design concepts and then to select the best concept though the use of a systematic design trade-off analysis.

Part V treats the very important topic of communications. A chapter on technical reports describes many aspects of technical writing. The most important lesson here is that a technical report is different than a paper for the History or English Departments. An effective professional report is written for a predefined audience with specific objectives. We describe the technical writing process and give many suggestions to facilitate composing, revision, editing and proofreading. The final chapter in the book covers design briefings. We draw a distinction between speeches, presentations and group discussions. Then we focus on the technical presentation and indicate the importance of preparing excellent visual aids. We make extensive use of PowerPoint in illustrating the types of visual aids to be employed in a design briefing. Finally, we include a discussion of the delivery of the presentation and the need for extensive rehearsing.

Part VI is new to Book 2. It contains three chapters dealing with engineering and society. Chapter 15, provides a historical perspective on the role engineering played in developing civilization and on improving the lives of the masses. In this chapter, we move from the past into the present and indicate the current relationship between business, consumers and society. Safety, risk and performance are covered in Chapter 16. Here we discuss failure, and it implications on the safety and well being of those using our products. Theoretical methods are introduced to determine both component and system reliability. Finally, in Chapter 17, we discuss ethics, character and engineering. A large number of topics are covered so the instructor can select from among them. The code of ethics recommended by the Accreditation Board for Engineering and Technology (ABET) is included. A description of the Challenger accident is also covered because it is an excellent case history covering safety related conflicts between management and engineers.

In the next few years, it is imperative that we think very seriously about ABET's new criteria for program accreditation. The new Criteria 2000 is very different than the current criteria for accreditation. We currently accredit our engineering programs based on input, and many pages of instructions are provide by ABET to guide in the assessment of that input. However, Criteria 2000 is based only on educational outcomes. Very little information given in Criteria 2000 is about program input. The guide provided by ABET for assessment of educational outcomes is very terse. It appears that it will be up to each program in each College of Engineering to specify the intended educational outcomes for their program, to define the metrics used to measure these outcomes, and to plan and place into effect methods of measurement of these metrics over extended periods of time.

We have considered Criteria 2000 in this writing this book, and believe a first course in engineering design should have the expected educational outcomes listed below:

Communication Skills

1. Engineering Graphics

   - Understand the role of graphics in engineering design.
   - Understand orthographic projection in producing multi-view drawings.
   - Understand three-dimensional representation with pictorial drawings.
   - Understand dimensioning and section views.
   - Demonstrate capability of preparing drawings using both manual and CAD techniques.
   - Demonstrate understanding by incorporating appropriate high-quality drawings in the design documentation.

2. Design Briefings:

   - Within a team format, present a design review for the class using appropriate visual aids.
   - Each team member participates in the briefing.

3. Design Reports:

   - The team's design is documented in a professional style report incorporating time schedules, costs, parts list, drawings and an analysis.

Team Experience

- Develop an awareness of the challenges occurring in teamwork.
- Demonstrate teamwork in the product realization process through a systematic design concept selection process involving participation of all the team members.
- Demonstrate planning from conceptualization to the evaluation of the prototype.
- Understand and demonstrate share responsibility among team members.
- Demonstrate teamwork in preparing design reports and presenting design briefings.

Software Applications

- Demonstrate entry level skills in using spreadsheets for calculations and data analysis.
- Show a capability to prepare graphs and charts with a spreadsheet.
- Show a capability to prepare professional quality visual aids.
- Understand entry level skills in employing a CAD program.

- Demonstrate these computer skills in preparing appropriate materials for design briefings and design reports.

Design Project:

- The design project is the over arching theme of the course.
- Utilize all the skills listed above to assist in the product development process.
- Demonstrate competence in defining design objectives.
- Generate design concepts that meet the design objectives.
- Understand the basis for design for manufacturing, assembly and maintenance.
- Manage the team and the project effectively.

Acknowledgments are always necessary in preparing a textbook because so many people and organizations are involved. First, we want to thank The National Science Foundation for their support. Their funding was important, but more critical than money was our need for credibility. Without the Engineering Education Coalition Program, the need for curriculum reform would not have received adequate attention from the college administrators and our colleagues. The NSF basically called for a reform of the curriculum, and with generous funding gave it the required status.

Second, we need to recognize the contributions of many of our colleagues at the University of Maryland. In 1990 the author taught the pilot offering of the course with the assistance of Dr. Guangming Zhang. Guangming attended every class and prepared an excellent set of notes that were essential in the second series of pilot courses that were taught by Tom Regan and Isabel Lloyd. Guangming also taught the course, with slight modifications, to several classes of high school women as part of an early entrance summer program at the University of Maryland. His success in teaching engineering design to 16 year old women clearly showed his superior ability as a teacher, and the robust nature of the course material that we had developed. We want to thank Ms. Jane Fines who maintained contact with many of our students in several offerings of the course. She gave us very valuable feedback about the reactions of the students as we modified the course over the years.

As always, thanks are due to the administrators who encouraged and supported the development of this course. George Dieter and Bill Destler, the former and current Dean of Engineering, authorized the small class size essential for effectively teaching this course, and committed to significant long term expenditures necessary to support the faculty involved. They also publicly supported the efforts of the small group of instructors during the early years as we were institutionalizing the course. In fact Bill Destler, a Department Chair in 1992, took time from his very busy schedule and taught the course. Thanks are also due to Marilyn Berman, formerly an Associate Dean of Engineering, for her ability to cut yards of red tape. Without her help we would never have been able to schedule the pilot class. The Department Chairs also supported the concept of teaching product design and development in the first year. The author is particularly indebted to Bill Fourney and Bill Walston in the Mechanical Engineering Department for their support and encouragement. Special thanks go to Davinder Anand, the current Chair of Mechanical Engineering, for providing a leave permitting me to complete Book 1 and Book 2 in a timely manner.

James W. Dally
Knoxville, TN

DEDICATION
ABOUT THE AUTHORS
CODE
PREFACE

CONTENTS

PART I WEIGHING MACHINES

CHAPTER 1 INTRODUCTION

PURPOSE OF THE TEXTBOOK	1 - 01
THE PRODUCT DEVELOPMENT PROCESS	1 - 01
WINNING PRODUCTS	1 - 03
YOUR PRODUCT DEVELOPMENT	1 - 06
PROTOTYPE DEVELOPMENT	1 - 07
DESIGN CONCEPTS	1 - 07
PREPARING THE ASSEMBLY KIT	1 - 13
PROTOTYPE ASSEMBLY	1 - 13
PROTOTYPE EVALUATION	1 - 14
TEAMWORK	1 - 14
OTHER COURSE OBJECTIVES	1 - 15
REFERENCES	1 - 16
EXERCISES	1 - 16

CHAPTER 2 ANALOG AND DIGITAL WEIGHING MACHINES

MEASURING WEIGHT	2 - 01
SOME HISTORY ABOUT MEASURING WEIGHT	2 - 03
A SIMPLE BEAM BALANCE	2 - 03

THE BISMAR BALANCE .... 2 - 05

THE STEELYARD SCALE .... 2 - 06

SPRING SCALES .... 2 - 07

THE PLATFORM SCALE .... 2 - 09

INDICATOR SCALES .... 2 - 10

HYDRAULIC SCALES .... 2 - 11

ELECTRONIC SCALES .... 2 - 12

THE PRODUCT SPECIFICATION .... 2 - 12

REFERENCES .... 2 - 16

EXERCISES .... 2 - 16

CHAPTER 3 MECHANICAL ASPECTS OF WEIGHING MACHINES

INTRODUCTION .... 3 - 01

EQUILIBRIUM BASED WEIGHING MACHINES ...... 3 - 04

SPRING BASED WEIGHING MACHINES .... 3 - 10

HYDRAULIC BASED WEIGHING MACHINES .... 3 - 14

STRAIN GAGES .... 3 - 16

STRAIN GAGE TRANSDUCERS .... 3 - 19

REFERENCES .... 3 - 22

EXERCISES .... 3 - 22

CHAPTER 4 ELECTRICAL AND DIGITAL ASPECTS OF WEIGHING MACHINES

INTRODUCTION .... 4 - 01

BASIC ELECTRICAL CONCEPTS .... 4 - 02

THE BATTERY .... 4 - 02

KIRCHHOFF'S LAWS .... 4 - 06

THE POTENTIOMETER .... 4 - 10

THE WHEATSTONE BRIDGE .... 4 - 13

VOLTMETERS .... 4 - 15

AMPLIFIERS .... 4 - 17

DIGITAL VOLTMETERS .... 4 - 20

SUMMARY .... 4 - 22

REFERENCES .... 4 - 22

EXERCISES .... 4 - 23

PART II ENGINEERING GRAPHICS

CHAPTER 5 THREE-VIEW DRAWINGS

INTRODUCTION	5 - 01
VIEW DRAWINGS	5 - 02
LINE STYLES	5 - 04
REPRESENTING FEATURES	5 - 06
BLOCK WITH SLOT AND STEP	5 - 06
BLOCK WITH TAPERED SLOT	5 - 09
BLOCK WITH STEP AND HOLE	5 - 10
DIMENSIONING	5 - 12
DIMENSIONING HOLES AND CYLINDERS	5 - 13
DRAWING BLOCKS	5 - 15
ADDITIONAL VIEWS	5 - 17
SUMMARY, REFERENCES AND EXERCISES	5 - 19

CHAPTER 6 PICTORIAL DRAWING

INTRODUCTION	6 - 01
ISOMETRIC DRAWINGS	6 - 02
OBLIQUE DRAWINGS	6 - 06
PERSPECTIVE DRAWINGS	6 - 10
ONE POINT PERSPECTIVE	6 - 11
TWO POINT PERSPECTIVE	6 - 13
SHADING AND SHADOWING	6 - 16
SUMMARY, REFERENCES AND EXERCISES	6 - 18

CHAPTER 7 TABLES AND GRAPHS

INTRODUCTION	7 - 01
TABLES	7 - 02
GRAPHS	7 - 04
PIE CHART	7 - 04
BAR CHART	7 - 06
LINEAR X-Y GRAPHS	7 - 07
X-Y GRAPHS WITH SEMI-LOG SCALES	7 - 11
X-Y GRAPHS WITH LOG-LOG SCALES	7 - 12
SPECIAL GRAPHS AND SUMMARY	7 - 14
EXERCISES	7 - 15

PART III SOFTWARE APPLICATIONS

CHAPTER 8 CAD KEY Complete: for WINDOWS

INTRODUCTION	8 - 01
THE KEY CAD SCREEN	8 - 02
THE TOOLBOXES AND THE TOOLS	8 - 03
THE DRAW TOOLBOX	8 - 03
RECTANGLE/SQUARE TOOL	8 - 03
OVAL/ CIRCLE TOOL	8 - 04
POLYGON TOOL	8 - 04
MULTIGON TOOL	8 - 04
ARC TOOL	8 - 04
LINE TOOL	8 - 05
FREE-FORMED SPLINE TOOL	8 - 06
THE EDIT TOOLBOX	8 - 06
SELECTION (MOVE) TOOL	8 - 06
POINT SELECTION TOOL	8 - 07
RESIZE (SCALE) TOOL	8 - 07
SKEW TOOL	8 - 07
ROTATE TOOL	8 - 07
FILLET TOOL	8 - 07
CHAMFER TOOL	8 - 08
TRIM TOOL	8 - 08
ZOOM TOOL	8 - 08
TEXT TOOL	8 - 08
THE SNAP TO TOOLBOX	8 - 09
THE SNAP TO ICONS	8 - 10
ANY POINT	8 - 10
END POINT	8 - 10
POINT ON LINE	8 - 10
CENTER POINT	8 - 11
CORNER POINT	8 - 11
INTERSECTION POINT	8 - 11
PERPENDICULAR	8 - 11
LINE CENTER	8 - 11
PERCENTAGE	8 - 12
ABSOLUTE POINT	8 - 12
STARTING A DRAWING	8 - 12
THREE VIEW DRAWINGS	8 - 15

DIMENSIONING WITH KEY CAD   8 - 17
    LINEAR TOOL   8 - 18
    PARALLEL TOOL   8 - 18
    ANGULAR TOOL   8 - 18
    DIAMETER TOOL   8 - 18
    RADIUS TOOL   8 - 18
    LEADER LINE TOOL   8 - 19
A THREE-VIEW DRAWING WITH A HOLE AND ARC   8 - 21
ISOMETRIC DRAWINGS   8 - 22
SUMMARY   8 - 23
REFERENCES   8 - 24
EXERCISES   8 - 24

CHAPTER 9 MICROSOFT EXCEL

INTRODUCTION   9 - 01
THE SCREEN   9 - 01
    THE FIVE TOP ROWS   9 - 02
    THE SPREADSHEET   9 - 03
PREPARING A PARTS LIST   9 - 04
PERFORMING CALCULATIONS   9 - 06
    THE RICH UNCLE EXAMPLE   9 - 06
    STRAIN ON A SIMPLY SUPPORTED BEAM   9 - 09
GRAPHS WITH EXCEL   9 - 12
    PIE GRAPHS   9 - 12
    BAR CHARTS   9 - 14
    X-Y GRAPHS   9 - 15
SUMMARY   9 - 19
REFERENCES   9 - 19
EXERCISES   9 - 19

CHAPTER 10 MICROSOFT PowerPoint

INTRODUCTION   10 - 01
THE PowerPoint WINDOW   10 - 02
PLANNING AND ORGANIZING WITH PowerPoint   10 - 03
ENHANCING THE SLIDES   10 - 08
REHEARSING   10 - 10
COMPUTER PROJECTED SLIDES   10 - 10
SUMMARY   10 - 11
EXERCISES   10 - 12

PART IV PRODUCT DEVELOPMENT

CHAPTER 11 DEVELOPMENT TEAMS

INTRODUCTION	11 - 01
TEAM LEADERSHIP	11 - 02
TEAM MEMBER RESPONSIBILITIES	11 - 03
TEAM MEMBER TRAITS	11 - 03
EFFECTIVE TEAM MEETINGS	11 - 05
PREPARING FOR MEETINGS	11 - 07
POSITIVE AND NEGATIVE TEAM BEHAVIOR	11 - 09
REFERENCES	11 - 10
EXERCISES	11 - 10

CHAPTER 12 A PRODUCT DEVELOPMENT PROCESS

INTRODUCTION	12 - 01
THE CUSTOMER	12 - 02
THE PRODUCT SPECIFICATION	12 - 03
DEFINING ALTERNATIVE DESIGN CONCEPTS	12 - 03
SELECTING THE SUPERIOR CONCEPT	12 - 06
DESIGN FOR ????	12 - 11
DESIGN FOR MANUFACTURING	12 - 12
DESIGN FOR ASSEMBLY	12 - 13
DESIGN FOR MAINTENANCE	12 - 14
PROTOTYPE BUILD AND EVALUATION	12 - 15
TOOLING AND PRODUCTION	12 - 15
MANAGING THE PROJECT	12 - 16
SUMMARY	12 - 19
REFERENCES	12 - 19
EXERCISES	12 - 20

PART V COMMUNICATIONS

CHAPTER 13 TECHNICAL REPORTS

INTRODUCTION   13 - 01
APPROACH AND ORGANIZATION   13 - 02
KNOW YOUR READERS AND YOUR OBJECTIVE   13 - 05
THE TECHNICAL WRITING PROCESS   13 - 06
REVISING, EDITING AND PROOFREADING   13 - 07
WORD PROCESSING   13 - 09
SUMMARY   13 - 10
REFERENCES   13 - 10
EXERCISES   13 - 11

CHAPTER 14 DESIGN BRIEFINGS

INTRODUCTION   14 - 01
SPEECHES, PRSENTATIONS AND DISCUSSIONS   14 - 02
PREPARING FOR THE BRIEFING   14 - 03
PRESENTATION STRUCTURE   14 - 04
TYPE OF VISUAL AIDS   14 - 10
DELIVERY OF THE PRESENTATION   14 - 11
SUMMARY   14 - 14
REFERENCES   14 - 14
EXERCISES   14 - 15

PART VI ENGINEERING AND SOCIETY

CHAPTER 15 ENGINEERING AND SOCIETY

INTRODUCTION   15 - 01
ENGINEERING IN EARLY WESTERN HISTORY   15 - 03
ENGINEERING AND THE INDUSTRIAL REVOLUTION   15 - 06
ENGINEERING IN THE 19TH AND 20TH CENTURIES   15 - 07
NEW UNDERSTANDINGS   15 - 09

BUSINESS, CONSUMERS AND SOCIETY   15 - 10
CONCLUSIONS   15 - 14
REFERENCES   15 - 15
EXERCISES   15 - 16

CHAPTER 16 SAFETY, RISK AND PERFORMANCE

INTRODUCTION   16 - 01
MINIMIZING THE RISK   16 - 03
FAILURE RATE   16 - 06
COMPONENT RELIABILITY   16 - 09
SYSTEM RELIABILITY   16 - 10
    RELIABILITY OF SERIES CONNECTED SYSTEMS   16 - 11
    RELIABILITY OF SYSTEMS WITH
    PARALLEL CONNECTED COMPONENTS   16 - 12
EVALUATING THE RISK   16 - 14
HAZARDS   16 - 15
SUMMARY   16 - 18
REFERENCES   16 - 19
EXERCISES   16 - 20

CHAPTER 17 ETHICS, CHARACTER AND ENGINEERING

INTRODUCTION   17 - 01
CONFUSED ABOUT ETHICAL BEHAVIOR   17 - 02
WHITE, BLACK OR GRAY   17 - 02
IT IS THE LAW   17 - 04
ETHICAL BUSINESS PRACTICES   17 - 05
HONOR CODES   17 - 06
CHARACTER   17 - 07
ETHICS OF ENGINEERS   17 - 08
THE CHALLENGER INCIDENT   17 - 16
BACKGROUND INFORMATION ABOUT THE SHUTTLE   17 - 17
THE SOLID ROCKET BOOSTERS   17 - 18
FAILURE OF THE O-RING SEAL   17 - 19
IGNORING THE PROBLEM   17 - 21
RECOGNIZING THE INFLUENCE OF TEMPERATURE   17 - 22
FLY EVEN IF IT IS COLD: A MANAGEMENT DECISION   17 - 22
APPROVALS AT THE TOP   17 - 23
SUMMARY   17 - 23
REFERENCES   17 - 24
EXERCISES   17 - 25

NOTES

PART I

WEIGHING

MACHINES

CHAPTER 1

INTRODUCTION

PURPOSE OF THE TEXTBOOK

The principle purpose of this textbook is to assist you in learning about the product realization process, and to provide an initial experience in engineering design. The textbook is written to support your efforts as you work within a team structure to develop a product. We recognize that most of our readers will have little or no design experience, and for this reason, we have selected two relatively simple products for your team to develop --- both are bathroom, or personal, scales for measuring someone's weight. One of the scales will have an analog display to show the weight, and the other will have a digital display. We will ask you to understand the principles of operation of the type of scale that your team chooses to develop, and then to design it completely, including the preparation of high quality engineering drawings. After the design is complete and released through a peer review process, we will ask you to prepare an assembly kit with all of the parts that are necessary to actually build the scales. Finally, you will assemble the scales, and evaluate their performance by conducting a test to measure their accuracy.

Before we describe the requirements of the analog and digital scales, let's briefly discuss the product realization process.

THE PRODUCT DEVELOPMENT PROCESS

We use hundreds of products every day. We are surrounded by products. This morning I prepared breakfast by toasting bread, frying eggs, and making coffee. How many products did I use in this simple task? The toaster (Black & Decker), the frying pan (T-FAL), the tea kettle (Revere), and the electric range (General Electric) are all products that are designed, manufactured and sold to customers both here and abroad. Some products are relatively simple, like the frying pan with only a few parts, but some are much more complex, such as your automobile with several thousand parts.

Corporations world-wide continuously develop their product lines with minor improvements introduced every year or two with more major improvements every four or five years. Product development involves many engineering disciplines, and is a major responsibility of engineers entering the work place as is illustrated in Table 1.

TABLE 1

Responsibilities of Mechanical Engineers in Their First Position [1]

ASSIGNMENT		PERCENT OF TIME
Design Engineering		40
Product Design	24	
Systems Design	9	
Equipment Design	7	
Plant Engineering/ Operations/ Maintenance		13
Quality Control/ Reliability/ Standards		12
Production Engineering		12
Sales Engineering		5
Management		4
Engineering	3	
Corporate	1	
Computer Applications/ Systems Analysis		4
Basic Research and Development		3
Other Activities		7

An examination of Table 1, shows that designing, manufacturing, and selling product represents more than 80 % of the responsibilities of mechanical engineers in their initial position in industry. This distribution of activities is typical of many of the engineering disciplines.

The viability of many corporations depends on the introduction of a steady stream of successful products to the marketplace. A successful product must satisfy the customer by providing robust and reliable service at a competitive price that completely meets the customer needs.

Products are developed by interdisciplinary teams with representation from engineering, marketing, manufacturing, production, purchasing, etc. A typical organization chart for a team developing a relatively simple electro-mechanical component is shown in Fig. 1. The team leader coordinates and directs the activities of the designers (mechanical, electronic, and industrial, etc.), and is supported by a marketing specialist to ensure that the evolving product meets the needs of the customer. Assistance in finance, sales and legal issues is usually provided on as-needed basis by corporate staff or outside contractors.

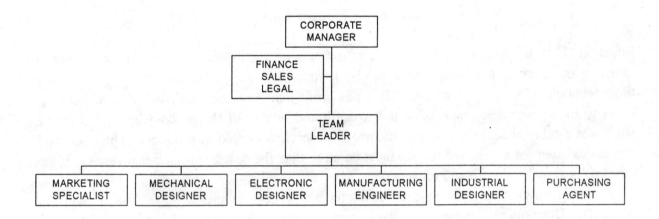

Fig. 1 Organization of a product development team for a relatively simple product.

The team leader is also supported by a purchasing agent who establishes close relations with the suppliers that provide materials and component parts used in the final assembly of the product. In the past decade, the relation between suppliers and end product producers has changed in a very significant manner. Suppliers become part of the development team. They often provide a significant level of engineering support that is external to the internal (core) development team. In some developments, these external (supplier funded) development teams, when totaled, are larger than the internal development team. For example, the internal development team for the Boeing 777 airliner involved 6,800 employees, and the external teams were estimated to include about 10,000 employees [2].

WINNING PRODUCTS

There are two primary aspects involved in developing successful products that win in the competitive market place. The first is the quality, performance, and price of the product. The second is the time and the cost of the development, and the cost to manufacture the product in production.

Let's discuss the product first. Is it attractive and easy to use? Is it durable and reliable? Is it effective? Does it meet the needs of the customer? Is it better than the products now available in the marketplace? If the answer to all of these questions is an unqualified YES, the customer may want to buy the product if the price is right. Next, you need to understand what is implied by product cost, and its relation to the price paid for the product. Cost and price are distinctly different quantities. Product cost clearly includes the cost of materials, components, manufacturing and assembly. The accountants also include other less obvious costs such as the prorated costs of capital equipment (the plant and its machinery), the tooling, the development cost, and even the inventory costs in determining the total cost of producing a unit of product.

Price is the amount of money that a customer pays to buy the product. The difference between the price and the cost of a product is the profit, which is usually expressed on a per unit basis.

$$\textbf{Profit = Product Price - Product Cost} \qquad \textbf{(1)}$$

Equation (1) is the most important relation in engineering and in any business. If a corporation cannot make a profit, it soon is forced into bankruptcy, it's employees lose their positions, and the stockholders lose their investment. It is this profit that everyone employed by a corporation seeks to maximize while maintaining the strength and vitality of the product lines. The same statement can be made for a business that provides services instead of products. The price paid by the customer for a specified service must be more than the cost to provide that service, if the business is to make a profit and prosper.

Let's now discuss the role of development process in producing a line of winning products. Developing a product involves many people with talent in different disciplines, it takes time, and it costs a lot of money. Let's first consider development time. Time, as it is used in this context, is time to market, i. e. the time from the kickoff initiating the product development process to the introduction of the product to the market. This is a very important target for a development team because of the many significant benefits that follow from being first to market. Many competitive advantages accrue from a fast development capability. First, the product's life is extended. For each month cut from the development schedule, a month is added to the life of the product in the marketplace with an additional month of sales revenue and profit. We show the benefits of being first to market on sales revenue in Fig. 2. The shaded region between the two curves to the left side of the graph is the enhanced revenue due to the longer sales.

A second benefit of early product release is increased market share. The first product to market has 100 % of market share in the absence of competing product. For products with periodic development of "new models," it is generally recognized that the earlier a product is introduced to compete with older models, without sacrificing quality and reliability, the better chance it has for acquiring and retaining a large share of the market. The effect of gaining a larger market share on sales revenue is illustrated in Fig. 2. The cross hatched region between the two curves at the top of the graph shows the enhanced sales revenue due to increased market share.

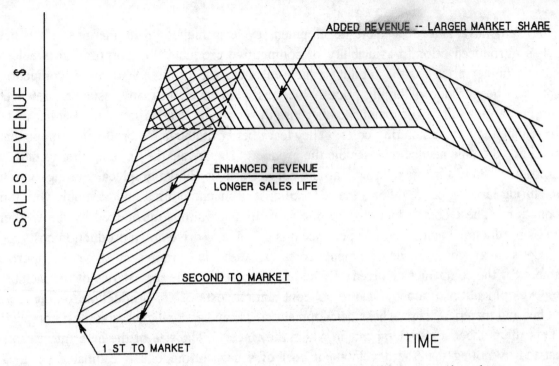

Fig. 2 Increased sales revenue due to extended market life and larger market share.

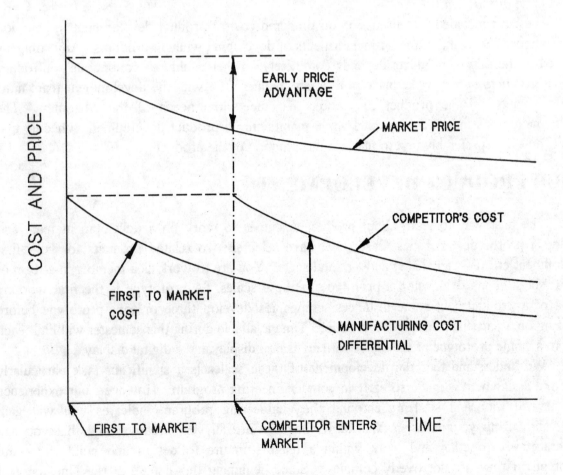

Fig. 3   The development team bringing the product to market first enjoys an initial price advantage and subsequent cost advantages from manufacturing efficiencies.

A third advantage of a short development cycle is higher profit margins. If a new product is introduced prior to availability of competitive products, the corporation is able to command a higher price for the product which enhances the profit. With time, competitive products will be introduced forcing price reductions. However, in many instances, relatively large profit margins still can be maintained because the company that is first to market has added time to reduce their manufacturing costs. They learn better methods for producing components and to reduce the time needed to assemble the product. The advantage of being first to market, with a product where a manufacturing learning curve exists, is shown graphically in Fig. 3. The manufacturing learning curve reflects the reduced cost of manufacturing and assembly with time in production. These cost reductions are due to many innovations introduced by the workers after mass production begins. With experience it is possible to drive down production costs.

Let's next consider development costs because they represent a very important investment for the companies involved. Development costs include the salaries of the members of the development team, money paid to subcontractors, costs of pre-production tooling, costs of supplies and materials, etc.. These development costs can be significant, and most companies must limit the number of developments in which they invest. The size of the investment can be appreciated by noting that the development cost of a new automobile is estimated [2] at $1 billion, with an additional investment of $500 to $700 million for the new tooling required for high-volume production.

We have included this discussion on time and cost of product development to help you begin to appreciate some of the business aspects of developing winning products. Any company involved in the sale of products depends completely on their ability to continuously introduce winning products in the marketplace in a timely manner. To win, the development team must bring a quality reliable product to market that meets the needs of the customer. The development costs must be minimized while maintaining a product development schedule that permits an early (preferably first to market) introduction of the product.

YOUR PRODUCT DEVELOPMENT

The best way to learn about product design is to work on a development team and develop a prototype. For this reason, we have selected two related products for your first development effort --- an analog and a digital scale. You are to work as a member in a team of five or six students to develop a prototype of these scales. A prototype is the first working model of a product. In some instances, companies develop three or four prototype before finalizing on a particular design of a product. Time available during this semester will limit each team to a single prototype of the scales with an analog display and a digital display.

We understand that the development of these scales is a significant task particularly when the assignment is made so early in your engineering program. However, our experience with several thousand students entering the engineering program indicates you will gain significantly from your efforts. We know that you are all very creative, and that you can cooperate well together and work within a team structure to design and build successful prototypes if they are not overly complex. Since beginning this course at the University of Maryland at College Park, in 1990, the participating students have developed play ground equipment, windmills for the generation of electricity, furniture, human power water pumps, wind powered vehicles, a solar cooking unit, and a solar desalination unit. Extensive surveys clearly indicate that you will spend a lot of time on the project, have fun doing so, learn a lot about engineering, and appreciate the lessons learned in working as a member of a development team.

PROTOTYPE DEVELOPMENT

In this course, we divide the prototype development process into three major phases which include:

1. Designing the prototype.
2. Preparing the assembly kit.
3. Prototype assembly and evaluation.

We will begin the process by providing you with a typical description of methods to measure weight. The description will initially be in general terms. Also you will be provided with design requirements pertaining to the input --- capacity of the scales and the accuracy of the measurement of a persons weight. The description will also include the size and location of the analog or digital display to insure that it can be read by the individual using the scale. We suggest that you make full use of the library to search the literature and the patent files. There is absolutely nothing to be gained by reinventing the wheel (or the scale in this case).

After you understand in general terms the requirements in the design of the scale, we will introduce the product specification [3]. The product specification is usually prepared in tabular format, and it tersely states design requirements and design limits. Remember we will always design with constraints and/or limits. These limits are stated in the product specification which is written very early in the development process. Also all of the performance criteria are included in the specification. Finally the specification provides design targets on weight, size, power, cost, etc.

We will place explicit design constraints which limit your flexibility in designing the prototype. First there is a limit of $25.00 (maximum) per team member for the cost of materials, supplies, and components purchased for the scales and the displays. The idea is to keep your out-of-pocket expenses at an affordable level, while giving the team adequate funds to build a good prototype. We will also limit the size of the scales, so that our laboratory facilities and workshops can handle the large number of scales that your teams will produce this semester.

DESIGN CONCEPTS

After you and each member of your development team understands the design specification and the design limitations, you are ready to begin to generate design concepts. (A much more complete description of the design requirements, constraints, etc. is given in Chapter 2 together with some initial design ideas for you to consider). Design concepts are simply ideas for performing each function involved in the operation of the scale(s) which will help you meet the product specifications. To illustrate design concepts, let's examine the functions (processes) involved in weighing a person, in say a bathroom, as depicted in Fig. 4.

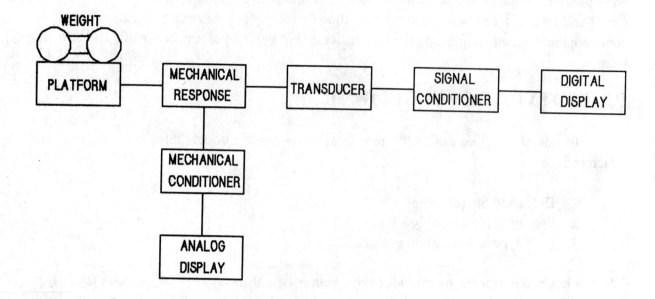

Fig. 4 Functions involved in the weighing process.

In a typical scale, we have some sort of a platform to accept the item that is to be weighed. When you step on the platform or place an item on the platform, it must be large enough to accommodate your feet or the item to be weighed. The platform must also be stable with no wobble or tilting. The customer would be very unhappy if he or she lost balance because of the platform movement and fell off your scale.

No problem ---- a block of wood would serve as a nice platform. It is solid and we can make it large enough for a person with a size 13 shoe. The platform must help us in the weighing process. It must respond somehow to the weight placed on it's surface. We need to fit the platform with some sort of mechanism that responds to the weight. But, what kind of a response would be helpful?

Recall some scales that you have seen. In the produce section of the local supermarket you have seen the scales that permit the shoppers to check the weight of their fruit or veggies. These are spring scales. When we place our apples in the pan (platform), a spring deflects with the applied weight, and a pointer attached to the spring indicates (displays) the weight. The platform response in this case was a displacement. The displacement of the spring was converted into a reading of the weight by measuring the displacement of the spring. The pointer rotation increases with the displacement and indicates the weight on a calibrated dial. The calibrated dial is an analog display. Unless you have lots of money, your bathroom scale at home is also a spring scale, but the mechanism is a bit more complex.

Examine Fig. 4 as it pertains to a spring type scale. The mechanical output in Fig. 4 is the displacement of the spring. We condition this mechanical output with a mechanism to convert linear motion (the stretch of the spring) to rotary motion (the turning of the shaft of the pointer). The mechanical display is the calibrated dial. The pointer is aligned with the calibrated dial markings to provide a measurement for the applied weight as shown in Fig. 5.

Fig. 5 An illustration of a simple spring scale.

Let's go back to the supermarket and move onto the delicatessen (delli) counter. On top of the counter, you will spot another scale. This one is a lot fancier than the spring scale which you noticed in the produce section of the store. When your sliced salami is placed on the platform, you as the customer can read the weight in red lighted numbers. This is a digital scale. In fact, it is a computer and a printer in addition to the scale. The delli clerk places the sliced salami on the platform and types in the per pound cost of it, and the computer calculates the cost of the quantity purchased and prints the label for the package together with a bar code for the automatic check out procedure.

The spring scales back in the produce department provided an analog display. They were totally mechanical. We converted the displacement of the spring to a pointer movement against the background of a calibrated dial as shown in Fig. 5 to give an analog readout of the weight.

The scales with digital readout have both mechanical and electrical components as illustrated in Fig. 4. The mechanical components are the platform, and some type of a mechanical member that undergoes a change (displacement, strain, pressure, etc.) in proportion to the imposed weight. Suppose the change we are monitoring is a quantity known as strain. The strain is sensed with strain gages (electrical resistors). The combination of the mechanical member undergoing the strain, and the strain gages comprise what is known as a transducer. The strain imposed on the strain gages produces a change in their resistance. The resistance change in turn is converted into a voltage output, V_0, with a signal conditioning circuit (Wheatstone bridge). The voltage out of the Wheatstone bridge is amplified and scaled and displayed as weight with a digital voltmeter. Clearly, in developing a scale with a digital display, you will be working with both mechanical and electrical components. An example of a digital scale is illustrated in Fig. 6.

Fig. 6 Photograph of a scale with a digital display.

We will discuss the different types of scales in Chapter 3, and give you some initial ideas to consider in the design of the mechanical components. The background for the digital scale will be covered in Chapter 4, where we will give you information on sensors, signal conditioning circuits and digital displays. Your digital and analog scales can have a common platform with interchangeable displays, or if your team prefers the scales can be totally independent.

You need to develop many different ideas for each subsystem included in the design of your scales. Each concept should be described in considerable detail prior to develop ing preliminary design proposals incorporating each of the design concepts. The more completely you develop the design proposals the better prepared you are to conduct what is termed a design-trade-off analysis, where you compare several design concepts and select the best one.

To perform a design-trade-off analysis, we consider each design proposal individually and list its strength and weaknesses. Factors usually considered in this analysis include size, weight, cost, ease of manufacturing, performance, appearance, etc. After the strengths and weaknesses of each design proposal has been listed, we can compare and evaluate each design proposal and select the most suitable for our prototype development.

We have briefly discussed design concepts, design proposals and design trade-off analyses for a single function that is involved in the process of weighing an individual. You have several functions to consider such as supporting the item, developing the mechanical response, mechanical conditioning for the analog system, mechanical display, the transducer, the signal conditioner and the digital display. The procedure in design is to generate ideas for performing each function, expanding these ideas with more detail until they can be treated as design proposals, and finally to perform a design trade-off analysis where the merits and faults of each proposal are evaluated.

When the design proposals are selected for each of the different functions involved in the weighing process, the winning concepts must then be integrated into a seamless system that accurately weighs an individual and displays the weight in an easy to read format.

We often refer to a part of the product that performs some function as a subsystem. The collection of all of the subsystems constitute a complete system, which when manufactured and assembled, become the prototype of the scale. The integration of the various subsystems is often difficult. One subsystem influences the design of the others. The manner in which the subsystems interact is defined as the interface between subsystems. In a mechanical application, the fit of a shaft in a bearing is an example of interface. Frequently, interfaces between the subsystems are troublesome and difficult to manage. It is vitally important that you control the interfaces to permit seamless integration of all of the subsystems without loss of effectiveness and efficiency.

The final step in designing the prototype is the preparation of the design package. The package, which is an extended engineering document, contains engineering sketches illustrating the concepts, engineering assembly drawings that describe how the various components fit together to form the complete system, and engineering drawings of all of the component parts in sufficient detail to permit the component to be manufactured by anyone capable of reading an engineering drawing. We recognize that many of you may need instruction in graphics, and we have included four chapters in this text to help you learn how to prepare three-view drawings, pictorial drawings and to make tables and graphs. One of the four chapters covers a computer aided drafting program, KEY CAD Complete, that will be of great assistance to you as you prepare the drawings for your design package.

The design package also contains a parts list that identifies every unique part employed in the assembly of the prototype. The quantity required of each component is also included on the line describing the component. For example, if you are going to use four No. 8 wood screws to fasten together the joints of your platform for the scale, you would:

1. Assign a part number on the parts list identifying the need for these screws.
2. List the quantity required as 4 in the quantity column.
3. Describe the screws in sufficient detail so that another team member can go to the store and procure exactly the type of screw that you want.

The description is brief, but complete and precise. (i. e. No. 8 wood screw, flat headed slotted, brass, 1 in. long). In this description, No. 8 gives the diameter; wood refers the type of application for the screw; flat headed describes its head; slotted indicates that you will use a flat blade screw driver in the installation; brass is the material from which the screw is fabricated; 1 in. is the length of the screw

The parts list contains all of the items that are to be purchased for your prototype. It also contains the components that must be manufactured. If the component is to be manufactured it is identified on the parts list by its name and a drawing number. We show an example of a parts list in Fig. 7.

PARTS LIST FOR A PLATFORM SCALE

NUMBER	NAME OF ITEM	DESCRIPTION OR DRAWING NO.	QUANTITY	PRICE
1	PLATFORM	WOODEN PLATE---- DWG. NO. 100-01	1	$ 1.20
2	BASE SUPPORT	WOODEN BOX -------DWG. NO. 100-02	1	$ 2.55
3	POST	PVC PIPE 2 IN. DIA. DWG. NO. 100-03	1	$ 3.50
4	LINKAGE	1/8 IN. DIA. WIRE CABLE --CUT TO LENGTH	1	$ 2.75
5	TOP LEVER	1/4 x 1 IN. AL. BAR --DWG. NO. 100-04	1	$ 3.00
6	MIDDLE LEVER	1/4 x 1 IN. AL. BAR --DWG. NO. 100-05	1	$ 5.00
7	LOWER LEVER	1/4 x 1 IN. AL. BAR --DWG. NO. 100-05	1	$ 3.00
8	BOLTS & NUTS	1/4- 20, 3/4 IN. LONG WITH LOCK NUTS	9	$ 2.70
9	COUNTER WEIGHT	AL BAR 1 IN. DIA. ---DWG. NO. 100-06	1	$ 2.75
10	TARE WEIGHT	AL BAR 1 IN. DIA. ---DWG. NO. 100-07	1	$ 2.75
11	TARE SCREW	NO. 8-32 BRASS THUMB SCREW 1/2 IN. LONG	1	$ 0.85
12	SUPPORT BARS	1/8 x 3/4 IN. STEEL BAR --- DWG. 100-08	4	$ 4.40

Fig. 7 Example of a partially complete parts list for a platform scale.

An engineering report is also included as part of the design package. This report supports the design by describing the key features of your scale. Your report should treat each function involved in the weighing process, and describe the design concepts that your team considered. A rationale for each design proposal that was adopted based on a systematic design trade-off studies is an essential element. Additionally, the design report contains theoretical analyses that you may have performed to predict in advance the performance of your prototype of the actual test to evaluate it's accuracy. More information on the preparation of an engineering report is given in a chapter included in this textbook on writing technical reports.

The design phase of the product development process is concluded with a final design presentation. This is a formal review of the design of the scale that your team presents to the class (peer review), to the instructor, and the teaching fellow. It is the team's responsibility to describe all of the unique features of the design and to predict the performance of the product. It is the responsibility of the class (peer group) and the instructors to question the feasibility of the design and the accuracy of the predicted performance. If you note a shortcoming or a flaw in the design, identify the problem to the team presenting their work. As a peer reviewer of another team's design, be tactful in your critique. Criticism is always difficult to offer to another. Offer your suggestions in good faith and in good taste. If you're on the receiving end of criticism, do not be defensive. The individual offering the critique is not attacking your capabilities. He or she is actually trying to give you a suggestion that might help you achieve a better design.

The purpose of the design review is to locate deficiencies and errors. It is better to correct errors in the paper stage of the process, not later in the hardware phase, when it is much more difficult and costly to fix the problems.

PREPARING THE ASSEMBLY KIT

The second major phase of the product development process is to prepare what is called an assembly kit. An assembly kit is a collection of all of the parts, in the appropriate quantities, that are required to completely build the prototype. The parts list is the essential document that directs our procurement and manufacture of the components needed to build the prototype. The parts list identifies everything that we need and indicates the precise number of each item.

An efficient method for collecting the required parts is to divide the list into three groups:

1. Those parts to be purchased.
2. The materials and supplies to be purchased.
3. The components to be manufactured.

One or two team members handle purchasing while the other team members work in the student workshop to fabricate the parts according to the detailed drawings prepared in the design phase. It is important to recognize that the drawing defines the part that is being manufactured. Always work from the drawing and not your memory or understanding of the part geometry.

It is suggested that one team member serve as the " inspector" to check the finished parts against the drawings for the parts that have been manufactured. If the parts have been purchased, the "inspector" should check the purchased items against the parts list to insure that the item exactly meets the description, and that the correct quantity has been procured.

PROTOTYPE ASSEMBLY

When the assembly kit is complete and checked against the parts list, we can begin the final phase --- prototype evaluation. The first part of the evaluation is to assemble or build the prototype. We often call this step the "first article build," because it is the first time that we have attempted to fit together all of the components needed to assemble the prototype. The "first article build" can go well if all of the parts are available, if they all fit together properly, if the tolerances on each part and each feature are correct, and if the surface finish on all of the parts is acceptable. Prototype design is often judged against the four Fs--- form, fit, finish and function. The "first article build" permits us to assess how well the team performed with regard to the first three of the Fs --- form, fit, and finish.

If the parts do not fit, modifications of one or more parts are required. These modifications require design change that is a dreaded and costly process in the real world. In fact one of the most important criterion used by management to judge the quality of a development team is the number of design changes required both before and after the introduction of a product to the market place. Clearly, we want to design each part in a product correctly during our first effort and minimize the number of design changes.

We anticipate that each team will make a few errors in the preparation of the detail design drawings and changes will be required. The natural tendency is to take the offending part to the model shop and to correct (modify) it, and get on with the "first article build." This behavior is acceptable only if the team revises the drawing of the offending part to reflect the modification made to correct the design deficiency. In the real world, the integrity of the drawing package is much more important than the prototype. The second and all subsequent assemblies will be fabricated from the details shown in your drawings, and not by examining the prototype. The prototype is often scrapped after it has been built and tested.

PROTOTYPE EVALUATION

After the assembly of the prototype is complete, it should be carefully inspected to insure it is safe. Fortunately, a scale is a relatively safe product since the parts do not undergo significant motion, and the hazards due to pressures and temperatures should not exist. However, make sure that you have eliminated all sharp edges and points on the prototype, and that you have avoided introducing pinch points.

The final step in this development process is to test your prototype to determine if it is functional and accurate. This is a big day and indeed a big hour. A one hour time slot has been scheduled for evaluating each team's prototype in the weight testing facility available in the ENES 100 Laboratory. Your prototype will be judged based on the following criteria:

1. Performance ---- accuracy over the entire specified range, and readability of the scales.
2. Cost ---- minimize.
3. Design innovation ---- be novel and be clever.
4. Quality of parts and assembly ---- good workman(women)ship.
5. Appearance ---- pleasing to the eye and in keeping with the decor in a master bathroom.

TEAMWORK

In this course, you are required to participate on a development team. There are three reasons for this requirement. First, the project is too ambitious for an individual to complete in the time available. You will need the collective efforts of the entire team to develop a pair of quality scales during the semester. We plan on pressing the development teams to complete the project on a prescribed schedule. Second, we want you to begin to learn teamwork skills. Experience has shown us that most students entering the engineering program have not developed these skills. From K-12, the educational process has focused on teaching you to work as an individual often in a setting where you competed against the other students in your class. We will demand that you begin functioning as a team member where cooperation, following, and listening are more important than individual effort. Leadership is important in a team setting, but following is also a critical element for successful team performance. The final reason is that you will probably find yourself on a development team early in your career if you take a position in industry. A recent study [4] by the American Society for Mechanical Engineers (ASME), the results of which are shown in Table 2, ranked teamwork as the most important skill to develop in an engineering program. Teamwork was also the first skill, in a list of 20, considered important by managers in industry. We are hopeful that this course will be instrumental in exposing you to team working skills so necessary for a successful career.

TABLE 2

Skills Considered Important for New Mechanical Engineers
with Bachelor Degrees
Priority Ranking

1. Teams and Teamwork
2. Communication
3. Design for Manufacture
4. CAD Systems
5. Professional Ethics
6. Creative Thinking
7. Design for Performance
8. Design for Reliability
9. Design for Safety
10. Concurrent Engineering
11. Sketching and Drawing
12. Design for Cost.
13. Application of Statistics
14. Reliability
15. Geometric Tolerancing
16. Value Engineering
17. Design Reviews
18. Manufacturing Processes
19. Systems Perspective
20. Design for Assembly

As you work within your team in the development of the analog and digital scales, you will be introduced to many of the 20 topics listed in Table 2. We trust that you will begin to develop an appreciation the need for these skills, and begin to enhance your level of understanding of the design process that is an inherent part of many of these skills.

OTHER COURSE OBJECTIVES

While your experience in developing the analog and digital scales is the primary objective of this course, there are several other related objectives. You will quickly recognize a critical need for graphics as you attempt to describe your design concepts to fellow team members. We have included four chapters on graphics to help you learn the basic skills required at this entry level. These chapters include materials on three view drawing, pictorial drawing, graphs and tables, and instructions for developing entry level skills in KEY CAD Complete, which is a computer aided drawing program.

You will also be required to become familiar and use three additional software application programs. These include a word processing, a spreadsheet, and a graphics presentation program. Our experience indicates that nearly all of you are already proficient with word processing; therefore, we will not cover this topic here. However, many of you are not familiar with a spreadsheet. Accordingly, we have included a chapter describing Microsoft Excel to help you prepare your parts list, to perform calculations and to prepare graphs. Spreadsheets represent a very powerful tool, that you will find useful in many different ways for the remainder of your life. We trust that you will take this opportunity to learn how to use this important application program. We have also included a description of Microsoft PowerPoint to aid you in the preparation of world class slides for your design briefings.

As you proceed with the development of the scales, we frequently will require you to communicate both orally and in writing. Design reviews before the class give you the opportunity to learn presentation skills such as style, timing and the preparation of world class visual aids. The design report will give you an experience in preparing complete, high-quality engineering drawings, and in writing a technical report containing text, figures, tables and graphs.

The final objective of the course is design analysis. We understand that your engineering analysis skills have yet to be developed. For this reason, we will present key equations that mathematically model the electrical and mechanical components used in the design of many of the subsystems that you employ in the design of the scales. We do not expect you to completely understand all of the theoretical aspects of the relatively complicated subsystems involved. Most of you will take several courses later in the curriculum dealing with these subjects in great detail However, at this stage of your career, we want you to begin to appreciate the inter relationship between analysis and design. To help you with the analysis we have included a chapter on the mechanical aspects of weighing machines and another chapter on electrical circuits and digital displays. The coverage is very brief and introductory, but it should give you a sample and a start in understanding both mechanical and electronic systems.

REFERENCES

1. Valenti, M. "Teaching Tomorrow's Engineers," Special Report, Mechanical Engineering, Vol. 118, No. 7, July 1996.
2. Ulrich, K. T., S. D. Eppinger, Product Design and Development, McGraw-Hill, New York, NY, 1995, p. 6.
3. Cross, N., Engineering Design Methods, 2nd Edition, Wiley, New York, NY, 1994, pp. 77-81.
4. Anon, "Integrating the Product Realization Process into the Undergraduate Curriculum," ASME report to the National Science Foundation, 1995.
5. Macaulay, D., The Way Things Work, Houghton Mifflin Co., Boston, 1988.

EXERCISES

1. List ten products that you have used today and the companies that manufactured and marketed them.
2. Suppose that you do not want to work on product development, manufacturing or sales. What opportunities remain for you in Engineering?
3. Write an Engineering brief describing the characteristics of a winning product.

4. Write the most important equation in engineering or business.
5. Why is Ford Motor Company reluctant to develop a brand new model of an automobile?
6. What is a design concept? How many concepts will your team generate in developing the scale this semester?
7. What are the differences between analog and digital scales? Define the meaning of the word analog and relate it to an analog device. Define the meaning of the word digital and relate it to a digital device.
8. Why is it important to prepare a parts list during a development program?
9. What is an assembly kit and why do we prepare one in the development of the first prototype?
10. Why do we invest scarce company funds in assembling a prototype?
11. What are the safety considerations with which you must be concerned in testing the analog and digital scales?
12. Why do you believe industry representatives ranked teams and teamwork as the most important skill for new Engineers?

NOTES

CHAPTER 2

ANALOG AND DIGITAL WEIGHING MACHINES

MEASURING WEIGHT

Analog and digital weighing machines are the products selected for the 1997-98 class of ENES 100 to design, build and evaluate. We have divided the class into development teams each with five or six members. Each team will have the responsibility of developing two weighing machines --- one with an analog readout and the other with a digital display.

Before describing the weighing machines in detail, let's briefly discuss the meaning of weight, and then consider the methods of measuring weight that have been used for many thousands of years.

Our weight is the force that is caused by the earth's gravitational pull on our bodies. Our bodies, or any body for that matter, has a mass which we will define as m_b. The earth has a mass m_e. When two masses are in close proximity, there is an attractive force F which develops between the two bodies. This force is given by:

$$F = C \, m_b \, m_e \, /r^2 \qquad\qquad (1)$$

where C is called the universal constant of gravitation.

r is the distance between the two bodies.

Earth bound bodies, either those of people or objects, are very small compared to the radius of the earth R which is equal to 3960 miles or 6.37×10^6 meters. Because of the large size of the earth, we can, without introducing significant error, set:

$$r = R \qquad (2)$$

Next we collect all of the quantities that have anything to do with the earth in Eqs. (1) and (2), and group them together to give:

$$g = Cm_e /R^2 \qquad (3)$$

where g is the gravitation constant for the earth. We treat g as a constant equal to 32.17 ft/s^2 or 9.806 m/s^2, although strictly speaking g varies a bit because the earth is not a true sphere, and R does not remain constant when we move from Pikes Peak to Death Valley. Rewriting Eq. (1) gives:

$$F = m_b\, g \qquad (4)$$

The force F in Eq. (4) is the weight of a body on Earth with a mass m_b.

If you examine the units of g (ft/s^2 or m/s^2), it is evident that it is an acceleration. The gravitational field of the earth produces an acceleration that acts on all the bodies on or near the surface of the earth. That is the reason we fall down a set of stairs, or out of a tree.

Let's look at one more relation that helps us describe the meaning of weight --- Newton's second law of motion:

$$F = ma \qquad (5)$$

where a is the acceleration.

   m is the mass of some body.

Sir Isaac Newton gave us this relation between force F and acceleration. Let's apply Eq.(5) to what we know about the earth. The gravitational field causes an acceleration a = g, where g is the earth's gravitational constant. The force F acting on the body in the gravitational field is the weight W. We can then write:

$$W = F = ma = m_b\, g \qquad (6)$$

If we travel from College Park to Chicago, our weight remains essentially constant. So we get confused and think of our weight as being the constant in Eq. (6). Not true. The term that is the constant in Eq. (6) is the mass m_b. Prove it! Go jump on the moon. We all know that we weigh much less on the moon, about one sixth as much as here on earth. The reason is that the gravitational acceleration on the moon is about g/6 because the mass of the moon is much smaller than the mass of the earth. Our mass m_b is the same whether we are on the moon, mars, or the earth.

OK! Now that we understand about mass being constant and that weight is a force due to the earth's gravitational field, let's take a look at the methods that have been developed over time to measure weight.

SOME HISTORY ABOUT MEASURING WEIGHT

Sometimes we are in such a hurry that we fail to look back. This statement is also true in design. As we approach the millennium year 2000, remember that we have some records of what folks did 8000 years ago. Our forebears were not lazy or dumb. They invented a lot of very neat products. True, the invention of steam power came late, electricity even later, and microelectronics in the past 35 years, but we had many very important developments long before our generation. The analog weighing machines have a very long history. The digital weighing machines were introduced in the market after microelectronics developed, and are so new that we have very little to report that is not extremely contemporary.

A SIMPLE BEAM BALANCE

The first weighing machines go back to the ancient Egypt and Babylon [1 - 4]. These weighing machines were beam balances as illustrated in Fig. 1. These were very simple devices consisting only of a beam (usually wooden) and some loops of string. The beam was hung from its center loop, and an unknown weight was placed in the right hand string loop. Reference weights were added in the left hand string loop until the beam balanced (became horizontal). At this position of the beam, the unknown weight was equal to the reference weight and the measurement of weight of the unknown mass was complete. You could weigh any item that could be supported by the beam and the strings, providing you had a good (and acceptable) set of reference weights. In the very early days, the reference weights were usually made of stone or glass, although when metals became more common brass or bronze was sometimes employed.

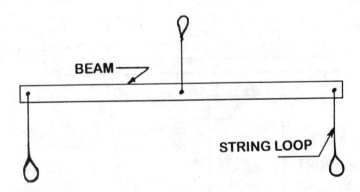

Fig. 1 The very early beam balance with string attachments for support, the unknown item, and the reference weights.

Over the centuries, the beam balances were improved with many design modifications. The first improvement was the additions of pans and hooks to support either the reference or unknown weights as shown in Fig. 2. Later when metals became available for the construction of the beams, pins were placed through the beams for attachment to clevis attachments for the pans and the center support.

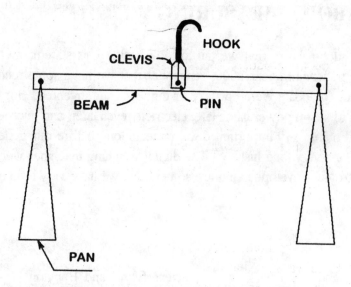

Fig. 2 Beam balance with pans for supporting the weights and a pinned central support.

When the accuracy of the weight measurements became more important, there was concern about the effects of friction at the pin associated with the center support. With friction, slight inequalities could exist between the unknown and the reference weights with the beam in apparent balance. To eliminate this friction, the center pin in the beam was replaced with a knife edge support as shown in Fig. 3. The knife edge eliminated friction, improving the accuracy of the measurements, but the sharp edge was fragile. Weighing machines were fitted with mechanisms for lifting the beam off of the knife edges as weights were added to the pans to protect the knife edges. When the weights were placed in both pans the assembly was lowered slowly onto the knife edges to check for balance.

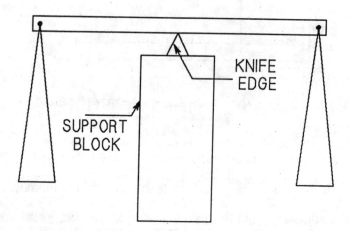

Fig. 3 Beam balance with pans and knife edge support to minimize friction.

The beam balances became more and more accurate. They were placed into glass cases to prevent air currents from disturbing the balance operations. Until recently, they could be found in chemistry laboratories. They were equipped with indicators to guide the operator in selecting weights to add to the reference pan. They were also fitted with dampers to reduce the number of oscillations required before the beam reached equilibrium.

Beam balances were constructed in many different sizes. Large balances were used to weigh large quantities of grain or other market produce. Medium size balances were employed to measure spices or other more valuable merchandise. Very small balances were used for very valuable items like gold or precious stones. It was recognized even in the time of the Egyptians that the accuracy of the beam balance depended on its capacity. The large beam balances were less accurate than the smaller balances. This fact is still true today for all weighing machines, electronic or mechanical, analog or digital. Accuracy and capacity (range) are inversely related.

THE BISMAR BALANCE

The balance, with it many modifications, served well for many centuries (in fact it is still in use today) before a new type of balance was introduced with unequal arms for the beam. There are two types of unequal arm balances --- the bismar and the steel yard. With the bismar balance, illustrated in Fig. 4, the beam is equipped with a large fixed weight (called a counterpoise) on the left end. The unknown mass is connected to the hook attached to the right end of the beam. The center support is a moveable loop of cord, rope or a metal band. The center support is moved toward one end of the beam or the other until the beam is positioned horizontally. In this horizontal position, the unknown weight of an item can be determined in terms of the weight of the counterpoise. The analysis that gives the relation for the unknown weight W_u in terms of the known weight of the counterpoise W_{cp} is based on the equilibrium of moments. To perform this analysis we start with a free-body diagram of the bismar balance illustrated in Fig. 5.

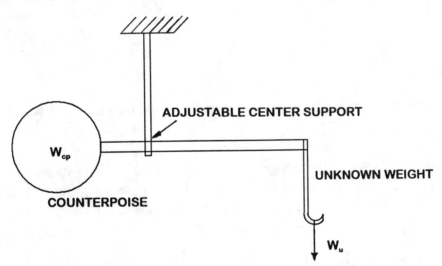

Fig. 4 Sketch illustrating the adjustable middle support and fixed weight locations on a bismar balance.

Since the beam is horizontal and at rest, it is in equilibrium. We can write the relation for equilibrium of moments about point O, which gives:

$$\sum M_O = 0 \qquad (7)$$

The two weights W_u and W_{cp} both produce a moment about point O. The moment is simply the weight times the perpendicular distance from the line of action of this force to the point O. Using this simple rule we can rewrite Eq. (7) as:

$$(W_u)d_u = (W_{cp})d_{cp}$$

Solving this relation for W_u yields:

$$W_u = (d_{cp} / d_u)W_{cp} \qquad (8)$$

Since we know the weight of the counterpoise, we can measure W_u by determining the two distances d_u and d_{cp}. Great! We have converted a difficult weight measurement into a simple task of measuring two lengths.

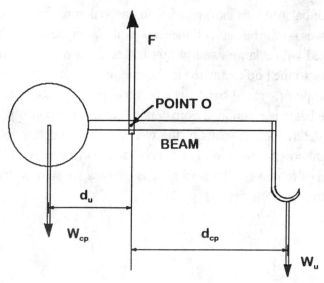

Fig. 5 Free-body diagram showing the forces and the distances involved in the equilibrium of a bismar balance.

Weight machines where the weight measurement is somehow transformed into a measurement of lengths or distances are usually called scales. The word scales derives from the instrument used to measure length, namely a scale. For the remainder of this chapter, we will usually use the word scales instead of weighing machines because it is shorter and in more common usage.

THE STEELYARD SCALE

Another 18th century scale, very similar to the bismar, was the steelyard. The design of the steelyard scale fixes the pivot point on the beam at an off-center position as indicated in Fig. 6. The position of the pan (or hook) supporting the unknown mass is fixed at the short end of the beam. The counterpoise is the moveable component in the design of a steelyard scale. The analysis of the free-body diagram of the steelyard scale is identical to that of the bismar scale, and we rewrite Eq. (8) to give:

$$W_u = (d_{cp} / d_u)W_{cp} \qquad\qquad (8 \text{ bis})$$

However, with the steelyard scale both W_{cp} and d_u are fixed and known quantities. We can group them together, and set them equal to a constant $C = d_u / W_{cp}$. Simplifying Eq. (8) gives the relation for W_u as:

$$W_u = Cd_{cp} \qquad\qquad (9)$$

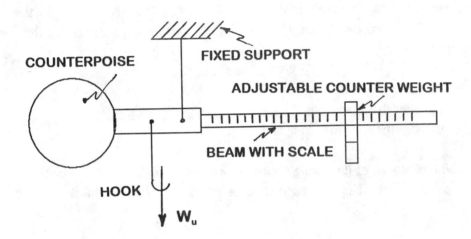

Fig. 6 Schematic illustration of a steelyard scale with an adjustable counterweight to achieve balance.

The steelyard scale has the advantage over the bismar scale in that only one measurement of distance d_{cp} is required. Again we have converted a difficult weight measurement into a relatively simple determination of a single length.

Have you been to your doctor's office lately? The nurse usually weighs you before you get a chance to see the doctor. Does the scale in your doctor's office resemble the steelyard scale? Did the nurse move a weight (maybe two)? Are these weights similar to the counterpoise? The scale in the doctor's office is a platform scale that we will discuss later; however, it uses the same principle of equilibrium of moments.

SPRING SCALES

Let's consider a different type of scale that does not use a balance beam. One that does not use reference weights (the counterpoise). Do you recognize a helical spring like the one shown in Fig. 7.

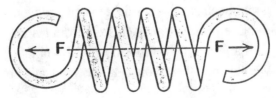

Fig. 7 Illustration of a helical spring.

Helical springs are wound from wire to form the helix that serves as the spring element. When springs are subjected to either tension or compression forces, they stretch or compress, and when the force is removed they return to their original shape. Springs are employed for three major reasons in mechanical design. First, they will return some object to its original position like the bolt in a lock on a door. Second, they store energy in the deflected state which can be returned to the system as the spring resumes its original shape. Finally, springs can be used to measure force. We are interested in springs because we know they can be employed in our weighing machine to convert a force measurement into a length measurement.

Springs have been used in the design of scales for a few hundred years, and they are still utilized in most of the lower cost mechanical scales on the market today. There are three essential components in a spring scale:

- The spring.
- A pointer that is activated by the deflection of the spring.
- An indicator/scale for displaying the weight.

We will discuss the mechanism of the spring scales in more detail in Chapter 3. For the time being, let's consider the equation describing the response of a helical spring under the action of an applied axial force F.

$$F = k\,d \qquad\qquad (10)$$

where d is the axial deflection of the spring in in. or m.

k is the spring constant (or spring rate) in lb/in. or N/m.

Consider an example. Let's take a spring with a spring constant of 500 lb/in. which means that it will deflect one inch if we apply a load of 500 lb to the spring. However, we will learn later that our product specification requires us to measure a weight that varies up to a maximum of 250 lb. Can we still use this spring? How much will its free end deflect? Let's analyze the spring, and then decide if it might help us in the design of a prototype?

Let F = 250 lb which is the required capacity of a scale that your team is to build. Take k = 500 lb/in.[1], and find the deflection d of the spring. Using Eq. (10) gives:

$$d = F/k = 250/500 = 0.5 \text{ in.}$$

The spring deflects 0.5 in. when we apply a force equal to 250 lb to one end and constrained at the other end.

We note that the spring is a mechanical transducer that converts a force (weight) measurement into a measurement of displacement. To make a scale from a spring, you need to track and display the spring displacement with a pointer. The position of the pointer is noted on a scale that is calibrated in terms of weight. We illustrate these concepts schematically in Fig. 8.

---

[1] To give you a sense of size, a spring wound with round edge flat wire 9/32 by 5/32 inch, that has a free length of 3 in., with an outside diameter of 1.25 in. and an inside diameter of 5/8 in., has a spring constant of 520 lb/in.

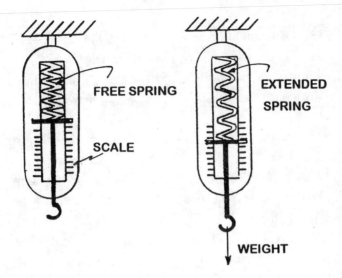

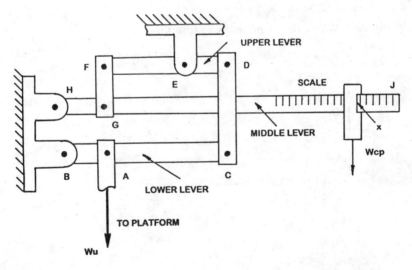

Fig. 8 Schematic illustration of a fish scale showing the features involved in a spring scale.

THE PLATFORM SCALE

The platform scale was introduced in the early 18th century when it became necessary to weigh carts and carriages for toll roads in England. Platform scales are really complex balances that employ compound levers. The pan for the unknown mass is a large platform that can accommodate a large object like a cart or a carriage. The force is transmitted to a series of compound levers as shown in Fig. 9.

Fig. 9 Platform scales employ compound levers to weigh very large masses.

The platform, shown in Fig. 9, is attached to the lower lever at point A. The lower lever arm moves downward at point A, rotating about point B and deflecting the pin at point C downward. This deflection is transmitted from the lower lever to the upper lever with the vertical link which connects points C and D. The downward deflection of point D causes the upper lever to rotate about point E. This clockwise rotation about point E produces an upward deflection of the pin at point F. This upward deflection is transmitted from the upper lever to the middle lever by a second vertical link that connects points F and G. Since point G is moving upward, the middle lever will rotate about the pin at point H and tilt upward. On the right side of the middle lever is a small weight that can be adjusted to bring this lever to a horizontal position. The right side of the middle lever is imprinted with a scale that is calibrated to indicate at the position of W_{cp} the weight of the object on the platform.

Whew! Compared to the simple balances introduced earlier, the platform scales are very complex. Why do we introduce this complexity? Two reasons. First, to limit the vertical movement of the platform. The pins and load application points are arranged to greatly amplify the motion of the middle lever at the end J. This large movement makes the scale sensitive, while the very limited motion of the platform makes it more stable.

Second, the use of three levers with cleverly arranged pin positions gives a very large leverage, and consequently a very small weight W_{cp} can be used to balance a very massive object placed on the platform. The small counterpoise W_{cp} is a very significant advantage, because it can be moved easily along the scale inscribed on the middle lever to accurately balance the large weight on the platform.

INDICATOR SCALES

Indicator scales have features similar to the simple balance, bismar and the steelyard scales. The indicator scale pivots about a fixed central support as shown in Fig. 10. However, the counterpoise is configured as a pendulum that dominates the right side of the mechanism. The unknown mass is suspended from the hook (or pan) positioned on the left side of the indicator scale. Depending on the weight of the unknown mass the entire system rotates changing the distances d_u and d_{cp} so as to automatically balance the moments about point O.

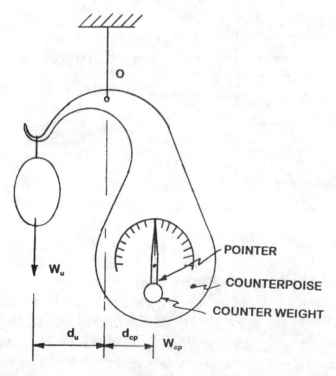

Fig 10 Sketch of an indicator scale showing the pendulum weight with the self aligning pointer.

With the indicator scale, unlike the bismar or steelyard scales, it is not necessary to measure the distances d_u and d_{cp} to determine the unknown weight. The length measurements are avoided by placing an indicator on the pendulum counterpoise as indicated in Fig. 10. The indicator is a circular scale and a vertical pointer. The circular scale is inscribed on the pendulum, and calibrated to read the weight of the unknown mass directly. A self-aligning pointer (it has a counter weight at its lower end to keep the pointer oriented in the vertical direction) points to the correct indication of the weight. The pointer always remains in the vertical orientation. As the unknown mass is changed, the pendulum and the circular scale both rotate below the pointer. This is a very clever design for an automatic indication of the weight of an unknown mass.

HYDRAULIC SCALES

Let's consider still another concept much different than balances and spring type scales, namely a weighing machine based on hydraulic principles. Suppose we place our unknown mass on a platform, and connect this platform to a piston and cylinder arrangement as illustrated in Fig. 11.

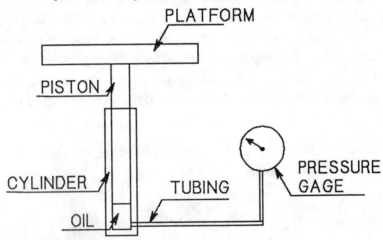

Fig. 11 Sketch illustrating the platform and piston system used to develop pressure in a close hydraulic system.

When the force W_u is applied to the platform, the piston is driven downward into the cylinder. This downward motion is resisted by the oil that is contained in the cavity of the cylinder. A pressure p is developed in the cylinder which is expressed as:

$$p = W_u / A_p \tag{11}$$

Where p is the hydraulic pressure in psi or Pa.
A_p is the cross-sectional area of the piston in in.^{2} or in m^2.

Clearly the weight and the pressure are related in a linear manner as indicated by Eq. (11). The piston-cylinder arrangement has changed our problem from measuring weight to one of measuring pressure. Is that an advantage? Yes! For large masses balances become extremely cumbersome, and springs get bulky and costly. We can handle high pressures (up to 3000 psi) with little difficulty and relatively accurate bourdon tube pressure gages are available at low cost.

But how do we convert pressure to weight? It is easy. As you might expect the pressure gage is a circular case with a pointer and a circular scale. The circular scale is calibrated to read the pressure. It is a simple matter to replace this pressure scale with one calibrated to read-out in terms of weight.

If you search the historical literature, you will not find the hydraulic scale (at least I did not). We were relatively late (18th century) in discovering pressure, and how to harness it to produce work.

ELECTRONIC SCALES

Electronic scales are even more contemporary than hydraulic scales. They have become common only in the past 25 years. There are two reasons for their late arrival in the market place. First, the technology (stable amplifiers and digital displays) was not available until very recently. Second, the cost of the electronic scales was much higher than the mechanical scales (this is still a fact today).

Most electronic scales are related to spring scales in that a mechanical member in the scale deforms under the action of the applied weight. The deformation of this spring element produces a high strain at one or more points on the spring element. Strain gages, illustrated in Fig. 12, which are variable electrical resistors, are bonded to the spring element at these highly strained points. The changes in the resistance of the strain gages is converted into a voltage change that is proportional to the weight on the scale. The voltage change can be monitored on a analog voltmeter or a digital voltmeter. If a digital voltmeter is employed the readout on the display is digital (lighted numbers).

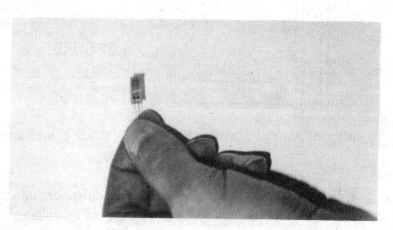

Fig. 12 Photograph of a bonded metal film resistor used to measure strain.

This subsection on electronic scales is only to mention that this "new" scale exists, and that they are very contemporary in the historical development of weighing machines. We will devote Chapter 4 to a complete and detailed treatment of the technology used in designing digital scales.

THE PRODUCT SPECIFICATION

We seek a pair of scales, which can be considered weighing machines, one with an analog readout and the other with a digital display. You will be expected to participate as a member, or possibly the leader, of a student development team in the design, manufacture, assembly and evaluation of the two prototypes. A computer laboratory with software for word processing, computer aided drafting (CAD), spreadsheets, and a graphics presentation program will be important in this development activity. A model shop and/or a work shop will also be helpful in fabricating the required components, and for fitting and assembly of each of the scales.

The evaluation of each of the scales will be performed by adding calibration weights to the platform of your weighing machines. These weights, shown in Fig. 13, are made from cast iron. They are fabricated with a handle so they can be lifted easily and placed on the scale during the calibration process. Ten weights, each 25 lb, will be available for your use. You will add the weights, one at a time, and record the measurement after each application of load. We anticipate that you will be able to complete the verification test to determine the accuracy of a scale in five minutes or less.

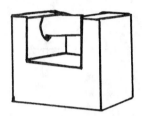

Fig. 13 Illustration of weights used to verify the accuracy of the scales.

The visibility of the displays, both analog and digital, will be checked by your instructor. Your instructor, with 20/20 vision, will verify the readings of the weight displayed from a distance of six feet.

Criteria that will be used to judge the merits of your team's development effort include:

1. Performance

   - Accuracy over the entire range 0 to 250 lb.
   - Resolution of the measurement.
   - Visibility of the display.

2. Cost

   - Minimize the cost of materials and components used in the prototype.
   - Minimize cost consistent with maintaining a quality product.

3. Design innovation

   - Simplicity.
   - Novel concepts.
   - Ease in manufacturing and assembly.

4. Quality

   - Reliability of the measurements.

- Close fitting parts.
- Minimum friction.
- Superior workman(women)ship.
- Alignment.

5. Appearance

- Pleasing to the eye.
- Decor matching a modern master bathroom.

6. Package size

- See extended description below.

7. Safety

- Avoid all hazards[2].

The criterion --- package size --- requires more explanation because it is more involved than the other criteria used in determining the adequacy of the scales which your team develops. The package size is a design constraint. We are limiting the size of each weighing machine by requiring that all parts fit into a cardboard box which normally contains ten packages (500 sheets each) of 8.5 by 11 in. copy paper.

You will also be required to assemble your prototypes using simple hand tools such as a drill, screw drivers, hammer, pliers, soldering iron, etc. You will be allotted 50 minutes immediately prior to the class period scheduled for the evaluation test to assemble both of the prototypes. You will also be required to disassemble both scales, placing all of the parts back into the copy paper box, in the 15 minute period immediately following the evaluation tests. Unless you receive instructions to the contrary from your instructor, you will have only one opportunity to "officially" determine the accuracy of your scales. However, the calibration weights will be available six weeks prior to the scheduled evaluation test, and you are encouraged to check your design concepts well in advance of your final class. When you have completed the evaluation test, pack the parts from both scales in the copy paper box, and dispose of it in an environmentally sensitive manner.

The scales will have a capacity of 250 lb, with a resolution of 5 lb. (We will discuss the meaning of resolution in more detail in Chapters 3 and 4). The accuracy requirements depend on the weight placed on the scale. From 25 to 100 lb, the weight measurement should be accurate to ± 5 %. From 100 to 250 lb, the accuracy should be ± 2 %. We show the allowable error bands in Fig. 14.

The weighing machines will be equipped with a control and/or mechanism for zeroing the scale. This zero adjustment is an important feature for both the analog and digital systems. The mechanical components in the display mechanism wear or change dimensions with temperature causing the zero position of the indicator to vary from zero. For electrical systems, the output

---

[2] See Chapter on Safety, Risk and Performance.

voltage drifts with time and temperature and adjustments are necessary to null out these parasitic voltages.

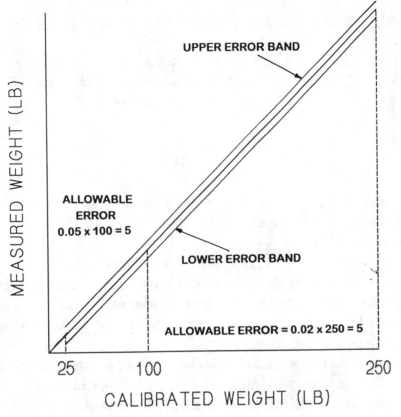

Fig. 14. Allowable error bands for the evaluation of the accuracy of the scales.

We anticipate that our customers will purchase these scales to be used on a daily basis in a master bathroom. Accordingly, the scale should be small to conserve floor space. It should also be attractive with rounded corners, flowing curves and smooth finishes. The colors employed should be consistent with the decor of a master bathroom.

The platform on the weighing machine should be easily accessible, and it should be stable. We recognize that the platform of a scale must move (a little), but the displacement in both the horizontal and vertical directions should be minimized. The platform should also be large enough to accommodate male feet fitted with a size 13 E shoe. Finally, the platform should not be slippery even when its surface is wet.

The displays must be visible to the individual using the scales. You should anticipate that he or she will stand on the platform of the scale, and attempt to read his or her weight from the display. Reading the display should be possible from the standing position. You may design a large display, magnify the display to enlarge the numerals, or you may choose to move the display to a location more suitable for reading.

While 120 V ac power is available in the modern master bathroom, we believe that it represents a safety hazard if used in your early prototype. Accordingly, **the use of the 120 Volt ac utility line power supply is prohibited**. We recognize that an electrical supply is essential for the digital display and/or scale. We anticipate that you will employ batteries to provide the necessary power. **Do not employ more than nine volt batteries without permission in writing from your instructor.**

REFERENCES

1. Kisch, B., Scales and Weights: A Historical Outline, Yale University Press, New Haven, CT, 1965.
2. Klein. H. A., The World of Measurement, Simon and Schuster, New York, 1974.
3. Lieberg, O. S., Wonders of Measurement, Dodd, Mead & Company, New York,1972.
4. Macauly, D., The Way Things Work, Houghton Mifflin Co., Boston, 1988.

EXERCISES

1. If the gravitational field on the moon produces a gravitational constant 1/6th as large as the gravitational constant on earth, what would you weigh on the moon? What is your mass on earth? What is your mass on the moon?
2. Discuss the advantages and disadvantages of the simple beam balance.
3. Describe the differences between the simple beam balance, the bismar scale and the steelyard scale. Also describe the similarities between these three weighing machines.
4. We have described the helical spring in this chapter. Define two other types of structural elements that are sometimes employed as springs.
5. What are the advantages of the platform scale? What is its most significant disadvantage?
6. While the indicator scale is very clever in design, why is it rare in comparison to the spring scales.
7. We have cited the advantage of the hydraulic weighing machine as it ability to determine the weight of very large masses. What are its two most important disadvantages?
8. The change of resistance of a strain gage (which is a variable resistor) is extremely small. What well known electrical circuit is used to convert this small resistance change into a voltage change?
9. Define the relation used to compute the error bands in Fig. 14.
10. Describe the difference between resolution and accuracy.
11. Define friction. What is the equation used to determine the friction forces? What are the factors that affect the magnitude of the friction force? What can you do as a designer to minimize friction?

CHAPTER 3

MECHANICAL ASPECTS OF WEIGHING MACHINES

INTRODUCTION

All weighing machines, analog or digital, incorporate mechanical elements in their design. Even the digital scales with their numeric displays of weight begin with mechanical components. The conversion from the initial mechanical response to the weight placed on the scale, to an electrical response, and then to a digital output is an important part of the task of designing a digital scale.

Before we get too deeply involved in the mechanical aspects of weighing machines, let's worry about the meanings of four words that we have used in the above paragraph:

- Mechanical
- Electrical
- Analog
- Digital

The word mechanical refers to components that we will be using in our scales such as platforms, springs, pins, levers, shafts, bearings, pointers, scales, bolts, nuts, etc. Electrical refers to components such as resistors, capacitors, inductors, chips, circuits and circuit boards. In performing mechanical design, we will be concerned with motion, equilibrium, strength, and friction. With electrical design we are concerned with voltage, current, signal amplification, and electrical noise.

The words analog and digital both refer to the way that we display the weight measurement. To illustrate the difference, examine the two voltmeters illustrated in Fig. 1. The meter to the left is an analog multi-meter. It measures voltage, resistance and current, and displays this reading with a pointer that sweeps over several a calibrated scales. It provides a continuous reading of the quantity being measured as indicated in Fig. 2. If the input voltage increases by a very small amount the output voltage that is displayed will also increase by this amount.

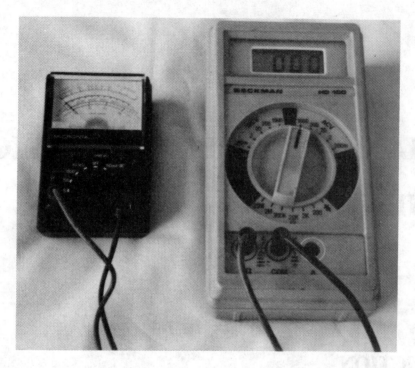

Fig. 1 Photographs of analog and digital multi-meters.

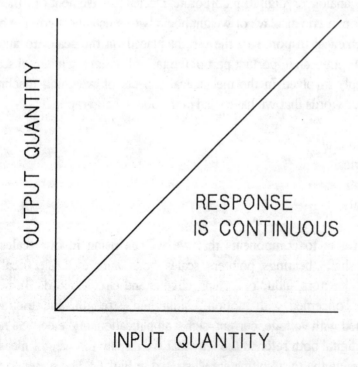

Fig. 2 An analog response is a continuous function.

The output of a digital display is not continuous. Inspection of Fig. 3 shows that the output corresponding to an input voltage that is a linear ramp resembles a staircase. As we increase the input voltage to the digital voltmeter, the output voltage that is displayed remains constant until we reach a threshold voltage. At the threshold, the digital voltage displayed increases as a step function and is actually is larger that the input voltage. From Fig. 3, it is clear that the digital voltmeter is in error almost all of the time. Not to worry! We keep the error within allowable bounds by maintaining very small step sizes when the input signal is digitized.

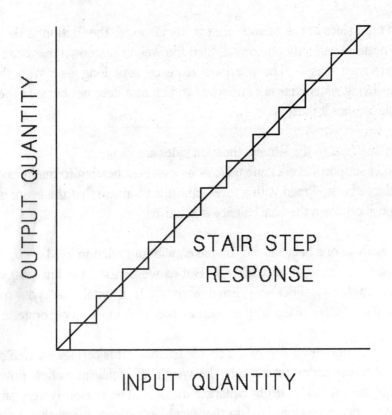

Fig. 3 The digital response to a linear ramp input voltage is a staircase output.

We will discuss digital displays and the electrical aspects of our weighing machines in more detail in Chapter 4. In this chapter let's examine in some detail weighing machines that are based on:

- Equilibrium or $\sum M_O = 0$.
- Springs that convert force into displacement.
- Hydraulic cylinders that convert force into pressure.
- Strain gage based transducers.

We have selected these four topics because they all are viable design concepts that your team may wish to consider in designing your prototype. In this Chapter, we will introduce the analysis methods that you will need to evaluate the merits of the concepts, and size the components to be used in your design. We will also try to illustrate some of the design ideas and/or features that have been employed to develop scales that are currently available on the market.

EQUILIBRIUM BASED WEIGHING MACHINES

The simple balance described in Chapter 2 is an equilibrium based weighing machine. The weight of the reference mass in the left hand pan produces a moment about the center support that is exactly canceled out by the weight of the unknown mass in the right hand pan (see Fig 3 of Chapter 2). The beam of the balance is horizontal when the two moments are equal. If the moments are not in balance, the beam tilts and the motion of the beam indicates that the selection of the reference mass was not correct.

If the left and right sides of the balance are exactly identical, the friction at the center support is negligible, and the beam is perfectly horizontal, then the weight unknown mass can be equated to the weight of the reference mass. The previous sentence is a long one with three qualifying statements. These qualifying statements are important to you as a designer because they indicate that if you develop a simple balance it must be:

- Perfectly symmetrical so the left and the right sides are identical.
- That the center support have a knife edge or an excellent bearing to minimize friction .
- That the balance be equipped with a level indicator to insure that the beam is in a perfectly horizontal position when the final balance is achieved.

When we perform an analysis of a balance, we use the equilibrium relation $\sum M_O = 0$, and we explore the conditions required for balance to be achieved. But as we perform the analysis, we make some assumptions that lead to the qualifications listed above. It is vital that you recognize these qualifications because they dictate many of the features that you must incorporate in the design of your balance.

Let's continue with our discussion of the simple balance. It is certainly a viable concept for a scale. In fact, it is still in use today to measure the weight of small quantities; however, it is not commonly employed to measure the weight of larger masses. The reason is very simple. We get tired of moving the reference masses around in the weighing process when they become large. It takes too long and requires too much energy. The scale is too cumbersome and too delicate to be used with objects weighing more than a few pounds.

So the simple balance is not a good idea. Do we abandon the concept of a weighing machine based on the equilibrium principle? Not if there is a way to eliminate the killer disadvantage with a different design while retaining the advantages of cost, simplicity, and accuracy.

We know from Chapter 2 that the platform scale, shown in Fig 9 of Chapter 2, was developed to weigh wagons and carriages for the toll roads operated in the 18th century. We also know that they are still widely used today by physicians for weighing patients. The platform scale uses a series of levers, links and pins to produce a mechanical advantage so that large weights placed on the platform can be balanced with small weights that are moved along a lever arm. This three lever mechanism eliminates the killer disadvantage, but is it suitable for your prototype?

Before we decide that a platform balance is viable or not, let's conduct an analysis of the mechanism. This analysis has three purposes:

1. It permits us to size the levers.
2. It provides information concerning the scale that is to be imprinted on the balancing lever.
3. It aids in the listing of the qualifications that must be placed on the weighing machine.

To begin the analysis we need to examine the drawing of the three lever mechanism for a platform scale that is shown in Fig. 4. There are three levers, the lower lever B-C, the upper lever F-D, and the middle lever H- J. The lower lever is connected to the upper lever through the pins and the link C-D. The upper lever is connected to the middle lever through the pins and link F-G. All three levers are fixed at a point, and they all will rotate about their respective fixed point. The platform is connected to the three lever mechanism by a vertical link that connects with the lower lever at point A.

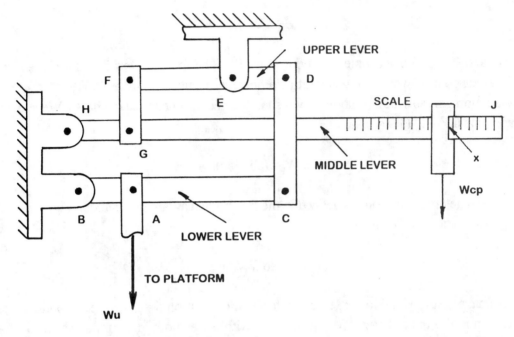

Fig. 4 The three lever mechanism of a platform scale.

The analysis is simple, but it is a bit tedious because of the numbers of levers involved. Let's start with the lower lever where the unknown load is imposed by the vertical linkage that extends from the platform to point A. We isolate the lower lever by removing it from the mechanism, and by representing it with a free body diagram as illustrated in Fig. 5.

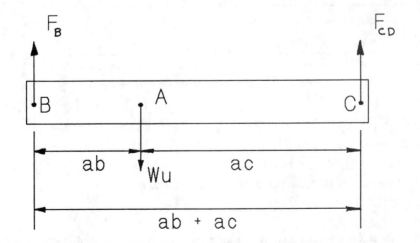

Fig. 5 A free body diagram of the lower lever arm from the platform scale.

In drawing the free body diagram we removed the pins, links, and clevis at points A, B and C and replaced them with forces at each point. The force at point A is identified as the weight of the unknown mass W_u. The forces at points B and C are not known, but we represent them with vectors (arrows) labeled F_B and F_{CD}. All of the forces are in the vertical direction because the links were all oriented in this direction.

We next write the equilibrium relation for moments and apply this relation to the fixed point on the lower lever, namely point B.

$$\sum M_B = 0 \tag{1}$$

Both W_u and F_{CD} produce moments[1] about point B. The force W_u produces a clockwise moment $(ab)W_u$ about point B which we will treat as a positive quantity. The force F_{CD} produces a counter clockwise moment $(ab + ac)F_{CD}$ about B which is treaded as a negative quantity. We substitute these two moments into Eq. (1) to obtain:

$$abW_u - (ab + ac)F_{CD} = 0$$

where ab and ac are dimensions as defined in the free body diagram presented in Fig. 5. Solving this relation for F_{CD} gives:

$$F_{CD} = abW_u /(ab + ac) \tag{2}$$

OK! We have analyzed the lower lever and have an expression for F_{CD}. Let's continue the analysis by considering the upper lever. We again begin with a free body diagram as shown in Fig. 6, and note that all three of the forces at points D, E and F are in the vertical direction.

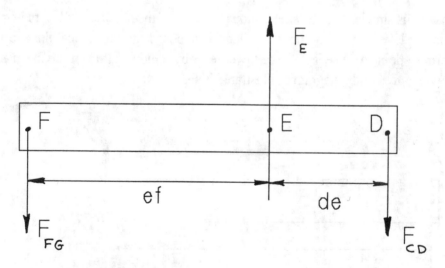

Fig. 6 Free body diagram for the upper lever of the platform scale.

---

[1] A moment about a point is the product of a force times the perpendicular distance from the point to the line of action of the force.

We apply the equilibrium relation for moments, selecting in this instance the fixed point E, by writing:

$$\sum M_E = 0 \tag{3}$$

Determining the moments about the point E that are produced by forces F_{CD} and F_{FG} and substituting in Eq. (3) yields:

$$de \, F_{CD} - ef \, F_{FG} = 0$$

Solving this relation for F_{FG} gives:

$$F_{FG} = (de/ef)F_{CD} \tag{4}$$

Substituting Eq. (2) into Eq. (4) gives:

$$F_{FG} = (de/ef)[ab/(ac + ab)]W_u \tag{5}$$

The equations are getting longer but the analysis has provided us with expressions for two of the forces acting on the levers in terms of the dimensions locating the pins and the unknown weight W_u. We will conclude the analysis by considering the free body of the middle lever that is presented in Fig. 7.

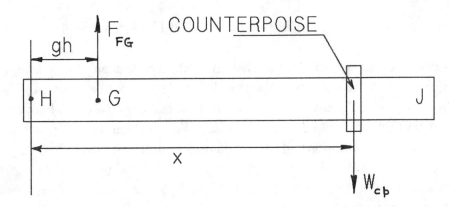

Fig. 7 Free body diagram of the middle lever of the platform scale.

In this application of the moment equation of equilibrium, we select the fixed point H, and write:

$$\sum M_H = 0 \tag{6}$$

The weight of the counterpoise W_{cp} produces a clockwise moment, $(x) W_{cp}$ about point B. The force F_{FG} produces a counterclockwise moment, $(gh)F_{FG}$, about point H. Summing these moments according to Eq. (6) gives:

$$x \, W_{cp} - gh \, F_{FG} = 0$$

Solving this relation for the distance x, substituting the results from Eq. (5), and simplifying gives:

$$x = [ab/(ab + ac)][ed/ef](gh)(W_u / W_{cp}) \qquad (7)$$

The first term in the [ ] brackets is determined by the selection of dimensions for the pins in the lower lever. The second term in the [ ] brackets is determined by the dimensions for the pins in the upper lever. The term in the parenthesizes (gh) is a dimension between pins on the middle lever. The last term W_u / W_{cp} is a ratio of the weights. As a designer you select all of these dimensions and the ratio of the weights. Equation (7) is the relation that guides you in making your selections.

Are you ready for an example to show you the importance of Eq. (7) in the design of a platform scale? Let's suppose that we are to design a platform scale that will weight big fat Jack who is pushing 300 lb. Where do we start? In design, you start by selecting dimensions for the location of the pins, and checking to determine if these are reasonable choices by using Eq. (7). Referring to Figs. 5, 6 and 7, we first select:

ab = 2 in. and ac = 10 in. for the lower lever.
ed = 2 in. and ef = 6 in. for the upper lever.
gh = 2 in. and W_{cp} = 2 lb. for the middle lever.
W_u = 300 lb. on the platform.

Substituting these initial selections in Eq. (7) gives:

$$x = [2/(2 + 10)][2/6](2)(300/2) = 16.67 \text{ in.}$$

This answer for x indicates that we will have to slide the counterpoise along the middle lever to a position where x = 16.67 in. to balance the scale when big fat Jack is on the platform. This means that the middle lever is a reasonable length, and the scale will not take up too much floor space. Our initial selections of dimensions appear to be a good choice. We can layout the levers and pins in our three lever mechanism to these dimensions. If you encounter any difficulties in the layout and need to change dimensions, it is an easy matter to employ Eq. (7) to check the feasibility of the new dimensions. Remember none of the dimensions are fixed, and you have many different choices that will give a mechanism of the proper size for a bathroom scale.

Clearly, the analysis of the three lever mechanism helped us in the selection of dimensions and in determining the weight of the counterpoise. But the analysis is useful for another reason. It permits us to examine the assumptions we made when we wrote the relations leading to Eq. (7). Although not explicitly stated in the analysis, we made two very important assumptions. First, we assumed that there was no friction at the joints, and that the levers would rotate freely at the pins located at points B, E and H. Friction free pins do not just happen by accident. As a designer you must specify a very low friction roller bearing at each of these points.

The second assumption pertained to the weight of the bars and the linkage. We neglected these weights as we wrote the equilibrium relation for the moments acting on all three of the levers. Neglecting these weights was a good assumption for the lower lever and even for the upper lever. However, it was not a good assumption for the middle lever. If the middle lever is fabricated from an aluminum bar ¼ in. thick, 1 in wide, and 20 in. long, it will weigh about 0.5 lb. This weight is not negligible in comparison to the weight of the counterpoise (2 lb. in our example), and it must be accounted for in the analysis.

The fact that we have neglected the weight of the middle lever does not invalidate the approach. However, we do need to modify the analysis to accommodate for the weight of the middle lever. We modify the free body diagram in Fig. 7 and add another term to the equilibrium relation. An exercise is given at the end of this chapter to allow you to make the modification necessary to account for the weight of the middle lever in deriving Eq. (7).

Let's next consider the scale imprinted on the middle lever. When we weigh big fat Jack, we move the counterpoise along the middle lever until the arm moves off the stops and comes into balance in a horizontal position. The indication of weight is provided by the position x of the counterpoise in the balance position. The reading of the weight is made from a calibrated scale imprinted on the middle lever.

When you build your prototype scale there are two approaches that you can follow in preparing a calibrated scale for the middle lever. You can modify Eq. (7) and use the results from this relation to calculate the position x for say 50, 100, 150, 200 and 250 lb. These distances are marked on the middle lever together with corresponding weights.

Another approach for preparing the calibrated scale is experimental. You can add calibrated weights to the platform in 25 or 50 lb. increments, and mark the middle lever at the balance positions as the weights are increased from 50 to 250 lb. A better approach is to use both methods for laying out the scale on the middle lever. Use the theoretical approach [Eq. (7)] for the initial marking of the scale followed by the experiment to check the accuracy of your solution.

The scale should be linear with equal distance between major marks indicating say 10 or 20 lb. increments as shown in Fig. 8. Minor marks are made between the major marks dividing the scale into say 2 or 5 lb. increments. These minor marks determine the resolution of the weight measurement. The resolution of the weighing machine is equal to the incremental weight between the minor marks on the calibrated scale.

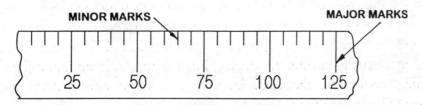

Fig. 8 A calibrated scale showing linearity, major markings and minor markings.

Resolution should not be confused with accuracy. You can improve the resolution by making many minor marks on the scale and reducing the increment between the marks to a fraction of a pound. However, if the joints in the lever mechanism exhibit friction the accuracy of the scale will be poor regardless of the resolution of your scale. The error E is determine by placing reference (standard) weights on your scale and noting the difference between the measured weight and the standard weight. The error E is:

$$E = (W_m - W_R)/W_R \qquad (8)$$

where W_m and W_R are the measured and reference weights respectively.

Of course, you will seek to minimize the error in the measurement of weight over the entire range from 25 to 250 lb.

SPRING BASED WEIGHING MACHINES

The spring based weighing machines are generally much smaller than the platform scales and are much less expensive. A platform scale with a 300 lb. capacity and a resolution of ¼ lb. costs about $250. A spring type bathroom scale typically sells for about $1/10^{th}$ this price. The idea with a spring type scale is to convert the force measurement into a measurement of deflection of the free end of a spring. Of course the deflection must be large enough so that it can be scaled with sufficient resolution and can be easily read from a suitable distance.

The most common type of spring used in mechanical design is a helical spring as depicted in Fig. 9. The coil is formed by winding a small diameter wire about a mandrel to form a larger diameter helix. There are many forms of helical springs including, compression, tension, conical and torsion springs all of which are illustrated in Fig. 9.

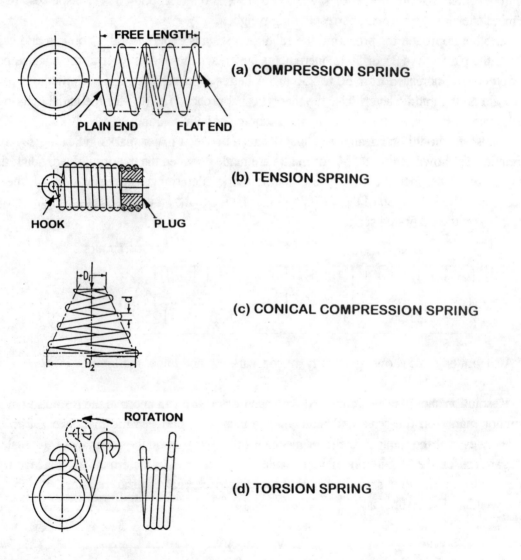

Fig. 9 Examples of helical springs.

The relation between the axial force F applied to the free end of the spring and its deflection d is:

$$F = kd \qquad\qquad (9)$$

where k is the spring rate or spring constant specified in lb./in. or N/m.

The spring constant depends on the material used in fabricating the spring, the diameter of the wire from which the spring was wound, the diameter of the helix, and the number of coils of wire. Designers do not normally fabricate springs because they are commercially available, competitively priced and marketed in all sizes shapes and forms by a number of different companies that are in business to manufacture springs.

You have all seen spring scales. They are hanging at many locations in the produce section of the typical supermarket. Low cost postal scales like the one presented in Fig. 10 use springs for weighing envelops and small packages. The spring scales all have several common features which include:

- A platform to support the unknown mass.
- A link to connect the platform to the free end of the spring.
- A support for the fixed end of the spring.
- A pointer to indicate the deflection of the spring.
- A calibrated scale to convert the deflection into a measurement of weight.
- An adjusting screw to located the pointer at the zero mark on the scale.

We have located these features for you on the photograph of the mechanism of an old fashion postage scale in Fig. 10.

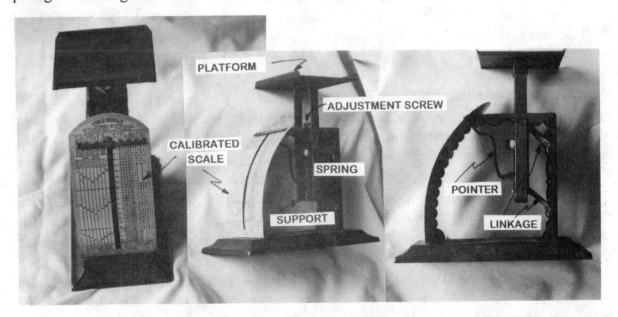

Fig. 10 Postal scale with a soft spring giving a large displacement to indicate the required postage[2].

---

[2] If you note the postage indicated on the scale you will recognize that is an antique. The 3 cent charge for a one ounce letter dates back to the 1940s.

But we are to design a scale for a bathroom not the post office. How does a bathroom scale differ from a postage scale? The bathroom scale sits on the floor and is small and unobtrusive. The platform is stable when we step on it to measure our weight. We can read the display from our standing position. The platform on my postage scale deflects about 1 inch when a packaging weighing two pounds is placed on the platform. That deflection is much too large for a bathroom scale. If the bathroom scale deflected very much, we might lose our balance and fall. A lawsuit would surely follow.

How are we going to limit the deflection of the platform of the bathroom scale? That question is easy to answer. Use a stiff spring (one with a high spring constant k). That solves the stability problem, but if the deflection of the spring is small how are we going to measure such a small deflection and convert it to a weight scale with a reasonable resolution. One problem leads to another and then still another.

Let's look at a scale that already exists. In design it is OK to copy if we do not infringe on a patent. In fact we have coined a word for copying when it comes to design. It is called reverse engineering. If we are designing a new model of a product, we purchase the latest model of all of the competitive products in the market. We test them (this is called benchmarking), and we take them apart and thoroughly inspect the mechanism to gain a complete understanding of all of the features incorporated in the product by our competitors. If you are worried about a patent infringement, conduct a patent search. Remember, patents protect a novel idea but only for 17 years. Spring scales were introduced in the 19th century, and most of the basic patents expired many decades ago.

Do we expect that you will somehow get your hands on a bathroom scale, take it apart and inspect the mechanism? Dissection of mechanisms is a very well established learning technique. We will be very disappointed in you if you miss this opportunity to learn.

Let's assume that you have procured a bathroom scale. (I bought one at a discount store for $9.99 plus tax). When you open the bathroom scale by loosening the two springs that hold together the top and bottom assemblies, you will find a very neat mechanism. First note the very large diameter dial and the four bars that support the platform. These bars transmit the force (weight) applied to the platform to a small spring plate shown in Fig. 11. The spring plate is connected to the free end of a relatively stiff spring. When the weight is applied to the platform the four bars transmit it to the spring plate, the spring extends a small amount, and the plate moves downward. The downward motion of the plate is converted into a horizontal motion and is amplified by a crank shown in Fig. 12. Attached to the crank is a rod, a rack and another small helical spring. When the crank rotates to track the downward motion of the plate, it pulls the rack which engages the pinion and rotates the dial upon which the scale is imprinted. The pointer is fixed to the body of the scale and the dial with its calibrated scale moves below the pointer. A thumb wheel is aligned with the spring and spring plate to permit adjustment of the tension on the spring to zero the scale.

This mechanism converts the very small downward extension of the spring into an amplified horizontal movement. It then converts the horizontal motion into a rotary motion with a rack and pinion gear. By making the diameter of the dial much larger that the diameter of the pinion gear [3], the

---

[3] On my bathroom scale the pinion diameter was only 3/16 in. and the diameter of the dial was 6 in. The ratio 6/(3/16) = 32 is a mechanical amplification of the motion. Note also that the crank affords another opportunity for mechanical amplification.

scale becomes much larger. The large diameter dial affords higher resolution and provides for better visibility.

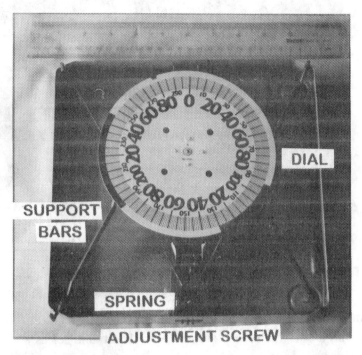

Fig. 11 The mechanism contained in a compact spring type bathroom scale.

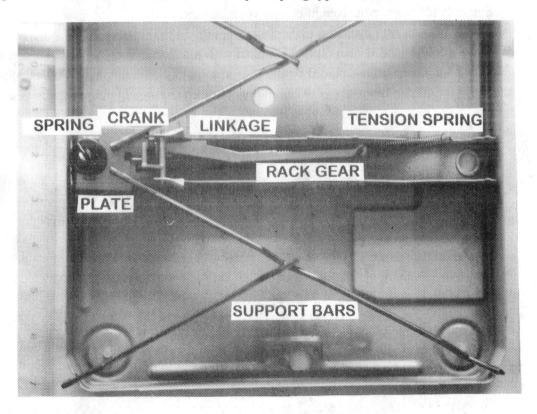

Fig. 12 A rack and pinion gear and a crank used to magnify the displacement of the spring element in a bathroom scale.

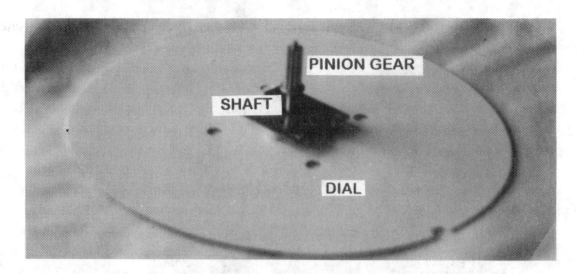

Fig. 13 The pinion gear attached to the dial.

HYDRAULIC BASED WEIGHING MACHINES

Hydraulic weighing machines are based on the idea of converting weight into pressure, and then measuring that pressure with a bourdon tube pressure gage. Hydraulic scales are not found in bathrooms because they are more expensive than the spring type scales. They are not found in physicians offices because they are usually less accurate than platform scales. However, they can be employed very effectively to measure the weight of very large masses.

A schematic illustration of a hydraulic scale, shown in Fig. 14, indicates the simplicity of the mechanism. We have a cylinder filled with oil that is closed with a piston which IN TURN supports the platform. When an unknown mass is placed on the platform, the piston is driven into the cylinder compressing the oil, and creating a pressure p. The relation between the weight W_u and the pressure is given by:

$$W_u = pA_p \qquad\qquad (10)$$

where A_p is the cross sectional area of the piston.

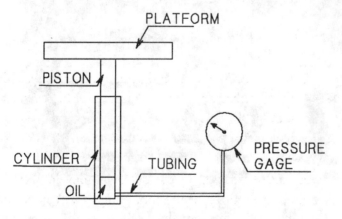

Fig. 14 The primary components used in a hydraulic weight measuring system.

The pressure is transmitted by the oil through a small diameter tube to a convenient location for the convenient location of the pressure gage. At the end of the tube, we place a bourdon type pressure gage which is illustrated in Fig. 15. The pressure gage contains a C shaped bourdon tube that is fixed at one end and free at the other. When the bourdon tube is pressurized it tends to straighten, and its free end moves outward. The distance moved by the free end is proportional to the pressure applied to the C shaped tube. A linkage mechanism with a rack and pinion gear is connected with linkages to the free end of the tube. When the bourdon tube straightens, the shaft of the pinion gear is rotated by the movement of the rack. The pinion is integral with a shaft which is attached to a pointer that sweeps over a dial to indicate the magnitude of the pressure. The dial is fixed to the cylindrical housing of the pressure gage. The dial is calibrated in pressure, however, if the hydraulic system was employed as a weighing machine, the dial would be marked to readout directly in terms of pounds or Newtons.

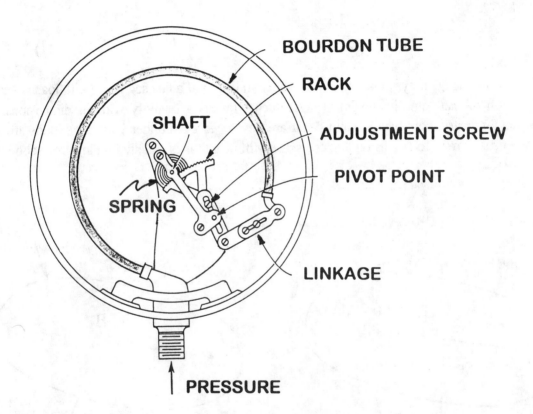

Fig. 15 A view of the mechanism used in a bourdon tube pressure gage.

There is only one difficult problem to overcome in designing a hydraulic weighing machine, namely to effect a low friction seal at the interface between the piston and the cylinder. Common seals like O-rings or chevron packing prevent leakage, but they also exhibit relatively high friction. Perhaps your team can develop a tight low friction method of sealing in the development of a hydraulic weighing machines.

Bourdon tube pressure gages can be adapted to the prototype development at low cost. They are sometimes employed on tire pressure gages or tire foot pumps that can be found in most auto supply stores.

STRAIN GAGES

In the past decade or so, many electronic scales have been introduced to the market. These are premium scales in that they are accurate, usually incorporate a digital display, and are relatively costly. Most of these electronic scales are based on the measurement of a quantity called strain with a strain gage sensor.

Let's introduce the concept of strain, describe how it is measured, and finally indicate a method for incorporating a strain gage in a weighing machine. Let's consider some solid body like the potato shown in Fig. 16. Suppose we place two well defined marks on the potato at points A and B a distance L_0 apart. Now apply some system of forces to the potato and watch it deform. As the potato deforms, the points A and B move and the distance between the points changes to L_f. Applying the forces to the potato has caused it to be strained (it deformed). The magnitude of that strain ε is given by:

$$\varepsilon = (L_f - L_0)/L_0 = \Delta L/L_0 \qquad (11)$$

Strain is a geometric quantity. It is simply the change in length of a line segment (with load) over the original length of that same line segment. As such strain is a quantity without dimensions. In components fabricated from metals, strains that are developed even under very large loads are very small. We usually refer to strains on a micro scale with magnitudes usually ranging from about 100 to 3000 x 10^{-6}.

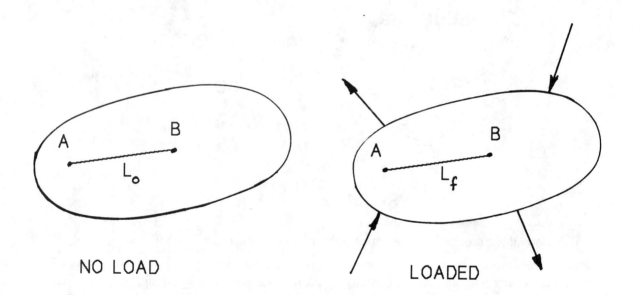

NO LOAD   LOADED

Fig. 16 A solid body deforms under the action of applied loads.

We measure strain by using a metal foil resistor called a strain gage that is shown in Fig. 17. A grid configuration is photoetched from the metallic foil to increase the resistance of the gage, and to give the gage a high directional sensitivity. The foil is very thin (about 200 μin.) and fragile; hence, it requires support. This support is provided by a thin plastic carrier which facilitates handling the gage.

Fig. 17 A metal foil electrical resistance strain gage supported by a thin plastic carrier.

To use the strain gage, we bond it to a mechanical member using an adhesive. The adhesive serves a vital function in the strain measuring system since it must transmit the strain from the mechanical component to the gage sensing element (the foil) without distortion.. The singularly unimpressive feat of adhesively bonding a strain gage to a mechanical component is one of the most critical steps in the entire process of measuring strain with a bonded resistance strain gage.

When bonding a strain gage to a mechanical component, it is important to carefully prepare the surface where the gage is to be mounted. The preparation includes sanding the surface until it is free of all rust or paint, cleaning it with a solvent to remove all traces of oil or grease, and finally to treat it with a basic solution to give the surface the proper chemical affinity for the adhesive.

The bonding process, illustrated in Fig. 18, utilizes a rigid transparent tape for handling and positioning the gage. When the gage is bonded and the adhesive has hardened, the tape is stripped away leaving the gage fixed to the surface of the mechanical component.

The strain gage acts as a variable resistor when subjected to a surface strain in the direction of its grid elements. The change in resistance ΔR is with the strain ε is given by:

$$\Delta R/R_0 = (R_f - R_0)/R_0 = (GF) \varepsilon \qquad (12)$$

where R_0 is the initial resistance of the gage usually 120 or 350 Ω (ohms).

R_f is the final resistance of the gage after the application of a strain ε.

GF is the gage factor or calibration constant for the gage.

ε is the strain in the direction of the gage grid lines.

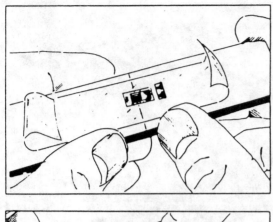

(a) Capture strain gage and solder tab on cellophane tape and position the gage on the beam.

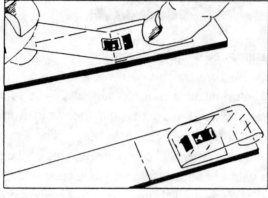

(b) Peel back and fold over the cellophane tape, but maintain its bond on one end.

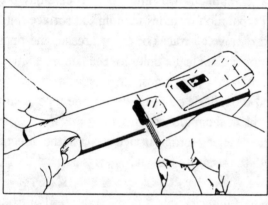

(c) Apply a thin layer of adhesive.

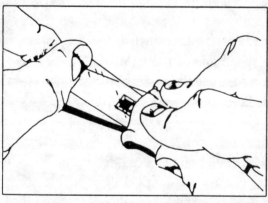

(d) Rotate tape down to reposition the gage and wipe out excessive adhesive.

Fig. 18 Recommended procedures for adhesively bonding electrical resistance strain gages (Courtesy of Micro Measurements)

The strain gage is a sensor that is used to measure strain and to convert that measurement into a change in the electrical resistance. Let's consider an example to show the magnitude of the resistance change that the gage produces under the action of an applied strain. Suppose we have a mechanical component that undergoes a strain of 1200×10^{-6} when subjected to a system of applied forces. If the strain gage has a gage factor GF = 2.00, what is the response of the gage in terms of $\Delta R/R_0$? We use Eq. (12) to determine:

$$\Delta R/R_0 = (GF)\,\varepsilon = (2.00)\,(1200 \times 10^{-6}) = 2.4 \times 10^{-3}\,\Omega.$$

The strain of 1200×10^{-6} is a reasonably large strain, but the response of the gage is very small (only $2.4 \times 10^{-3}\,\Omega$). How do we measure such a small change in resistance? Our ohm meter is not nearly sensitive enough to measure with any accuracy such a small quantity. We use a Wheatstone bridge circuit that was developed by Lord Kelvin way back in 1856 [1]. The Wheatstone bridge is not a structure, but is a simple electrical circuit used to convert very small changes of resistance into changes in voltage. We will describe this circuit in more detail in Chapter 4 when some introductory principles of electrical circuits are introduced.

STRAIN GAGE TRANSDUCERS

Let's next explore techniques for using strain gages to produce transducers. Transducers are mechanical components that incorporate a sensor. The transducer measures a mechanical quantity like force, torque, pressure, velocity, acceleration, etc., and converts that quantity into an electrical quantity. A strain gage transducer incorporates a bonded electrical resistance strain gage as the sensor. Let's explore how to design a strain gage transducer to measure weight and to convert that weight into an electrical output.

A complete transducer system contains five principle components as indicated in Fig. 19. We begin with a mechanical input of some quantity (in our case it is the weight of big fat Jack). The first component is a mechanical element that takes the weight and converts that weight to strain. The second component is a sensor such as a strain gage that converts the strain into some electrical quantity such as $\Delta R/R_0$. The third component is a signal conditioning circuit such as a Wheatstone bridge that converts the resistance change into a voltage change. The fourth component, if needed, is an amplifier to increase the magnitude of the voltage output from the bridge so that it can be displayed on a suitable voltage measuring instrument. The final component is the display which is usually a voltmeter, either digital or analog.

In this section, we will describe techniques to follow in the design of the first two components of the transducer system shown in Fig. 19. The last three components will be described in Chapter 4.

Let's consider the design of a weighing machine similar in size and shape to the spring type scale depicted in Fig. 11. We have a platform that presses down on a four bar support mechanism. These four bars are in turn supported by a simply supported beam as illustrated in Fig. 20. The beam has a span S, and a cross section h high and b wide. The unknown weight W_u is applied at the center of the beam. A strain gage sensor is bonded to the underside of the beam directly beneath the loading point.

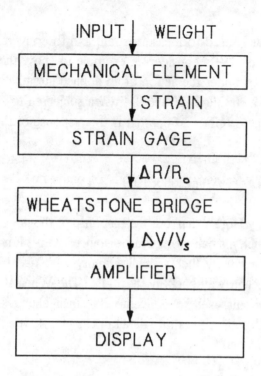

Fig. 19 Block diagram showing the five components of a transducer system.

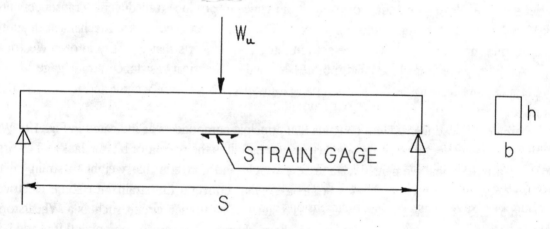

Fig. 20 A simply supported beam with a strain gage sensor.

The weight W_u produces a bending stress σ in the beam that is given by:

$$\sigma = (3\ W_u\ S)/(2\ b\ h^2) \qquad (13)$$

From Eq. (13), we can determine the bending stresses σ, but we are more interested in strain. How do we find the strain if we know the stress? Use Hooke's law which states for a uniaxial member like a beam that:

$$\sigma = E\ \varepsilon \qquad (14)$$

Now if we combine the results of Eqs. (13) and (14) we obtain:

$$\varepsilon = \sigma / E = (3 \, W_u \, S)/(2 \, b \, h^2 \, E) \qquad (15)$$

where E is the elastic (Young's) modulus of the material from which the beam is fabricated. For aluminum $E = 10.6 \times 10^6$ psi or 73 GPa.

OK! We now have a relation for strain in terms of the geometry of the beam (S, b and h), the beam material (E), and the load W_u applied to the center point of the beam. As a designer, how do you make use of this interesting expression. You must make initial selections pertaining to the geometry and material of the beam. We know the weight of big fat Jack, and we know that we would like the strain to be large, but not so large as to cause the beam to fail by yielding. Let's make some initial estimates for the dimensions of our beam, and use Eq. (15) to determine their feasibility.

Select S = 6 in for the span of the beam. Take b = h = ¼ in., and make the beam from a 2024 T4 aluminum alloy with $E = 10.6 \times 10^6$ psi. Remember that big fat Jack weighs 300 lb. Substitute into Eq. (14) to obtain:

$$\varepsilon = (3)(300)(6)(4)^3 / (2)(10.6)(10^6) = 16,300 \times 10^{-6}$$

Let's interpret this result for the strain. We certainly wanted a very large strain, but $16,300 \times 10^{-6}$ may be too much strain for the beam to withstand. Let's check to determine the stress in the beam by using Hooke's law.

$$\sigma = E \, \varepsilon = 10.6 \times 10^6 \times 13,585 \times 10^{-6} = 144,000 \text{ psi}$$

If we check on the strength of our aluminum alloy 2024T4 in a materials handbook, we find that it will yield when the bending stress exceeds 47,000 psi. Our initial selection of the beam's dimensions will cause the beam in our transducer to yield before the maximum load is placed on the scale. We need to change the dimensions of the cross sectional area to reduce the stress and strain imposed on the beam so that it does not fail in service.

Suppose we increase the beam dimensions b = h = ½ in. Repeating the calculation for strain gives:

$$\varepsilon = (3)(300)(6)(2)^3 / (2)(10.6)(10^{-6}) = 2,040 \times 10^{-6}$$

By increasing the values of b and h from ¼ in. to ½ in., we have reduced both the strain and the stress by a factor of eight. The reduced stress is only 21,600 psi which is much lower than the yield strength of 47,000 psi. We have a safety factor (strength /stress) = (47,000/21,600) = 2.17 which is a reasonable value.

If we place a strain gage on the lower side of the beam directly under the point of the application of the load as shown in Fig. 20, we will record a strain of 2040×10^{-6} when the platform is loaded to the maximum capacity of 300 lb. What is the signal output in terms of resistance from this strain gage? From Eq. (12) we determine:

$$\Delta R / R_0 = (GF) \, \varepsilon = (2.00) \, 2040 \times 10^{-6} = 4.08 \times 10^{-3} \, \Omega$$

Clearly the change in resistance of a strain gage is very small. For a gage with an initial resistance R_0 = 350 Ω, the change in resistance ΔR = 1.42 Ω. This value is too small to be measured accurately with a typical ohmmeter. A special circuit called a Wheatstone bridge is employed to convert these small changes in resistance to a voltage that can be measured with a suitable voltmeter. We will return to this result in the next chapter in the section covering the Wheatstone bridge where you will be shown the methods for determining the voltage output from the bridge.

REFERENCES

1. Thomson, W. (Lord Kelvin): "On the Electrodynamic Qualities of Metals," Proceedings. Royal Society, 1856.
2. Beer, F. P. and E. R. Johnston, Jr., Mechanics of Materials, 2nd Edition, McGraw-Hill, NY, 1992.
3. Dally, J. W., Riley, W. F. and K. G. McConnell, Instrumentation for Engineering Measurements, 2nd Edition, John Wiley, New York, 1993.
4. Dally, J. W. and W. F. Riley, Experimental Stress Analysis, 3rd Edition, McGraw-Hill, New York, 1991.

EXERCISES

1. Describe the difference between analog and digital displays.
2. Why is the simple balance a poor choice as a design concept for a bathroom scale?
3. Suppose you design a platform scale with a mechanism with dimensions:

    ab = ed = gh = 1 in.
    ac = 8 in., ef = 5 in.

    Find the ratio of weights W_u / W_{cp} if x is to be equal to 14 in.
    If W_u is 220 lb. when x = 14 in., determine W_{cp}.

4. Draw a calibrated scale 300 mm long with a range from 0 to 1000 N. Use 100 N for the major marks on the scale, and 10 N for minor marks on the scale. What is the resolution of the scale?
5. You place a reference weight of 125 lb. on a platform scale, and measure a weight of 126.3 lb. Determine the error of the measurement.
6. You design a scale with a helical spring that is subjected to the maximum weight of 250 lb. If the spring has a spring constant of 1000 lb./in., what is its deflection?
7. A pinion gear has a 10 teeth. If the teeth on the mating rack gear have a pitch of 1 mm, determine the arc of rotation of the pinion gear if the rack moves through a distance of 6 mm.
8. For the rack, pinion dial mechanism of exercise 7, determine the mechanical amplification of the movement of the rack gear if the dial has diameter of 150 mm.
9. You decide to develop a hydraulic scale using a piston with a diameter of ¾ in. Determine the pressure in the oil if the platform is loaded with a weight of 300 lb.
10. A gage length L_0 = 20 mm is scribed on a component fabricated from an aluminum alloy 2024 T4. If this component is subjected to a system of forces and the new gage length becomes L_f = 20.006 mm, find the strain and the stress at the location of the scribed line.
11. Will the component described in exercise 10 fail by yielding?

12. What assumption did you make in exercise 10 when you converted strain to stress by using Hooke's law?

13. An electrical resistance strain gage with a resistance of 350.00 Ω, and gage factor of 2.06 is subject to a strain of 0.000900. Determine new resistance of the strain gage.

14. Describe the four components usually found in a transducer measuring system. Relate these four component to a measuring system on your automobile

15. For the beam shown in Fig. 20, find the stress and strain if W = 110 lb., b = h = 0.1875 in., and S = 5 in. Assume that the beam is fabricated from an aluminum alloy 2024 T4.

NOTES

CHAPTER 4

ELECTRICAL AND DIGITAL ASPECTS OF WEIGHING MACHINES

INTRODUCTION

In the previous chapter, we found that a spring could be used to convert an applied force (weight) into a displacement. The conversion was beneficial because we measure the displacement with a scale, and eliminate the reference weights that are necessary when we weigh objects with a balance. In this chapter, let's continue to employ a spring to convert force into displacement. However, instead of measuring the displacement with a scale, we will convert it again to a voltage and then measure it with a voltmeter.

The double conversion system from force to displacement to voltage is depicted in Fig. 1. Note, the system begins with a mechanical component (the spring), but the second component (electrical) changes the technology of the system from mechanical to electrical. Because the system described in the block diagram of Fig. 1 has both mechanical and electrical components, it is called a hybrid. We will describe this hybrid system in detail later in this chapter, and suggest some ideas to aid you in your efforts to design the prototype for a weighing machine.

Another hybrid system incorporates a strain gage as the electrical component to convert strain (a mechanical quantity) into a change in resistance (an electrical quantity). We have already covered this topic in Chapter 3, but we left you with a resistance change $\Delta R/R_0$ that was too small to measure with a common ohmmeter. To deal with a very small $\Delta R/R_0$, we employ a signal conditioning circuit called a Wheatstone bridge. It is a very useful circuit to convert small resistance changes into voltage changes. We will describe the Wheatstone bridge and show the relation between $\Delta R/R_0$ and the voltage change $\Delta V/V_s$.

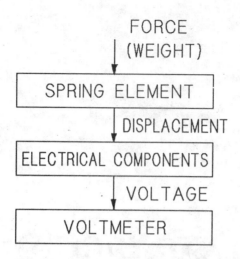

FORCE
(WEIGHT)
↓
SPRING ELEMENT
DISPLACEMENT
↓
ELECTRICAL COMPONENTS
VOLTAGE
↓
VOLTMETER

Fig. 1 Block diagram showing a hybrid system containing both mechanical and electrical components.

As we describe these electrical circuits, it will become necessary to introduce you to some basic concepts including voltage sources, voltage, current, and passive electrical components. We will deal only with direct current (dc) in introducing the simple laws governing current flow in circuits.

Finally, we will introduce some very elementary concepts in digital electronics by covering the binary code and the digitization process. We will introduce the analog to digital converter and describe simple digital displays.

BASIC ELECTRICAL CONCEPTS

Electricity involves charge carriers either electrons which are negative charge carriers or holes which are positive charge carriers. Since we cannot see, smell or touch an electron, electricity has always been somewhat mysterious to most folks. Holes, which are atoms with an electron missing from its outer ring, are even more obscure. Nevertheless, we will try to introduce many of the basic electrical concepts in a simple straight forward manner and avoid the mystery.

THE BATTERY

Let's start with a source of electricity, namely a battery which is shown in Fig. 2. This is a wet cell type battery which consists of two metal bars that are submerged in a liquid electrolyte. The electrolyte is an acid ($H_2 SO_4$) which attacks both of the metal bars. The zinc gives off positive ions (Zn^{++}) to the acid which leaves the metal electrode with a negative charge. The copper gives off negative ions (Cu^{-}) to the acid which in turn leaves it with a positive charge. If we attach a voltmeter[1] across the two metal bars (electrodes), we measure a voltage that corresponds to the

---

[1]The voltmeter must have a high input impedance to limit the current flow during the measurement, otherwise the voltage measurement will be in error.

difference in the charges that have developed on the Zn and the Cu electrodes. This voltage is 0.34 - (-0.76) = 1.10 V (volt). The electrode potential of Cu is 0.34 volts and that of Zn is -0.76 volts.

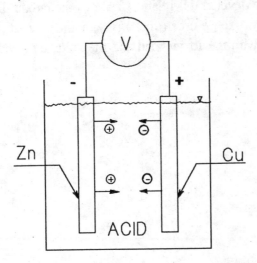

Fig. 2 A Cu - Zn wet cell (battery) that serves as a voltage source.

The battery is a voltage source. It converts chemical energy into electrical energy. We can generate higher voltages by connecting several individual battery cells in a series arrangement. We can also select other metals with higher electrode potentials in the electromotive series to use in the fabrication of the cells to generate higher voltages.

Let's change the connections on our battery, and add a resistor R in a circuit from the Zn electrode to the Cu electrode as shown in Fig. 3. What happens? We have completed the circuit. The battery with an open circuit was a voltage source, but the battery with a closed circuit becomes a current source. An electric current (I) flows from the positively charged Cu electrode through the resistor to the negatively charged Zn electrode. The electrons (charge carriers) flow from the negative charge to the positive charge as shown in the inset in Fig. 3.

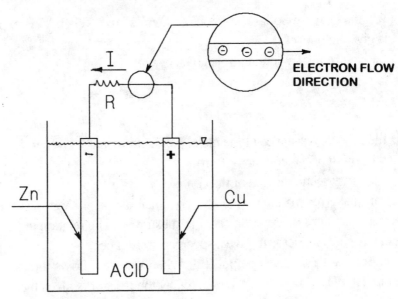

Fig. 3 A battery with a resistive load connection showing the opposite directions of current flow and electron flow.

OK! We understand that a battery is a voltage source which supplies current when connected to a closed circuit. Let's now examine the relation between the voltage provided by the battery and the current that flows in the circuit. To develop this relation, refer to the circuit diagram presented in Fig. 4. Note the symbols used for the voltage source (the battery) and the resistor R. We depict the current (I) flowing through the circuit with the arrow head and the voltage provide by the battery by the symbol (V_s).

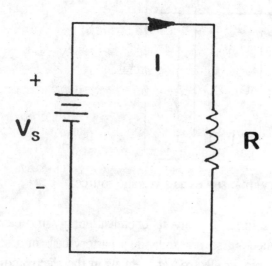

Fig. 4 A simple circuit loop containing a voltage source (battery) and a resistor.

The relation between the applied voltage and the current flow (I) is given by:

$$V_s = I R \qquad (1a)$$
$$I = V_s / R \qquad (1b)$$

Equations (1a) and (1b) are both expressions of Ohm's law. When V_s is expressed in volts, and I in amperes, R is given in ohms Ω.

The resistance R of a conductor is given by:

$$R = \rho L / A \qquad (2)$$

where ρ is the specific resistance of the metal conductor usually given in Ω-cm.
L is the length of the conductor (cm).
A is the cross-sectional area of the conductor (cm^2).

We can make resistors by winding very-small-diameter, high-resistance wire on coils or by evaporating very thin films of metal on ceramic substrates. The relation in Eq. (2) permits us to select the metal and size the conductor to give a specified resistance.

A resistor is a dissipative component. When a current flows through a resistor power is lost in the form of heat. The power that is dissipated by the resistor is given by:

$$P = V_d I \qquad (3)$$

where   P is the power expressed in Watts (W).

V$_d$ is the voltage drop across the resistor.

Substituting Eqs. (1a) and (1b) into Eq. (3), we obtain:

$$P = I^2 R = V_d^2 / R \qquad\qquad (4)$$

We have used subscript s with the voltage in Eqs. (1a) and (1b) and the subscript d with the voltage in Eq. (4). For the simple circuit shown in Fig 4 the supply voltage V$_s$ is equal to the voltage drop across the resistor V$_d$. In this case the subscripts can be interchanged. Clearly, it is easy for us to determine the power lost if we know the current flowing through a resistor or the voltage drop across the resistor.

Let's consider another passive electrical component, namely a capacitor as shown in Fig. 5. The capacitor consists of two flat, parallel plate electrodes separated by a dielectric that also serves as an electrical insulator. The capacitance C of a flat plate capacitor is given by:

$$C = \varepsilon \, A/h \qquad\qquad (5)$$

where ε is the dielectric constant.

A is the area of the plates.

h is the distance between the two plates as shown in Fig. 5.

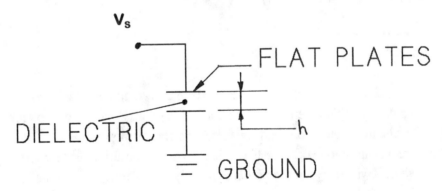

Fig. 5 A flat parallel plate capacitor.

When a voltage V is applied across the capacitor, it stores a charge q that is given by:

$$q = C \, V \qquad\qquad (6)$$

where C is the capacitance given in farads and denoted with the symbol F.

q is the charge on the capacitor given in coulombs.

As the capacitor is charged, it becomes a storage device. The electrical energy stored is:

$$\mathcal{E} = C \, V^2 / 2 \qquad\qquad (7)$$

where $\mathcal{E}$ is the energy in Joules.

When a voltage is first applied to the capacitor current flows and the capacitor becomes charged. However, when it is fully charged and maintained at a constant voltage, no current flows through the capacitor. The capacitor conducts current only when the voltage changes with respect to time.

The final passive component important in electrical systems is the inductor shown in Fig. 6. We make an inductor by winding a coil of small diameter insulated wire about a core. The inductance L is given by:

$$L = 4\pi N^2 \, A\mu/l \times 10^{-9} \tag{8}$$

where   L is the inductance in Henrys
       N is the number of turns of wire in the coil.
       μ is the permeability of the media within the coil.
       A is the cross sectional area of the coil.
       l is the length of the coil.

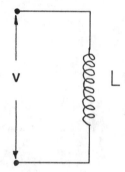

Fig. 6 An inductor coil.

A solenoid is a coil that is commonly used to develop a magnetic field that in turn is used to produce a force on a plunger that extends into the core of the coil. You have a solenoid in your auto that engages a relay which switches your battery across the starter motor.

In dc circuits an inductor coil acts like a small resistor and has little effect on the current flowing in the circuit. However, when the coil is exposed to a current that fluctuates with respect to time there is a significant voltage drop across the coil which is given by:

$$V_d = L \, (dI/dt) \tag{9}$$

where (dI/dt) is the rate of change of the current with respect to time.

KIRCHHOFF'S LAWS

We introduced Ohm's law with a simple circuit loop containing a voltage source and a resistor as shown in Fig. 4. This is a good beginning, but what happens when we begin to add components to the circuit? We need to develop a few more tools to use in analyzing slightly more complex circuits. Let's begin by adding one additional resistor to the simple circuit loop as shown in Fig. 7a, and determine the relation for the current flowing in the circuit.

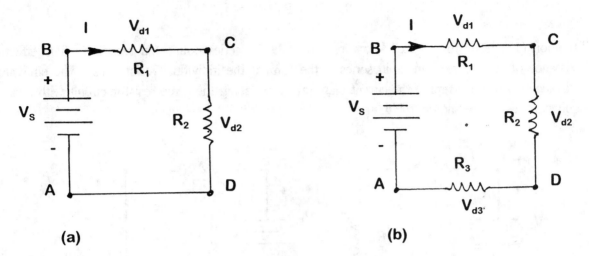

Fig. 7 Circuit loops containing two or three resistors.

To approach this problem, we need to use Kirchhoff's voltage law that states that the sum of the voltages changes about the circuit loop is zero. If we refer to the drawing of the circuit shown in Fig. 7a, we can write Kirchhoff's law in an equation format as:

$$+ V_{A-B} - V_{B-C} - V_{C-D} - V_{D-A} = 0 \qquad (10)$$

In the interpretation of Eq. (10), we move in the direction of the current flow (clockwise in this case). As we move from point (node) A to B, we pass through the battery which increases the voltage at B relative to A. This is the reason for the + sign shown on the term V_{A-B}. Continuing around the circuit loop, we proceed from node B to C and pass through resistor R_1. There is a voltage drop V_{d1} as the current passes through this resistor equal to $I\,R_1$. The fact that the voltage drops across R_1 is the reason for the minus sign on term V_{B-C}. Moving onto the leg of the circuit between node C and D, we encounter a second voltage drop $V_{d2} = I\,R_2$. Because the voltage decreases as we move from node C to D, we assign a minus sign to the term V_{C-D}. Finally, we consider the final leg of the circuit between nodes D - A. We have a perfect conductor between these two points with no loss or gain of voltage. For this reason $V_{D-A} = 0$. Substituting these four voltage changes into Eq. (10) gives:

$$V_s - I\,R_1 - IR_2 - 0 = 0$$
$$I = V_s / (R_1 + R_2) \qquad (11)$$

If we apply the same analysis techniques to the circuit with three resistors that is shown in Fig. 7b, it is easy to add another term and write:

$$V_s - I\,R_1 - IR_2 - IR_3 = 0$$
$$I = V_s / (R_1 + R_2 + R_3) \qquad (12)$$

Examine Eqs. (11) and (12), and note that the resistors in series in the circuit add together to give an equivalent resistance R_e given by:

$$R_e = R_1 + R_2 + R_3 = \sum R \qquad (13)$$

Equation (13) leads to the well know resistor rule for series connected resistors. The effective resistance of several resistances in series is the sum of the individual resistances. The equivalent resistance is used to simplify the circuit diagram by replacing the three resistor circuit with a single equivalent resistor as shown in Fig. 8.

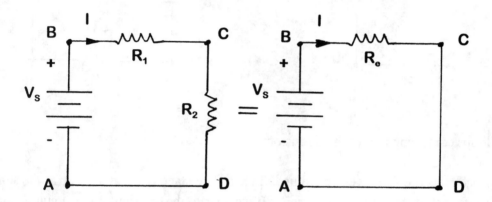

Fig. 8 An equivalent circuit with an equivalent resistor replaces a more complex circuit.

But what if the resistors are not all connected in a series arrangement? What if we have a parallel arrangement of the resistors such as that shown in Figs. 9a and 9b? To analyze this type of a circuit we need to introduce Kirchhoff's current law. This is a simple law that states that the sum of the currents flowing into a node must be equal to the sum of the currents flowing from that node. Let's refer to the node X presented in Fig. 10 with currents flowing into and out of this node. Kirchhoff's current law permits us to write:

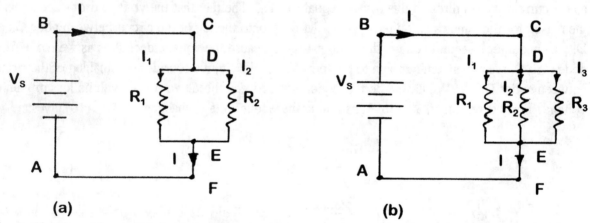

(a)   (b)

Fig. 9 Circuits with parallel arrangements of resistors.

$$I_1 + I_2 - I_3 - I_4 = 0 \qquad (14)$$

To deal with the resistors in parallel, as shown in Fig. 9a, we will apply Eq. (14) to the current flow at node D and write:

$$I - I_1 - I_2 = 0 \qquad (a)$$

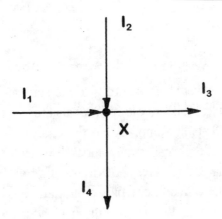

Fig. 10 Node X with current flows in and out.

Next consider Kirchhoff's voltage law Eq. (10), and apply it to the circuit in Fig. 9. It is clear then that the voltage gain due to the battery is offset by the voltage drop across the two parallel resistors between points D - E. We write this fact in an equation:

$$V_s - V_{D-E} = 0 \qquad\qquad (b)$$

We know that the current splits at node D, and I_1 flows through R_1 and I_2 flows through R_2 to produce the voltage drop V_{D-E}. This knowledge permits us to write:

$$V_{D-E} = I_1 R_1 = I_2 R_2 = V_s \qquad\qquad (c)$$

Equation (c) gives:

$$I_1 = V_s / R_1 \qquad and \qquad I_2 = V_s / R_2 \qquad\qquad (d)$$

Following the form of Eqs. (d), we write:

$$I = V_s / R_e \qquad\qquad (e)$$

where R_e is the equivalent resistor that replaces the two parallel resistors in Fig. 9a to give the same equivalent circuit as shown in Fig. 8.

Let's substitute Eqs. (d) and (e) into Eq. (a) to obtain:

$$V_s/R_e = V_s [(1/R_1) + (1/R_2)] \qquad\qquad (f)$$

Eliminating V_s from Eq. (f) gives:

$$1/R_e = (1/R_1) + (1/R_2) \qquad\qquad (15a)$$

or
$$R_e = (R_1 + R_2) / (R_1 R_2) \qquad\qquad (15b)$$

If we find three resistors in parallel, we can follow the same procedure in the analysis of the circuit and show that the equivalent resistance for the three parallel resistors is:

$$1/R_e = (1/R_1) + (1/R_2) + (1/R_3) \qquad (16)$$

While resistors connected in series were summed to give the equivalent resistance, resistors in parallel follow a different rule. For parallel resistors the reciprocals are summed and set equal to the reciprocal of the equivalent resistance. This fact implies that the equivalent resistance for a parallel arrangement of resistors is less than that of the smallest individual resistance.

OK! Hopefully you recognize that Ohm's law and Kirchhoff's two laws can be employed to analyze circuits. Also you should be able to take a complex circuit and use equivalent resistances to reduce the circuit to its most elementary form. We will use these analytical tools together with a few passive electrical components to develop a circuit that can be used to convert a displacement into a voltage.

THE POTENTIOMETER

A potentiometer is a variable resistor. The simplest way of making a potentiometer is to take a length l of a very high resistance wire and attach it across a voltage source such as a battery. We provide a wiper (an electrical contact) which slides along the length of the wire as shown in Fig. 11. The wiper divides the wire and the resistance of the wire R_w into two parts. The resistance R_x, defined in Fig. 11 is a function of the position x of the wiper, and is given by:

$$R_x = (x/l)\, R_w \qquad (17)$$

The current flowing through the potentiometer wire is determined from Eq. (1b) as:

$$I = V_s /R_w \qquad (a)$$

The voltage drop V_x across the resistance R_x is obtained from Eqs. (17), (1a) and (a) as:

$$V_d = (x/l)V_s = C_p\, x \qquad (18)$$

where $C_p = V_s/l$ is a constant depending on the slide wire length and the supply voltage V_s.

Examination of Eq.(18) show that we have a means for converting a displacement x into a voltage. This is an interesting fact and has application in the development of your weighing machine. Suppose you develop a spring element with a spring rate k that undergoes a displacement d so that:

$$W = kd \qquad (19)$$

where W is the weight producing the displacement d of the spring.

If you couple the free end of the spring to the wiper on the slide wire resistor together with a linkage, you have a relationship between the displacement of the spring d and the position of the wiper x. Let's design the linkage so that:

$$x = K_L\, d \qquad (20)$$

where K_L is the amplification of the motion provided by the linkage mechanism.

We have all the pieces in Eqs. (18), (19) and (20). Let's combine them to find:

$$V_d = (C_p K_L /k)W \qquad\qquad (21)$$

Equation (21) is your design relationship. It indicates that you can generate a voltage V_d that is linearly proportional to the weight W on the platform of our weighing machine. The constant of proportionality is the term $(C_p K_L /k)$. The factors C_p, K_L, and k in the constant of proportionality are all determined by your selections of components that you use in the design of the prototype of the weighing machine. You have a lot of decisions to make in the process of designing a scale based on a potentiometer sensor; however, Eq. (21) will guide you in the selection of the individual components and their sizes.

You have a voltage V_d that is proportional to the weight. What do you do with the voltage to give a measurement of weight? A voltmeter will measure the voltage. If you have an analog voltmeter, all you have to do is to remove the cover and modify the scale so that it reads out in weight instead of voltage.

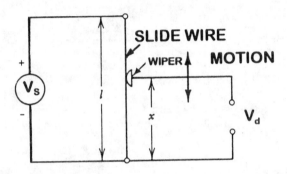

Fig. 11 A slide wire resistance potentiometer.

The simple slide wire resistor, shown in Fig. 11, is easy to understand. Unfortunately it is rarely used because the specific resistance of common metals is too low. For instance, if we used a constantan (45% Ni and 55% Cu) wire with a diameter of 0.010 in. for the potentiometer, the resistance per unit length of the wire would only be 0.25 Ω/in. A potentiometer made with a single strand of this wire would have a resistance much too low to be useful. If the resistance applied across the terminals of the battery is too low, the current drawn from the battery becomes excessively high, and the life of the battery is prohibitively short. To alleviate this difficulty, high-resistance, wire wound potentiometers are manufactured by winding the resistance wire around an insulating core as shown in Fig. 12. The potentiometer presented in Fig. 12a is used for linear displacement measurements.

Cylindrically shaped potentiometers, similar to the one illustrated in Fig. 12b, are used for angular measurements. Of course, cylindrical potentiometers can be used to measure linear displacements by converting their rotary motion to a linear motion. An example of a conversion method is illustrated in Fig. 13 where a pulley is attached to the shaft of the potentiometer. A small diameter wire is wrapped around the pulley and led back to a plate that is fixed to the free end of a spring. As the weight W deflects the spring, the linear motion is transmitted to the pulley to rotate the potentiometer.

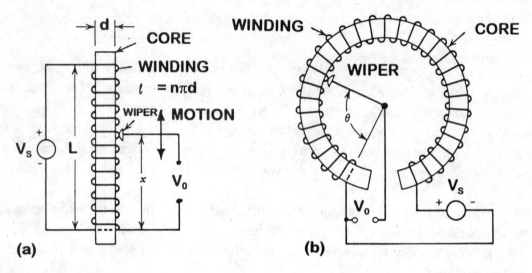

Fig. 12 Wire wrapped resistance potentiometers with (a) linear core and (b) cylindrical core.

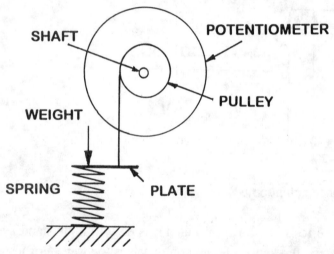

Fig. 13 A mechanism for converting a cylindrical potentiometer into a linear motion sensor.

The resistance of commercially available potentiometers range from 10 to about 10^6 Ω depending on the selection of the wire material, its diameter and the length of the coil. If we are to employ a potentiometer in the design of our weighing machine, we want the resistance of the potentiometer (its input impedance) to be relatively high to reduce the current drain on the battery. For example, suppose you have a potentiometer with a total resistance of 100,000 Ω, and a battery with a voltage of 10V. The current flowing through the potentiometer is determined from Eq. (1b) as:

$$I = V_s / R_p = 10/100,000 = 0.10 \text{ mA}$$

This result indicates that for every 1 Ampere-hour capacity of the battery we can anticipate a life of 10,000 hours. Selecting a high resistance potentiometer clearly enhances the battery life.

THE WHEATSTONE BRIDGE

   In chapter 3, we described the strain gage as a sensor for converting strain into a resistance change. We also indicated that the change in resistance of a metal foil strain gage was so small that it could not be measured with sufficient accuracy by using an ohmmeter. This problem was solved by Lord Kelvin when he invented the Wheatstone bridge for the measurement of small changes of resistance ΔR.

   Let's examine the Wheatstone bridge circuit that is shown in Fig. 14. The bridge consists of four resistors arranged in the shape of a diamond. We have identified these resistors as R_1, R_2, R_3 and R_4. A dc voltage supply is connected across node A and C with the positive terminal connected to point A. A high impedance voltmeter is connected across nodes B and D to read the output voltage V_0 from the bridge.

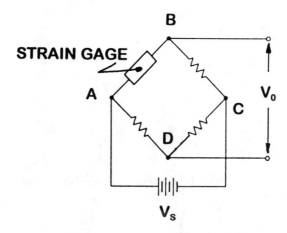

Fig. 14 A Wheatstone bridge circuit showing an active strain gage in arm R_1.

   If we select the four resistances such that:

$$R_1 R_3 = R_2 R_4 \tag{22}$$

The output voltage V_0 will be zero and the bridge is said to be in perfect balance. Equation (22) describes the condition of the resistors necessary to achieve this balance. This is an important relation because we will always initially balance the bridge to achieve a zero output voltage prior to applying a strain to the strain gage.

   Next, let's place a strain gage in arm R_1 of the bridge between nodes at A and B so that:

$$R_1 = R_g \tag{23}$$

where R_g is the resistance of the strain gage (usually 120 or 350 Ω).

   Suppose we load the mechanical member upon which the strain gage is bonded and it undergoes some strain ε. The strain gage responds by changing its resistance by an amount ΔR_g given by:

$$\Delta R_g / R_g = (GF) \, \varepsilon \tag{24}$$

Where GF is the gage factor or calibration constant for the strain gage(about 2.0 for most metal foil gages).

As the strain gage changes its resistance, the bridge goes out of balance and an output voltage V_0 occurs across the nodes B - D. This output voltage is given by:

$$V_0 = \frac{R_g R_2}{(R_1 + R_2)^2} \frac{\Delta R_g}{R_g} V_s \tag{25}$$

Let's simplify the design of the bridge by selecting all four of the resistors with the same value so that:

$$R_1 = R_g = R_2 = R_3 = R_4 \tag{26}$$

Substituting Eqs. (26) into Equation (25) gives:

$$V_s = \frac{V_s}{4} \frac{\Delta R_g}{R_g} \tag{27}$$

Finally, by substitute Eq. (24) into Eq. (27) to give a relation for the voltage output from the bridge in terms of the strain applied to the gage.

$$V_0 = (V_s /4)(GF)\varepsilon \tag{28}$$

OK! The Wheatstone bridge converted the resistance change from the strain gage to a voltage V_0. Let's consider an example and determine the output voltage from the bridge. Suppose we select a strain gage with a gage factor (calibration constant) equal to 2.00, and apply that gage to a beam that exhibits a strain of 0.001 when subjected to a load of 240 lbs. Moreover, we use a six volt battery to power the bridge. What output voltage will be measured if we attach a high impedance voltmeter across the terminals at nodes B - D of Fig. 14?

The solution is given by Eq. (27) as:

$$V = (6)(2.00)(0.001) = 12 \text{ mV}$$

The value of 12 millivolts for the full load applied to our weighing machine may be too small to provide an accurate measurement of the 240 pound weight. The adequacy of the output voltage depends on the sensitivity of the voltmeter. If the voltmeter can provide a measurement accurate to $\pm$ 0.1 mV, then an output voltage of 12 mV gives 120 distinct readings. For a 240 lb weight the increment between readings is (240 lb)/120 = 2 lb. The accuracy of $\pm$ 0.1 mV in the voltage measurement corresponds to an accuracy of $\pm$ 2 lbs in measuring the weight. In terms of percent, the accuracy A in the weight measurement is given by:

$$\mathcal{A} = (100)(W_m - W_T)/W_T \qquad (29)$$

where W_m is the measured weight.

 W_T is the true weight.

Applying Eq. (29) to our example gives the accuracy of the weight measurement as:

$$\mathcal{A} = (100)(\pm 2)/(240) = 0.83 \%$$

If we want to measure the weight with more accuracy we either increase the output voltage from the bridge or increase the sensitivity of the voltmeter. We can increase the sensitivity of the voltmeter, but higher sensitivity voltmeters cost more. We can increase the output from the bridge by increasing the battery voltage or by increasing the strain on the gage. However, the increases which can be achieved are usually relatively small. One must expect less than 100 mV or less output from strain gage bridges even under maximum load conditions. To increase the output voltage substantially requires the use of a voltage amplifier. We will discuss these approaches in more detail in the next section.

VOLTMETERS

There are two different types of voltmeters. The first is the analog meter with its calibrated scale and a pointer. The second is the digital meter with a digital readout of the voltage often on a liquid crystal display. Both of these meters are available commercially as multi-meters providing a means of measuring current and resistance in addition to ac and dc voltages.

Let's consider the analog meter which incorporates a D'Arsonval galvanometer as the sensor element as shown in Fig. 15. The galvanometer has a coil that is supported in a magnetic field with either jeweled bearings (low friction) or torsion springs. When a current flows through the coil, it rotates in the magnetic field until restrained by springs that are part of the suspension system. The coil reaches an equilibrium position and the pointer indicates the reading of the quantity being measured.

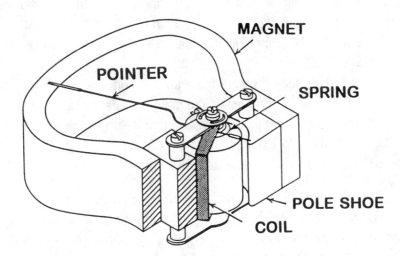

Fig. 15 The D'Arsonval galvanometer used as the sensing movement in analog voltmeters.

The galvanometer is converted into a voltmeter by placing a resistor in series with the coil to control the current flow. The range and sensitivity of the voltmeter is varied by changing this resistance. Typical analog voltmeters have several scales with full scale reading from 100 mV to 1000 V. These are low cost instruments ($10 to $50) with accuracies of ± 2 to 3 % of full scale. When operated on the lower voltage ranges, the input impedance of the voltmeter is relatively low and loading errors can occur because of the current that passes through the instrument. The input impedance for a typical analog voltmeter (without amplifier) in terms of ohms per volt full scale varies form about 2000 to 20,000 Ω/V.

What is a loading error? Do we have to deal with it? You do if you are going to use an analog meter and try to measure small (tens of millivolts) voltages. When we analyze a circuit like the Wheatstone bridge and develop equations for the output voltage, we assume that the voltmeter used to measure V_0 does not draw current from the bridge. Unfortunately the analog voltmeters with the D'Arsonval galvanometers require a significant current to activate the movement.

The error E due to the current draw through the meter is given by:

$$E = \frac{R_s / R_{Imp}}{1 + R_s / R_{Imp}} \qquad (30)$$

where R_{Imp} is the input impedance for the meter.

R_s is the impedance of the voltage source.

Let's consider an example to illustrate loading error. Suppose that the Wheatstone bridge is comprised of four resistors each 350 Ω, and we attempt to measure its output with a voltmeter with an input impedance of 3500 Ω. Since the bridge exhibits a source impedance $R = 350$ Ω, the ratio of the source resistance to the input impedance resistance of the voltmeter is $R_s / R_{Imp} = 350/3500 = 0.1$. From Eq. (30) we determine the loading error due to voltmeter as:

$$E = 0.1/(1 + 0.1) = 0.091 = 9.1 \%$$

This example shows that the voltmeter in this case introduces a large error. Too large for our project. How do we reduce the loading error due to the voltmeter? In our design of the circuits and the selection of the voltmeter, we must make certain that the ratio $R_s / R_{Imp} < 1/100$. In other words, if the input impedance is more than 100 times the impedance of the voltage source, then the loading error will be less that one percent.

Digital voltmeters differ in many respects from analog voltmeters. First, a low resistance galvanometer is not involved, and as a consequence the input impedance of the digital voltmeters is very high. The digital voltmeter shown in Fig. 16, purchased at Radio Shack for $24.95 (plus tax for the governor), has an input impedance of 10 $M\Omega$. This impedance is so large ($M\Omega$ is common) that loading errors are not an issue with modern digital instrumentation.

Fig. 16 Photograph of an inexpensive pocket size digital multi-meter.

AMPLIFIERS

Amplifiers are active devices which are used in conditioning electrical signals. An amplifier is employed in nearly every instrument system to increase small voltages from a transducer to a level sufficiently large for recording with a suitable voltmeter. The symbol for an amplifier is shown in Fig. 17. The input to the amplifier is denoted as V_I, and the output is V_0. The ratio of the V_0 / V_I is called the gain G of the amplifier. As the input voltage is increased the output voltage increases in the linear range of the amplifier according to:

$$G = V_0 / V_I \qquad (31)$$

Amplifiers with gains of 10, 100 or even 1000 are common and relatively inexpensive. However, an instrument amplifier with a gain of your choice probably will not be available at your local electronics supply store. They are available from more specialized suppliers such as Analog Devices of Norwood, MA. Operational amplifiers are available at most electronic supply outlets for very modest costs. They can be used in simple circuits to build an instrument amplifier with stable gains that will meet your specifications.

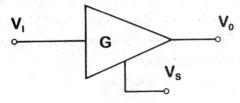

Fig. 17 Symbol representing an amplifier.

An operational amplifier (op-amp) is an integrated circuit with its components placed on a small silicon chip to form a complete amplifier circuit. The gain of an op-amp is extremely high (10^5

to 10^6), but unfortunately the gain is not stable. Also with such a large gain almost any input voltage will be so large as to saturate the amplifier and drive it into its non-linear region. Neither of these facts is good news if you try to use the op-amp directly as an instrument amplifier. Its gain is too high and it is unstable.

To avoid these difficulties we use the op-amp as a circuit element in an instrument amplifier. We can build many different electronic instruments with op-amps as the central element including, instrument amplifiers, differential amplifiers, inverting and non-inverting amplifiers, integrating amplifiers, filters, etc. Let's examine two of these amplifiers that you may find useful in developing the prototype of the weighing machine.

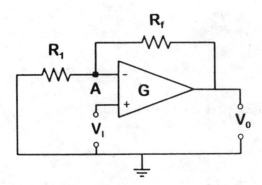

Fig. 18 Circuit diagram for a non-inverting instrument amplifier that utilizes an op-amp as the central electronic component.

The circuit diagram shown in Fig. 18 is that of a non-inverting amplifier. Non-inverting means that the output voltage V_0 is the same sign as the input voltage V_I. Two resistors are employed with this simple circuit. Resistor R_1 is on the lead to the negative pin of the op-amp, and R_f is a feedback resistor that returns current to the node A. An analysis of this circuit for the non-inverting amplifier indicates that the output voltage is given by:

$$V_0 = \left(1 + \frac{R_f}{R_1}\right)V_I = G_c V_I \tag{32}$$

Note that the gain G_c of the circuit is given by the selection of the resistors R_1 and R_f as:

$$G_c = [1 + (R_f/R_1)] \tag{33}$$

Suppose that you select $R_1 = 0.1$ M and $R_f = 1$ M, then the gain $G_c = 11$. The gain of the non-inverting amplifier will be stable even if the gain of the op-amp does fluctuate with time. The circuit (amplifier) gain depends only on the ratio of the two resistors, and is essentially independent of the gain of the op-amp.

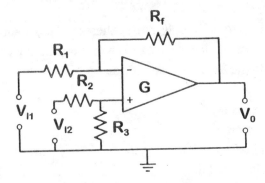

Fig. 19 Circuit diagram for a differential amplifier using both inputs to the op-amp.

A differential amplifier, presented in Fig. 19, is also very useful in making engineering measurements. In fact, it the preferred amplifier to employ with the Wheatstone bridge. It is called a differential amplifier because the instrument amplifies the difference between the two input voltages V_{I1}, and V_{I2}. The differential amplifier is only slightly more complex than the non-inverting amplifier in that it requires four resistors instead of two. The output voltage V_0 is a function of the values selected for these four resistors as indicated by:

$$V_0 = \frac{R_3}{R_2} \left(\frac{1 + \dfrac{R_f}{R_1}}{1 + \dfrac{R_3}{R_2}} \right) V_{I2} - \left(\frac{R_f}{R_1} \right) V_{I1} \tag{34}$$

If we select $R_f / R_1 = R_3 / R_2$, then Eq. (34) reduces to:

$$V_0 = (R_f / R_1)(V_{I2} - V_{I1}) \tag{35}$$

where the circuit gain G_c is again given by the ratio of R_f / R_1.

In connecting the Wheatstone bridge to a differential amplifier the circuit diagram shown in Fig. 20 should be used. Gains of 10 to 100 are usually sufficient to increase the output voltage to a value easily measured with very inexpensive voltmeters. The input impedance of an op-amp is so high (several MΩ) that loading errors are negligible.

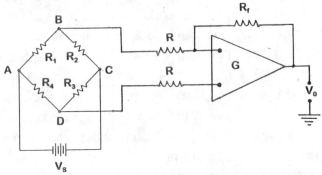

Fig. 20 Recommended circuit for connecting a differential amplifier to a Wheatstone bridge.

If you find a well stocked electronic supply store, there will be many different op-amps available and most will be cheap (less than a dollar each). Which one should you buy? The widest selection includes op-amps that include the following categories:

1. General purpose (lowest cost).
2. FET input, low bias current.
3. Electrometers (lowest bias current).
4. High accuracy, low differential drift.
5. Chopper amplifiers (lowest drift).
6. Fast with wideband.
7. Isolated.

In this project categories 1, 2 and 4 are probably the most suitable. Since drift is always a problem with op-amp circuits, we recommend the low differential drift op-amp.

It is a lot of fun to build an amplifier, particularly if it works well, and gives you the gain that is needed. Try building a low cost op-amp to increase the voltage from the Wheatstone bridge or to reduce significantly the error due to analog voltmeter loading of the measuring circuit.

DIGITAL VOLTMETERS

Digital voltmeters (DVM) are relatively new, and they have largely replaced the analog voltmeters because they are more accurate, exhibit extremely small loading errors, have better resolution, and are easier to read. The voltage is displayed with lighted numerals as shown in Fig. 16, rather than a pointer on a continuous scale as is the case with analog voltmeters.

The range of a DVM is determined by the number of full digits on the display. For example, a three digit DVM can record a count of 999. If the full scale or the DVM is set at 1, the count of 999 would be recorded as a voltage of 0.999V. Many DVMs are equipped with a partial digit to extend the range. The (1/2) partial digit can display the 0 and 1 numerals, and the (3/4) digit can display 0, 1, 2, 3 and 4. The value of the partial digit in extending the range of a DVM is apparent in the following example. Suppose that you have a 3 digit DVM and attempt to measure a voltage of 16.66 V. Your 3 digit voltmeter would record this measurement as 16.7 V. If you made the same measurement using a 3- ½ digit DVM, the reading would be 16.66 V, and you would not lose the last digit.

The resolution of a DVM is determined by the maximum count displayed. For example a 3 ½ digit DVM with a maximum count of 2,000 has a resolution of 1 part in 2,000. The sensitivity of a DVM is the smallest increment of voltage that can be detected. It is determined by multiplying the lowest full scale range by the resolution. For example, the 3 - ½ DVM with a 100 mV lowest (full scale) range has a sensitivity of 100 mV x 1/2000 = 0.05 mV.

The accuracy of a DVM is usually expressed as ± x % of full scale ± N digits. The specifications for our inexpensive Radio Shack DVM with 3 - ¾ digits indicates an accuracy ± 0.2 % of full scale plus ± 1 in the last digit. If we are operating this DVM on the 100 mV scale, the accuracy is ± (0.002 x 100 mV + 0.1 mV) = ± 0.3 mV.

The input to a multi-meter may be voltage dc or ac, current or resistance. In all instances, the input is immediately converted to a dc voltage as the signal enters the instrument. The voltage is amplified and then input to an analog to digital (A/D) converter. The A/D converter changes the dc voltage input to a clock count by using what is called the dual slope integration method illustrated in Fig. 21.

There are three different time controlled operations in the dual-slope integration technique.

1. Auto zero --- where the output voltage from the integrator is zeroed for a fixed time of the order of 100 ms.
2. The dc voltage input is integrated for a fixed time (about 100 ms). The output of the integrator is an increasing voltage with time that gives a ramp function as shown in Fig. 21.
3. The voltage on the integrator is discharged at a constant rate with respect to time to produce a linear run-down ramp also shown in Fig. 21.

Since both the ramp-up and ramp-down voltages exhibit slopes on the graph of voltage versus time, this A/D conversion technique is called dual slope integration.

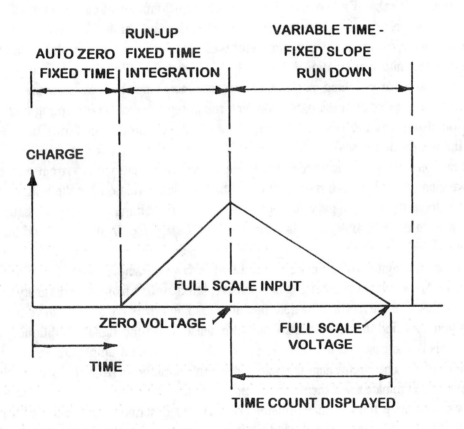

Fig. 21 Voltage time trace for a dual slope integrating DVM.

A counter (clock) is initiated at the beginning of the ramp-down; it runs until the voltage from the integrator goes to zero, and is stopped at this instant. This time interval is proportional to the dc voltage that was first applied to the integrator. The time count is presented on the digital display as the dc voltage.

SUMMARY

In developing a weighing machine with a digital display, we will have to convert from a mechanical response of a spring or strain element to an electrical output somewhere in the system. To aid you in dealing with the electrical aspects of your design, we have introduced some of the most elementary electrical concepts.

Batteries (wet cells) have been introduced to show you the method of converting chemical energy into electrical energy. This treatment also afforded the opportunity to acquaint you with voltage, and both electron and current flow.

We have also described Ohm's law, resistance, and the power dissipated by current flow through a resistor. We have also introduced the other common passive components, namely the capacitor and the inductor. You may not use the capacitor or the inductor in the design of the weighing machine, but it advantages to think of the resistor, capacitor and inductor together as the three passive components found in almost all electrical systems.

Circuit analysis is vitally important for all engineers regardless of their discipline. For this reason we have covered Kirchhoff's two laws, and showed two examples of circuit analysis that lead to the relations for the equivalent resistance for circuits with series and parallel connected resistors.

A variable resistor known as the potentiometer was introduced as a sensor that could be used to convert motion, either linear or rotational, into a change in voltage. For those of you planning to incorporate a spring into your weighing machine, the potentiometer may be the sensor that will prove to be the most suitable. In this discussion, we combine the potentiometer with a spring and develop the design relation that gives the output voltage from the potentiometer in terms of the design parameters and the weight on the scale.

The Wheatstone bridge is introduced for those of you intending to use strain gages as the sensor in your weighing machine. We start with the initial balance condition for the bridge and then show the output voltage for a single gage placed in one arm of the bridge. Finally, the equation for the voltage output from the bridge is given as a function of the gage factor, the strain and the supply voltage to the bridge.

Both analog and digital voltmeters are described in considerable detail. The older analog voltmeter with its D'Arsonval galvanometer is inexpensive, but it suffers from the possibility of severe loading errors. We describe the method for determining if loading errors are a factor in your design.

Some of you may find that the signals from your sensors and the signal conditioning circuits are too small. If this is the case the information provided on instrument amplifiers and operational amplifiers will be helpful. The circuit gain from an instrument amplifier of 10, 100 or even 1000 can enhance the output and eliminate the difficulties associate with a weak signal.

Finally, we introduce the digital voltmeter (DVM) and describe its characteristics such as resolution, sensitivity and accuracy. The dual slope integration technique which is used in many of the lower cost DVMs is described. It is interesting that the dual slope technique permits a simple digital clock to be used as a voltmeter.

REFERENCES

1. Dally, J. W., Riley, W. F. and K. G. McConnell, <u>Instrumentation for Engineering Measurements</u>, 2nd edition, John Wiley & Sons, New York, 1993.

2. Sheingold, D. H., editor, <u>Transducer Interfacing Handbook</u>, Analog Devices, Norwood, MA, 1981.

3. Dorf, R. C., <u>Introduction to Electric Circuits</u>, 2nd edition, John Wiley & Sons, New York, 1993.

EXERCISES

1. Write an engineering brief and describe a hybrid system (mechanical and electrical) giving an example from your own experience.

2. Draw the symbols used on drawings to represent a resistor, a capacitor and an inductor. Also, write the relations for the voltage across these components in terms to the current flow.

3. If you design a circuit with a 9 V battery containing an equivalent resistance of 2,000 Ω, find the current drained from the battery. If the battery has a capacity of 2A-h, determine the life of the battery. Assume that it cannot be recharged. What is the power dissipated by the resistor.

4. Write Kirchhoff's two laws. Use circuit diagrams to illustrate these two laws.

5. Draw a circuit with four series connected resistors and give the expression for the equivalent resistance. If these resistors have values of 1000, 400, 1500 and 10,000 Ω, find the equivalent resistance. Draw an circuit diagram which contains the equivalent resistor which is also equivalent to your initial circuit diagram.

6. Draw a circuit with four parallel connected resistors and give the expression for the equivalent resistance. If these resistors have values of 1000, 400, 1500 and 10,000 Ω, find the equivalent resistance. Draw an circuit diagram which contains the equivalent resistor which is also equivalent to your initial circuit diagram.

7. In adapting a potentiometer to a spring, we employ some type of a linkage which amplifies the small displacement of the free end of the spring. Sketch the design of such a linkage that amplifies the motion by a factor of three.

8. Suppose that your search of the electronic supply stores does not yield a linear motion potentiometer, but you locate several rotary models. Sketch a method showing how to use a rotary potentiometer to measure linear displacement.

9. If you employ a strain gage as a sensor with a gage factor of 2.05, a supply voltage of 9 V on the Wheatstone bridge circuit, and a strain of 1800×10^{-6}, determine the output voltage from the bridge. Comment on the magnitude of this voltage. Is it large enough for you to employ with an inexpensive analog voltmeter?

10. Prepare a table showing the loading error as a function of the ratio of the source impedance to the meter impedance. From this table establish a rule to maintain the loading error to less than two percent.

11. Draw a circuit of a non-inverting instrument amplifier with a gain of 100. Use an operational amplifier in this circuit and select the values of the resistors to be employed.

12. Draw a circuit of a differential inverting instrument amplifier with a gain of 1000. Use an operational amplifier in this circuit and select the values of the resistors to be employed.

13. Examine the specifications for a DVM and cite the statement for the accuracy. Convert this specification into a numerical value for the accuracy if you are using the most sensitive scale on the DVM. Perform the same conversion for the less sensitive scale.

PART II

ENGINEERING

GRAPHICS

CHAPTER 5

THREE-VIEW DRAWINGS

INTRODUCTION

Engineering graphics is a broad term that is used to describe a means of communication. We normally think of communication in terms of writing and speaking, because they are more commonly employed in the normal course of our life. However, when we try to communicate design ideas, we find writing and speaking insufficient to express our thoughts. A more visual way to communicate is needed. It more effective to present our ideas by means of drawings, sketches, pictures, graphs of many different types, etc.. Visuals aids, such as drawings, convey our ideas quickly and with remarkable precision. It is easier to transmit information, and it is easier to receive information if it is conveyed in the form of drawings, sketches or graphs.

The objective of this part of the textbook is to introduce you to methods of communication using visual aids. You should understand the advantages of presenting information in drawings, sketches and graphs, and that these are very common techniques used by engineers to communicate very complex ideas quickly and with precision. You will begin to learn how to prepare orthographic projections, isometric drawing, sketches and several different types of graphs. There are two general approaches used in preparing drawings and graphs. The first is a manual approach. We draw the visuals by hand using a few simple drawing instruments. The second is using a computer and suitable software to aid in the preparation of the drawing or graph. In this chapter we will cover the manual methods for preparing drawings and graphs. In separate chapters we will introduce KEY CAD Complete, which is a software program for preparing engineering drawings. In another chapter we will cover EXCEL, which is a spread sheet program that permits us to produce a several different types of different graphs.

VIEW DRAWINGS

Let's begin by considering the block like object shown in Fig. 1. We could describe this as a rectangular block with a slot cut out of its top. However, this written description is vague because we haven't conveyed the relative proportions of the block, the precise location of the slot, or the size of any of its features. We need to communicate proportion, exact location, and size of every feature so that the object can be reproduced by anyone capable of reading a drawing. We use pictorial drawings and/or multi-view drawings to quickly convey this information to the reader. The drawing shown in Fig. 1 is a pictorial, that we will initially use to help us describe multi-view drawings.

The pictorial (isometric) drawing in Fig. 1 is shown with three arrows. These arrows represent the directions of viewing, which isolate the front, top and side views. When taken individually, each view is a two dimensional rendering of the three dimensional object. Of course, two-dimensional drawings are incomplete, because they show only the information in one of many possible views. Nevertheless, the two-dimensional views are important because we can draw objects to true scale on a sheet of ordinary paper. We avoid the problem of incomplete information inherent in two-dimensional drawings, by presenting a sufficient number of views to completely locate and size all of the features of the object.

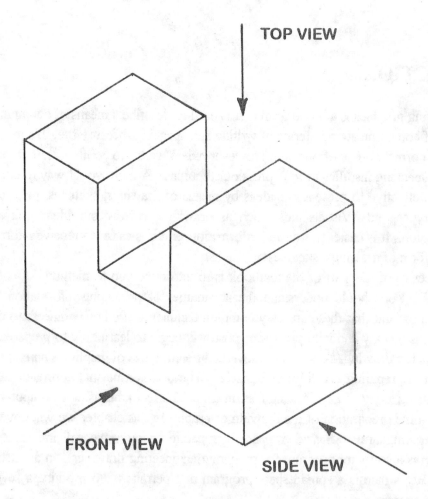

Fig. 1 Pictorial drawing showing a three dimensional rendering of a rectangular block with a slot. The three directions of viewing give the front, top and side views.

A three-view drawing of the block presented in Fig. 1 is illustrated in Fig. 2. The single pictorial drawing of the block is now represented with three different two-dimensional drawings representing the front, top and side views. These three drawings completely define the proportions of the block, its size and the location and size of the principle feature (the slot).

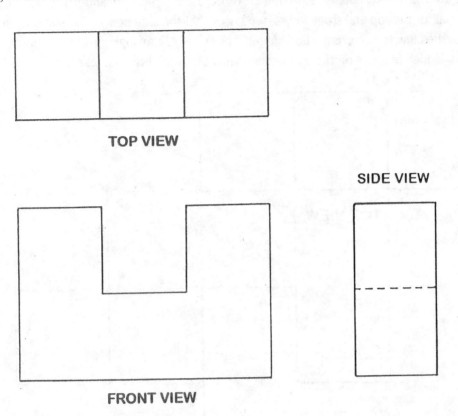

Fig. 2 A three-view drawing of the object shows the front, top and side views.

The arrangement of the views is important. The front view is placed in the lower left corner of the paper, the top view is directly above it, and the side view is directly to the right of the front view. This arrangement permits us to prepare the three-view drawing using orthographic projection. In orthographic projection, we project dimensions from one view to another. For example, we measure and draw the front view and project lines upward to the top view. The width of the object is shared in both the top and front views. We also project construction lines to the right from the front view, to give the height of the side view. The front and side views share the height dimensions of the block. We can also project from the top view to the side view if we draw a 45° construction line from the right hand corner of the front view, as shown in Fig. 3. We then project lines to the right from the top view until they intersect the 45° construction line, and then we turn them downward to show the depth of the object on the side view. All of the construction lines used to prepare Fig. 2 are illustrated in Fig. 3. Remember, orthographic projection will save you time in preparing your drawings. It reduces the number of time consuming measurements that you must make in preparing the different views necessary to completely describe any given part. The construction lines also help you in visualizing the feature locations on the adjacent views.

One more point to make before we leave Fig. 2. Did you notice the dashed line on the side view? Why were all of the lines solid except that one? We have a convention in engineering drawing for the use of line styles and line weight. Solid lines represent edges of features that we can see in a given view. As we look at the top of the block shown in Fig. 1, we can see the four sides of the rectangle, and we can see the two edges of the slot. We use solid lines then to represent all six edges that we can see in the top view. However, when we view the object from the side, we can only see the four edges outlining the rectangle. Looking from the right side, we cannot see the slot. If we look at the pictorial, or the top and front views, we know that the slot exists. To show it on the side view, we use a dashed line to represent a line hidden from view. Drawings differ from photographs in that we can show hidden features on the appropriate view by using dashed lines.

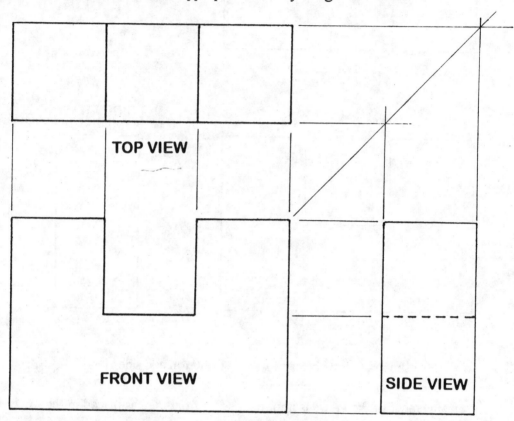

TOP VIEW

FRONT VIEW

SIDE VIEW

Fig. 3 Construction lines used to convey dimensions from one view to another in orthographic projection.

LINE STYLES

We use different line styles and different line weights (thickness and darkness of the lines) in preparing engineering drawings as illustrated in Fig. 4. We control the weight and the thickness of the lines by the selection of the pencil lead. The degree of hardness is governed by variable amounts of clay that is mixed with graphite to form pencil lead. Hard leads have little clay while soft leads have lots of clay. The scale from hard and light to soft and dark is given below:

Hard and Light $\Leftarrow$ 6H 5H 4H 3H 2H H F HB B 2B 3B 4B 5B 6B $\Rightarrow$ Soft and Dark

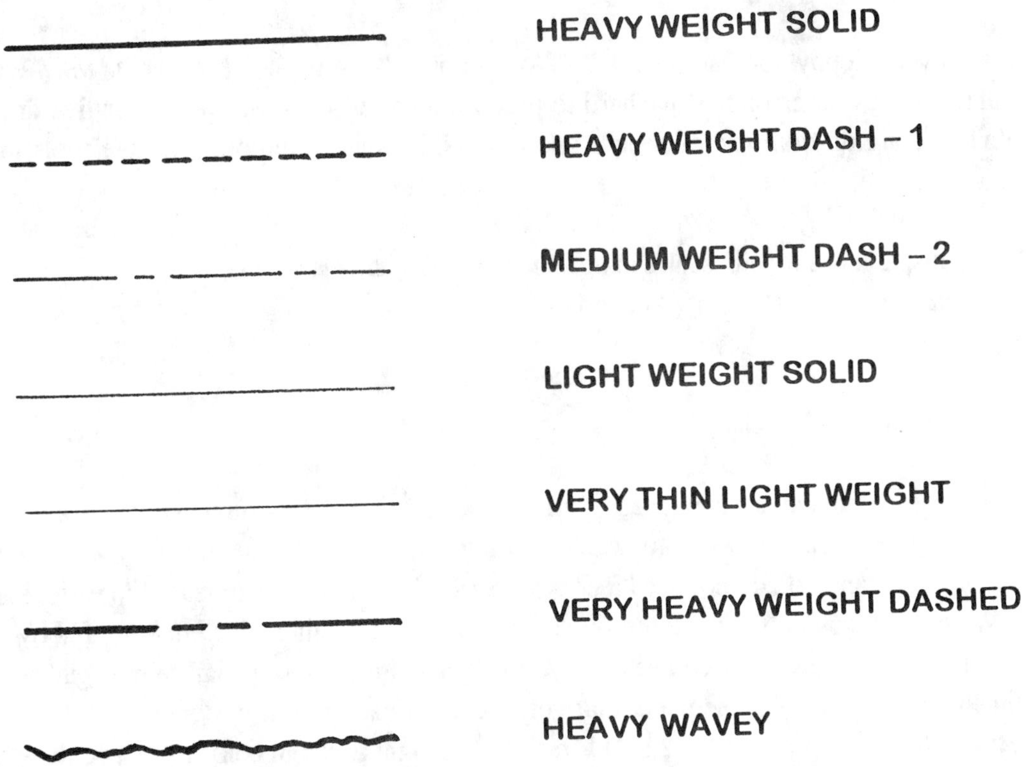

HEAVY WEIGHT SOLID

HEAVY WEIGHT DASH – 1

MEDIUM WEIGHT DASH – 2

LIGHT WEIGHT SOLID

VERY THIN LIGHT WEIGHT

VERY HEAVY WEIGHT DASHED

HEAVY WAVEY

Fig. 4 Different line styles used in view drawing.
    a. Heavy weight solid lines representing edges
    b. Heavy weight dashed lines representing hidden lines.
    c. Medium weight long dash then short dash lines for centerlines.
    d. Light weight solid for dimensioning.
    e. Very thin light lines for construction.
    f. Very heavy weight dashed lines for section cuts.
    g. Heavy wavy lines to indicate a break in the view.

We use 4H to 6H lead with a very sharp point to draw construction lines. These lines are so light and thin that they will not show when the drawing is reproduced. Light weight dimension lines are drawn so they will remain visible when the drawing is copied. We use an H or F lead to give the correct relative darkness, and use a sharp point with a very slightly rounded tip. Medium weight centerlines are drawn with a HB or B lead with a rounded point. Heavy weight lines, both solid and dashed, should be drawn with B or 2B lead with a rounded point.

The very soft leads --- 3B to 6B are usually employed by artists to prepare sketchings, but not by engineers preparing multi-view drawings. Engineers handle a completed drawing much more than an artist handles a completed sketch. In handling the soft leads smear, degrade the appearance of the drawing, and detract from the quality of the copies that are usually made from the originals.

We recognize that many of you will be short on pencils, but perhaps the team members can get together and share resources. Please do not use ball-point pens for drawing. There are two problems with ball-point pens. First, you have no control over the width of the line as it is determined by the diameter of the ball in the pen. Second, you will probably make several mistakes as you prepare the drawing. The lines drawn with a ball-point cannot be erased, and you eventually finish a drawing that is very sloppy and hard to read. Work in pencil and use a clean, white, vinyl eraser to eliminate all traces of your errors. Strive hard to produce an accurate, clean, and error-free drawing.

While we are on the topic of pencils and erasers, let's list some other simple tools that you will want to acquire.

1. Transparent ruler with scales in both inches and millimeters.
2. Triangles ---30/60 and 45/45 degrees.
3. Protractor
4. Compass
5. Templates circles, ellipses, etc.

This is a short list. We have not included a complete set of drawing instruments or a drafting board with T-square. Our purpose is not to teach drafting, but to introduce you to the essentials of engineering graphics, and you do not need a complete set of drafting tools to learn the early lessons in graphics. We encourage you to invest the small sum needed to acquire the items listed above. You will find them useful in many other courses that you will take during your program in engineering.

Finally, a suggestion for the paper that you use in preparing your drawings. We recommend National Brand engineering paper 8-1/2 by 11 in size. It is light green in color to relieve the strain on your eyes. It also has a grid (five squares to the inch) on the back side of each sheet. The grid shows through to the front side providing guidelines that are helpful in projecting the construction lines needed to prepare three-view drawings. The paper erases well, produces good copies, and is punched for a three ring binder.

Now that you are equipped with the proper pencils, eraser, tools and paper you are ready to begin to learn how to draw several different features normally encountered in preparing multi-view drawings of engineering components.

REPRESENTING FEATURES

We will demonstrate the techniques for drawing various features in three-view drawings. The examples selected include a block with a slot and a step, a block with a tapered slot, and a block with a step and a hole.

BLOCK WITH SLOT AND STEP

The rectangular block in Fig. 1 had one feature, namely a slot. We encounter many different geometric features in depicting parts that we design, including holes, curved boundaries, steps, and tapers. A pictorial of an object with both a slot and a step, presented in Fig. 5 serves as our next example. Let's prepare a three-view drawing of this object using the methods of orthographic projection. Look at the pictorial, and start with the front view.

As you examine the front view in Fig. 5, visualize the outline of the block. Even with the step and the slot, the outline of the block (the outside edges) is a rectangle. Draw that rectangle representing the outline of the front view in the lower left hand corner of your quadrille paper. We will not worry about exact dimensions of the rectangle at this stage; however, try to maintain the proportions shown in Fig. 5. When drawing the rectangle in the front view, extend the vertical lines

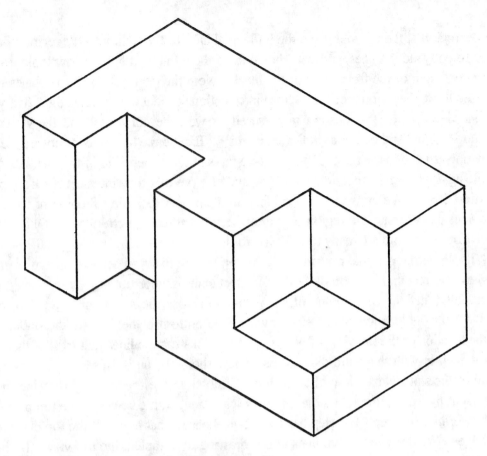

Fig. 5 A pictorial drawing of a block with a slot and a step.

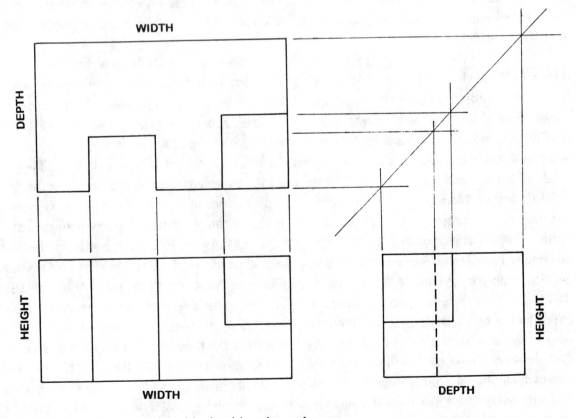

Fig. 6 A three-view drawing of the block with a slot and a step.

upward into the region of the top view, and the horizontal lines to the right into the region of the side view. These are construction lines, and you should keep them thin and light. Now look at the step that is in the upper right of the front view of the block. Note the intersection of the planes defining the step with the front plane of the block. Those intersections produce two new edges that you can see in the front view. Draw the location of these edges with the horizontal and the vertical lines shown in the upper right hand corner as indicated in Fig. 6. Extend the lines used to locate the step in the front view upward and to the right. Finally, locate the slot on the left side of the pictorial drawing in Fig. 5. Note again the planes that define the slot, and observe that they intersect the front plane to form two vertical edges. We draw two vertical lines in the proper location to represent these edges. Project these lines upward into the region of the top view using construction lines. OK! We have completed the front view, and are ready to move on to the top view.

Look down on the pictorial drawing. Again, the outline that you see is rectangular in shape, but in this view the rectangle is not perfect. The rectangle is interrupted by the slot. Looking downward, the edge defining the outline of the rectangle across the slot is missing; We can see through the block at the location of the slot down to the bottom plane. Draw the outline of the rectangle in the location of the top view of Fig. 6. Use the construction lines that exist at the location of the top view to help with the width dimensions. There should be an open section in the rectangle at the location of the slot. Next, show the depth of the notch in the top view by drawing the three lines that represent the edges formed by the three vertical planes of the slot with the top plane of the block. Finally, examine the step. The step has two vertical planes that intersect the top plane to form the defining edges. We draw those two lines in our drawing to complete the top view. If you have made full use of the construction lines projected from the front view into the region of the top view, it was easy to draw the top view. You only had to locate the position of three horizontal lines that defined the depth of the slot, step and block. We carry these depth dimensions to the right with construction lines.

The final view is the (right) side view. We could also do a left side view, but convention calls for us to draw the right view. After having completed the front and top views, including the orthographic projection of construction lines, the side view is easy to prepare. Draw a construction line at 45° so that it cuts the horizontal construction lines from the top view. Then draw a second set of construction lines downward from these intersection points into the side view. The side view now contains seven construction lines. The three horizontal lines projected from the front view give the height of the step and the block. The four vertical lines projected from the top view give the depth of the slot, step and block. We have the location and dimensions of all of the lines in the side view from our orthographic projections. All that we need to do to complete the side view is to darken select portions of the construction lines that represent edges. Again, look at the side of the pictorial drawing in Fig. 5 and observe that the rectangular outline of the block is completed. Also note the edges produced by the intersections of the planes forming the step with the plane of the right side of the block. Draw these two edges in the upper left corner of the side view. From a visual perspective, our drawing of the right side is complete; however, a multi-view drawing often carries more information than what we can visualize. We know from the front and top views that a slot exists. Our projection lines from the top view to the side view show the depth of the slot. Even though we cannot "see" the slot in the side view, we represent its presence and its location with a dashed (hidden) line. Our three-view drawing of a block containing a slot and a step is complete as illustrated in Fig. 6.

BLOCK WITH TAPERED SLOT

Let's consider drawing still another feature, namely a block that incorporates a tapered slot as illustrated pictorially in Fig. 7. Begin a three-view drawing with the front view in the lower left corner of your drawing paper as shown in Fig.8. Look at the front view in Fig. 7 and observe the edges produced by all of the planes that intersect the front plane. Drawing these edges gives us four vertical lines and three horizontal lines. The tapered surface intersects the front plane giving an edge shown in the front view. The top surface is represented by still another edge visible in the front view. We project all of these lines into the remaining two views with construction lines.

Examine the top view and observe that the rectangular outline of the block remains intact. Observe three edges inside the rectangular outline. The edge due to the intersection of the taper plane with the top surface produces a horizontal line, and the two vertical planes intersect the top surface to produce two vertical lines in the top view. Project the horizontal lines in the top view to the right so that they intersect the 45° construction line shown in Fig. 8. Then project these three lines downward from the intersection points on the 45° line into the side view.

The six construction lines projected into the side view give us most of the information needed to complete this view. Clearly, it is evident from Fig. 7 that the rectangle outlining the side view is intact. In fact as we view the right side, a simple rectangle formed by the four edges is all that is visible. The information regarding the tapered surface that is evident in the pictorial drawing, and the front and side view is not visible in the side view. We add information showing the location of the tapered surface on the side view by using the dashed (hidden) lines.

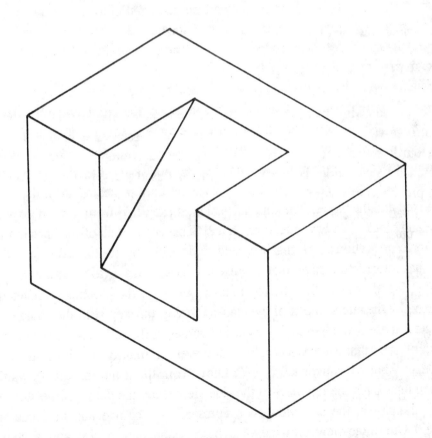

Fig. 7 A pictorial drawing of a block with a tapered slot.

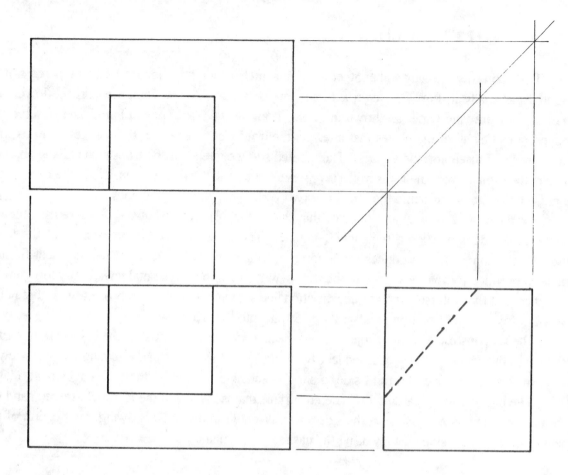

Fig. 8 A three-view drawing of the block with a tapered slot.

BLOCK WITH STEP AND HOLE

A pictorial drawing of a block with a step and a hole is illustrated in Fig. 9. We have prepared the three-view drawing of this object and have presented it in Fig. 10. The purpose in showing the fourth example of a three-view drawing is to illustrate how holes and curved boundaries are represented in engineering drawings. Assume in this discussion that you understand enough based on the previous examples to draw the outlines of the three-views. The only question concerns drawing the curved boundary that outlines the step and the hole that is drilled through the step.

The circular boundary outlining the step is evident only in the top view. The front and side views give no evidence of the presence of a curved boundary, because the vertical curved surface does not produce intersections (edges) on the front or side planes in either of these two views. Produce the curved surface in the top view using a compass with its point located in the center of the hole. The radius is set on the compass, so that the circular arc is tangent to the horizontal lines forming the outline of the block in the top view.

The compass is also employed to draw the hole in the top view. Note the two orthogonal lines that define the center of the hole. These are center lines that locate the position of the hole relative to some other feature on the block. The presence of the hole is also depicted in the front and side views even though it is not visible in either view. Use dashed lines (hidden) to locate the edges of the holes. Also locate the centerline of the hole in the front and side views. Note the long dash, short dash line used to denote the centerlines in all three-views. The centerlines are very important when we dimension the drawings, because we dimension holes to their centerlines and not to their edges.

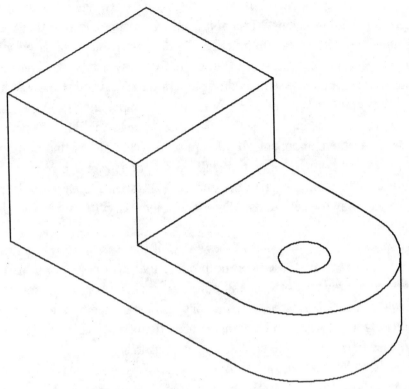

Fig. 9 A pictorial drawing of a block with a step and a hole.

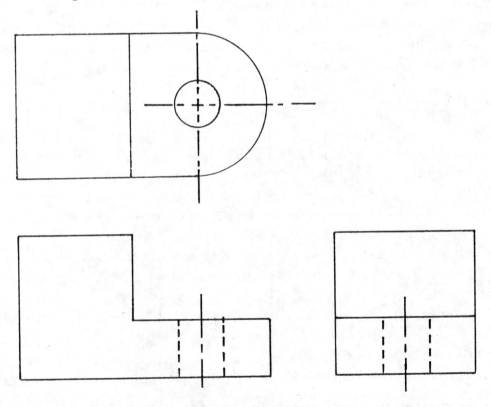

Fig. 10 A three-view drawing of the block with a step and a hole.

DIMENSIONING

In the previous section, we prepared our three-view drawings maintaining proportionality, but without concern for the dimensions. This approach was a simplification that we took to clarify the discussion of three-view drawings. In practice, the dimensions are critical because they control the size of the component and the location of all of its features. Let's examine dimensions from two different points of view. First as the individual preparing the drawing, you must know (or decide) on all of the dimensions required to completely define the component. If you are a designer of a new component, you start with a blank sheet of paper and assign the dimensions that you believe will optimize the design of the component. If you are redesigning an existing component, you will start with the drawing of that component and modify existing dimensions to refine the design. The point here is that you (as the person preparing the drawing) know the dimensions. It is your responsibility to include on your drawing all of the dimensions necessary for someone else to fabricate the component.

Next consider the second point of view, that of the person using the drawing. This person may manufacture the part, may assemble the product that incorporates the part, or may repair the product. Engineering drawings serve many purposes and are used by several other people that are not involved with design. The three-view drawing defines this component without ambiguity. The dimensions precisely give its size. The component as defined by the drawing, and its dimension can be made by anyone in the world with complete interchangability.

Let's begin to learn about dimensioning by adding dimensions to the drawing of the block with a slot that is shown in Fig. 2. We have copied this drawing, and have added dimensions as presented in Fig. 11. Examine the front view of this drawing noting that we have given the width of the block as 3. This dimension is inserted in a break in the dimension line. No units are given except for a notation in the drawing block that all dimensions are in inches, mm, ft, etc.. The dimension line is terminated by arrowheads that point to the two extension lines. The arrowhead is long (about 1/8 in.) and thin. The extension lines are separated from the view drawings by a small gap (about 1/16 in.).

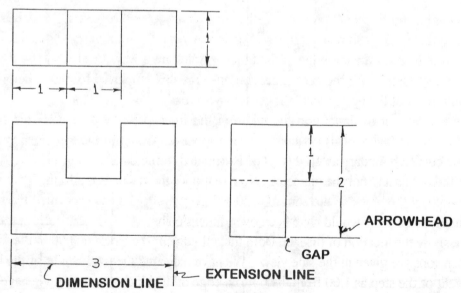

Fig. 11 Dimensioning example showing the dimension line, extension line, arrowhead and gap.

In the example presented in Fig. 11, we have dimensioned the block using inches as the unit of measure. The fact that we have defined the width as 3, indicates to the person manufacturing the block that the width can be 3 ± 1/64 in. The tolerance on the width is implied or is given in the drawing block. If you, as the designer, want to specify a block manufactured with more precision, you would specify the dimension using decimals. For example if you wrote 3.00, the person manufacturing the part would understand that the width of the block would have to be between 3.00 ± 0.01 in. If you need still more precision specify the dimension as 3.000 and the block must be produced with a width between 3.000 ± 0.005 in.. Remember there is a great difference between 3, 3.00 and 3.000 in the precision and tooling required to manufacture the components. There is also a great difference in the cost. As we tighten the tolerances, the cost of manufacturing a part increases significantly.

Let's return to Fig. 11 and examine the dimensioning of the top and side views. We have already shown the width on the front view, but we need dimensions for the depth and height of the block and the location and size of the slot. Remember that the top view is used to give dimensions for the width and depth, and the side view is used to give dimensions for the height and depth. In the top view of Fig. 11, we have specified the depth as 1. We have also specified the location of the slot and its width on this view. The height of the block and the height of the slot are given on the side view.

We have completed the dimensioning of the block and the slot. Every dimension necessary to manufacture the component has been specified in the drawing. Also the drawing has not been over dimensioned. (We have not specified the same dimension twice). The choice of where we place the dimensions is somewhat arbitrary. We can dimension the width on either the front or top views, the depth on the top or side views and the height on the front or side views. Two choices for each dimension are possible. We usually select one view or the other for a given dimension to keep the drawing clear and uncluttered..

DIMENSIONING HOLES AND CYLINDERS

To demonstrate techniques for dimensioning either holes or cylindrical boundaries, let's add dimensions to the drawing shown in Fig. 10. Again start with the front view, and locate the centerline of the hole as 3.20 from the left edge of the block. Holes are always located by the dimension from an edge to a centerline. Use the centerlines, because the drill employed to make the hole is inserted into the component at the point where the center lines cross.

We choose not to dimension the width in the front view, because we want to specify the radius of the curved (cylindrical) boundary in the top view. Going to the top view, we indicate the radius of the curved boundary as 1.20 R. The R is added to the dimension to indicate that it is given in terms of the radius and not the diameter. Note that once the radius R is specified, we have in effect given the width of the block as the sum of 1.20 + 3.20 = 4.40. To have specified the total width of the block in the front view would have been over dimensioning.

We show the location of the step from the left edge of the block as 2.00 in the top view. The height dimensions are given in the side view. We have indicated the total height of the block as 2.50 and the height of the step as 1.00 from the bottom edge of the block. Note that we have dimensioned the diameter of the hole in the side view as 0.75 DIA. We dimension holes in terms of diameter because the drill sizes used in manufacturing the holes are given in terms of their diameter, and not their radius.

Have we completed the dimensions as shown in Fig. 12? You might question if we have specified the depth of the block since it has not been indicated in the side view. Well, we specified the depth when we called out the radius of the curved boundary in the top view. The depth is equal to two times the radius R or 2.40.

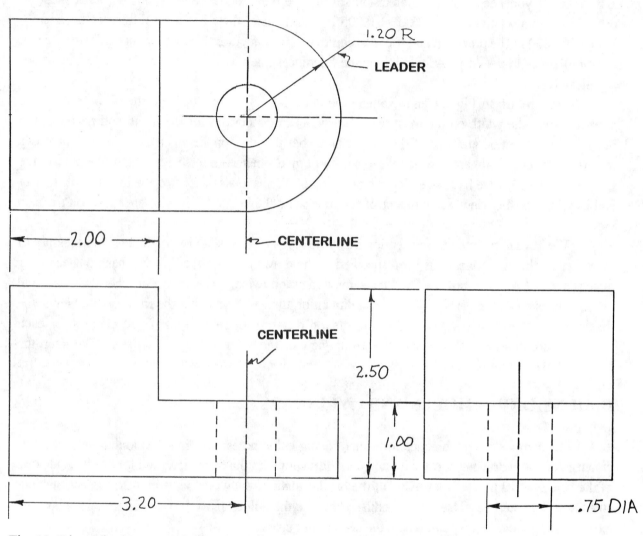

Fig. 12 Dimensioning example showing centerlines, leader, and radius and diameter indicators.

Again we have not shown units in the dimensioning. The units are defined in the drawing block along with other common information pertaining to the component. The use of two standards for units, the U. S. Customary and the SI systems, cause some difficulty in preparing and reading drawings. Some drawings are prepared using U. S. Customary units of in. or ft, and others are prepared using the SI system with units given in mm or m. Sometimes when the drawing is to be used by many people from different countries, both systems are employed in the dimensioning. An example of dual dimensioning is given in Fig. 13. In this case we do not break the dimension line because the U. S. Customary unit is placed above the line, and the SI unit is placed below the line and enclosed with parentheses.

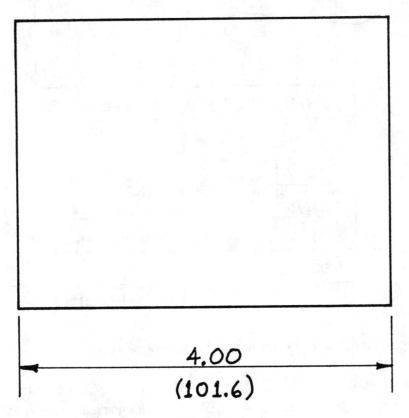

4.00

(101.6)

Fig. 13 Illustration of a convention used for dual dimensioning.

DRAWING BLOCKS

It is time to discuss the drawing block, because it serves a very important function on an engineering drawing. In the previous section we referred to the fact that the units for the dimensions were specified in the drawing block. The unit used in the drawing is only one of the many facts presented in the drawing block. As illustrated in Fig. 14, the drawing block is located in the lower right hand corner of the drawing, just inside the border. A typical drawing block conveys important information, some of which are common to all drawings produced by a certain company, and some which are unique to the individual drawing. Information commonly shown in the drawing block include:

1. Name of the company issuing the drawing.
2. Name of the part that the drawing defines.
3. Scale used in preparing the drawing.
4. Tolerances to be employed in manufacturing the part.
5. Date of the completion or the release of the drawing.
6. Material to be used in manufacturing the part.
7. Heat treatment of the part after manufacturing.
8. Units of measurement to be used in manufacturing the part.
9. Initials of the individual preparing the drawing.
10. Initials of the individual checking the drawing.
11. A drawing number which uniquely identifies the drawing.

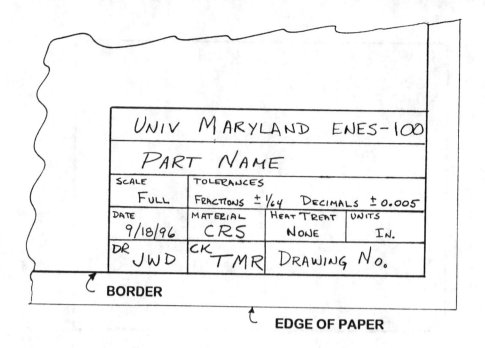

Fig. 14 An example of the information presented in a drawing block.

Let's examine each of these items. The name of the company is self evident. We illustrate it in Fig. 14 with the University of Maryland --- ENES 100. In designing a part, you will identify it with some name such as bracket, support, shaft, etc., and include this name on your drawing. The part name should be brief (no more than three or four words).

Next you will indicate the scale used in preparing the drawing. The example shown in Fig. 14 indicates that we used full scale in drawing the three-views of the part. The scale that we select depends on the size of the part and the size of the paper that is used for the drawing. In this class we are constrained to standard size copy paper (8-1/2 by 11 inches), because our laser printers are limited to this size. Suppose the part that you are designing is 12 inches wide, 6 inches high and 1 inch deep. Will the three-view drawing of this part fit on this standard size paper? The answer to this question is a clear NO. We need to scale down the part dimensions on the drawing so that it will fit on a single sheet of paper. In this case you would probably use ¼ scale. With ¼ scale the front view of the part would be 3 inches wide by 1 ½ inches high. You would dimension the drawing using the actual sizes of the part (i. e. you would show the width as 12 inches, although it is only 3 inches on the scaled down drawing). You indicate to those reading the drawing that you have reduced the size of each view by a scaling factor of ¼.

Tolerances that are to be employed in manufacturing the part are called out in the drawing block. Many companies have standardized their tolerances, and these limits are preprinted in drawing blocks that are incorporated on the company's drawing paper. The date of completion of the drawing is also shown. In some instances the drawing release date is used instead of the completion date. The release date identifies when the design engineers turn over the ownership of the design of a product to the operations (production) function which is responsible for producing the product.

The material from which the part is to be fabricated is often identified in the drawing block. In the example shown, the abbreviation CRS indicates that cold rolled steel will be used in manufacturing the component. Sometimes the heat treatment of the component, if required, is identified in the drawing block. Heat treatment is a multi-step process employed to enhance the strength and hardness of a component. Usually heat treatment follows the machining of a part, and we indicate whether or not it is required to assist those folks in charge of operations in controlling the flow of parts in the production process.

The units used in the drawing are given in the drawing block. We can use U. S. Customary units, SI units, or dual units. We cannot use mixed units (Note the difference between mixed and dual units). U. S. Customary units are in terms of inch (in.) or foot (ft). SI units are in terms of millimeter (mm) or meter (m).

Those responsible for the drawing are identified. The individual, who has prepared the drawing, signs the block with his or her initials. The drawing is checked for completeness and accuracy, and the individual checking the details also initials the drawing block. Finally, the drawing is given a number. In a large company, which may release thousands of drawings each year, the drawing numbers are controlled by the engineering records department. This department maintains a numbering system which insures that each part or component has a unique drawing number. The engineering records department also organizes the numbering system to group together all of the drawings needed to build a specific product.

ADDITIONAL VIEWS

Some of the parts that we design are complex and addition views may be needed to ensure that the drawings are properly interpreted. We are not restricted to three-view drawings. For very simple parts, we can adequately describe the object with one or two views. There is no need to waste time drawing additional views. For complex parts that are more difficult to visualize, we can provide additional views clarify the drawing. These extra views can show either external surfaces or internal sections. If it helps to visualize the object, we can add back, bottom and left side views to the more commonly employed front, top and right side views. To draw these additional external views, we simply rearrange the layout of the views on the drawing as shown in Fig. 15. This drawing shows five views of a multiply tapered block, illustrates the proper arrangement to show additional external views, and maintains the advantages of orthographic projection. Clearly, the extra views help us visualize the complex shape of this block.

An addition internal (sectional) view is often more useful in clarifying a drawing than an additional external view. The concept of a sectional view requires us to mentally cut the object with a cutting plane. We then divide the part into two and view the internal section revealed by the cut. An example of the technique used is illustrated in Fig. 16 where we have modified Fig. 8 by adding a sectional view. On the front view drawing in Fig. 16, we show a very heavy dashed line to indicate the location of the cut. The cut line is turned through 90° at both ends, and arrowheads are drawn to indicate the direction of the view. The arrowheads are labeled with the letter A. In this case, we are viewing the section cut from the right side. Accordingly, we expect the section view to closely resemble the right side view. We then draw the section that is revealed by the cut.

The section view may be placed at any convenient location on the drawing. We have placed it in the open space in the upper right hand corner of the drawing. The section view is usually encircled with a wavy line to distinguish it from the usual three-views. We also match the section view with the cut by using the letters A-A for identification. In some drawings, we might make several cuts each revealing different sections with each sectional view identified by A-A, B-B, C-C, etc. In the sectional view, we show the view revealed by the cut with cross hatching. The remaining part of the view (outside the cut section) is drawn without cross hatching. Try to prepare a section view of the component shown in Fig. 12. Make the section cut along the centerline through the hole in the top view, and then look at the section from the right side.

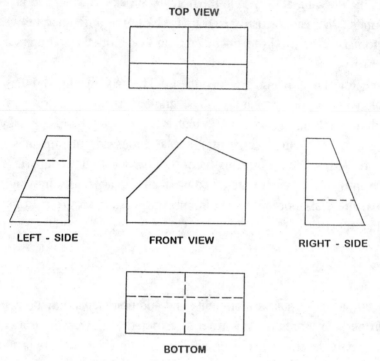

Fig. 15 A five view drawing of a complex block.

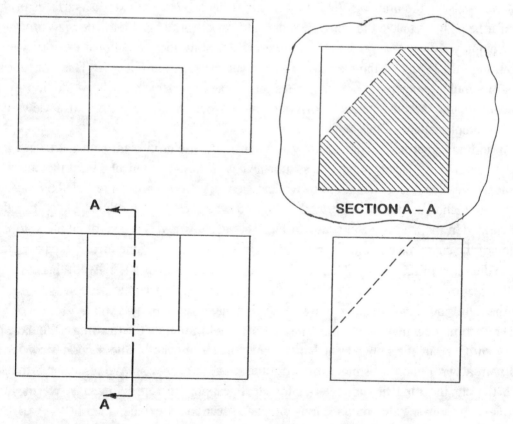

Fig. 16 A section cut and the corresponding section view.

SUMMARY

We have introduced engineering graphics by describing techniques for preparing multi-view drawings. Multi-view (usually three-view) drawings are used to communicate the relative proportions of a component and the exact size of all of its features. The drawing techniques are covered in detail with a few simple examples. We hope that you will take the time necessary to learn these techniques, because learning to draw and read three-view drawings is extremely important in engineering.

Dimensioning is a significant part of preparing an error free drawing. All of the dimensions must be included so that the part can be manufactured from the drawing. Often the part is manufactured in another city, or state or even another country. The person manufacturing the part does not know you and cannot ask you questions to clarify the ambiguities in your drawing. The drawing and all of its dimensions must completely define the component. An important implication of the dimensioning are the tolerances required in fabricating the part. As you write the numbers for the dimensions, you implicitly assign the tolerances. There is a great difference between 3, and 3.000 in the precision required in manufacturing and in the cost of the component.

We introduced drawing blocks and indicated the type of information that they normally convey. Remember drawing blocks are not unique and you will see differences in the information presented by different companies. We also introduced the concept of scale in this section. The scale used in preparing a drawing is your decision. Select the scale factor so that the three-views fit the paper without crowding. Sometimes you will scale down the view drawings so they will fit on the sheet. Other times you will scale up the views of a very small part, so that you can see its very small features.

The coverage of three-view drawings presented in this chapter is very brief. If you need to learn more, we recommend the very complete textbook by James Earle [1] for your additional reading.

REFERENCES

1. Earle, J. H., Engineering Design Graphics, 4th edition, Addison Wesley, Reading, MA 1983.

EXERCISES

1. Prepare a three-view drawing similar to the one shown in Fig. 2 except change the width of the notch from 1 inch to 1 ½ inch.
2. Take a piece of clay or foam plastic and use a razor knife to manufacture a block with the shape shown in Fig. 6.
3. Prepare a three-view drawing of a block with a taper notch like that shown in Fig. 8. Select the dimensions yourself, but be consistent from one view to another with these dimensions.
4. Prepare a three-view drawing similar to that shown in Fig. 10 except increase the diameter of the hole to 1 ¼ inch.
5. Dimension the drawing that you prepared for exercise 1.
6. Dimension the drawing that you prepared for exercise 3.
7. Dimension the drawing that you prepared for exercise 4.

8. Design a drawing block that your team can employ with their drawings of the scale.
9. Prepare a five view drawing of the object shown in Fig. 9.
10. Prepare a drawing with a section view for the block defined in Fig. 1.

CHAPTER 6

PICTORIAL DRAWINGS

INTRODUCTION

Pictorial drawings are three-dimensional illustrations of a component or an object. For a person trying to visualizing an object, the pictorial drawing is the most effective means to convey its size and shape. Some folks have difficulty putting the three standard (front, top and side) views together to "see" the object. Pictorials drawings assemble the three-view s on a single sheet giving a three-dimensional view, that facilitates visualization. Because pictorials are so easy to visualize, they are often used for catalogs, maintenance manuals, and assembly instructions.

Three different types of pictorials are in common usage:
- Isometric
- Oblique
- Perspective

A simple cube with three different types of pictorials is illustrated in Fig. 1. The isometric pictorial is drawn with its three axes spaced 120° apart. The term isometric means "equal measurement" indicating that the sides are all scaled by the same factor relative to their true length. Parallel lines defining edges on the object are also parallel on the isometric drawing. Drawing paper with isometric axes is available in good office and drafting supply stores. We encourage you to use it, because it greatly facilitates the preparation of an isometric pictorial.

Oblique pictorials are drawn with the front view in the x-y plane. We project oblique lines, which represent the z axis, at some angle often 45°. The angle used for the oblique lines can vary from 0 to 90°. Parallel lines defining edges on the object are also parallel on the oblique drawing. If the true length of the lines is employed in scaling all three sides, we call it a cavalier oblique pictorial. The cavalier oblique style is frequently used, but the resulting pictorial is a distorted. The distortion is due to the depth dimension which appears to be too long.

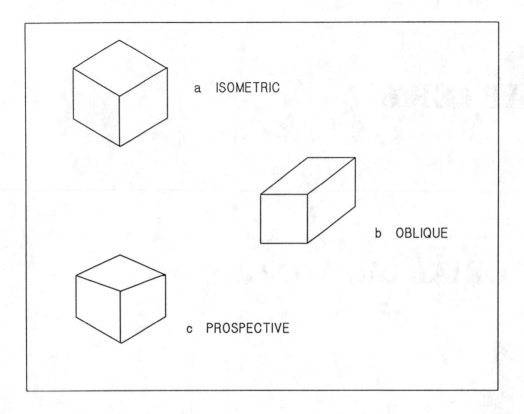

Fig. 1 A cube represented with isometric, cavalier oblique, and perspective pictorials.

The perspective is a pictorial drawing that represents what we actually see. Artists draw or paint using perspective style. Engineers sometimes represent their designs in this style; however, this is the most difficult of the three types of pictorials to master. In perspective drawing, we do not have a well defined coordinate system. Parallel lines tend to converge to a vanishing point as they recede from the observer. Scales on the different axes are different in order to shorten the lines located some distance from the picture plane. The use of converging lines instead of parallel lines and the foreshortening of select dimensions gives the drawing perspective. The prospective drawing looks like the object that we see.

ISOMETRIC DRAWINGS

In our discussion of the three forms of pictorial drawings, we will introduce the axes used to frame the drawing, the direction of viewing the three-dimensional object, and the dimensions used for the width, height and depth. Let's begin with the isometric pictorial, shown in Fig. 2, that employs axes which make 120° with each other. The axes divide the paper into three zones that are utilized to present three-views. If the axes make a Y with the vertical line directed downward from the two branches, we are looking downward at the object. From this perspective, we visualize the top view in the region between the branches of the Y as shown in Fig. 2. The front view is displayed in the region to the right of the vertical axis, and the left-side view is drawn in the region to the left of this axis.

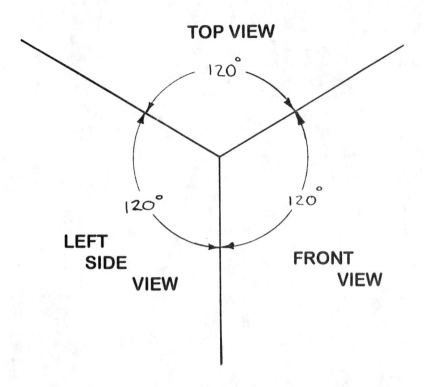

Fig. 2 Isometric axes at 120° angles give three regions for the front, top and left-side views.

OK! We have defined the axes and the direction of viewing the object in Fig. 2. Let's discuss the dimensions that we use on the drawing for the width, height and depth. Isometric means equal measurement, so it's clear that we will use the same scale along all three axes. To illustrate the techniques used to prepare isometric drawings, consider a simple rectangular block with width W, height H, and depth D. We have prepared an isometric pictorial of this block in Fig. 3. To draw this isometric pictorial:

1. Use a 30° triangle to draw the isometric axes identified with the numbers 1, 2, and 3 as shown in Fig. 3.
2. Use your scale and measure down the vertical a length equal to H, and establish point A.
3. From point A draw two more lines (numbers 4 and 5) that are parallel to lines number 1 and 2.
4. Along line 5, measure the width W, locating point B. Similarly measure the depth D along line 4 to give point C.
5. From points B and C, draw the vertical lines 6 and 7 that intersect lines 1 and 2, and locate points E and F.
6. From point F, draw line 8 parallel to line 2. From point E draw line 9 parallel to line 1.
7. Lines 8 and 9 that intersect at point G complete the isometric drawing.

The isometric pictorial that we have drawn shows the left-side view, the top view, and the front view, because we are viewing the object from above looking from the left to the right. The origin of the isometric coordinates is positioned at the upper left hand corner of the rectangular block. The procedure for preparing isometric drawings is easy to follow. All of the lines are parallel to the isometric axes, and all of the measurements are to the same scale.

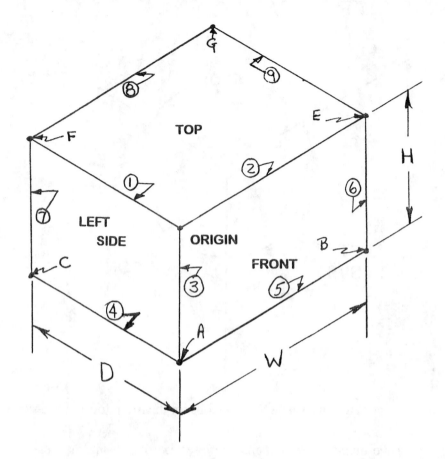

Fig. 3 Isometric pictorial of a rectangular block.

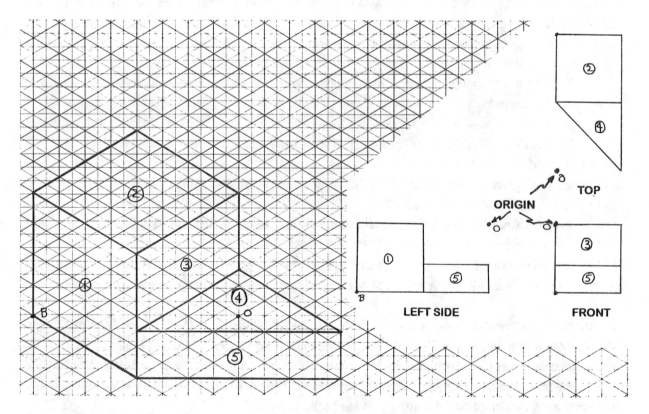

Fig. 4 Isometric pictorial of a block with a step and a tapered side.

The rectangular block was too easy. Let's try a more complex geometry as shown in Fig. 4. Look first to the inset of Fig. 4, that shows a three-view drawing of an odd shaped block. It is an unusual three-view drawing in that it presents the left-side view rather than the right-side view. The three-view drawing is presented in this manner, because the isometric drawing gives the same three-view s.

For the simple rectangular block, we placed the origin of the isometric axes at the upper left hand corner of the block. However, in this case that location is not recommended as a point to begin our pictorial, because it is located out in space in our three-dimensional representation. This fact is evident, if we locate the origin by the point O in the three-view insert of Fig. 4. In this example, it is easier to work from some other point. Let's try point B located in the left rear corner as a point to start your drawing.

The isometric pictorial, presented in Fig. 4, is drawn on isometric paper with only a straight edge to guide the lines. If you have a steady hand, you can sketch the lines without the straight edge. Isometric paper gives many evenly spaced lines parallel to the isometric axes eliminating the need for the 30° triangle and the scale. We begin at point B and draw plane 1 in the region of the left view. We do not draw the entire left-side view, because there is a taper to the block that complicates this view in the isometric representation. We leave the left-side view incomplete, and move to the top view and draw plane 2. Both planes 1 and 2 are clearly defined in the three-view drawing and are easy to construct on the isometric pictorial. Again we do not complete the top view, because the step and the taper complicate the geometry. We move next to the front view and add plane 3 which defines the depth of the step. We can now go back to the top view and draw plane 4 as shown in Fig. 4. Now that the top view is complete in the isometric drawing, and we know the height of the step, it is easy to draw plane 5. Note that plane 5 is on the surface formed by the taper. It does not lie in a plane formed by the isometric axes. We locate this non-isometric plane by first drawing all four of the isometric planes on the pictorial. It is then clear from the location of planes 1 and 4, where plane 5 is positioned.

You now understand how to draw simple rectangular blocks, and more complex blocks with both a step and a taper.

Let's next consider a cylinder of diameter D and height H, and illustrate it with an isometric pictorial. To draw the cylinder, lay out two isometric axes located a distance H apart using light construction lines as indicated in Fig. 5. On the upper set of axes, draw a square in the top plane with the length of the sides of the square equal to the diameter of the cylinder. Then select an isometric ellipse (35° - 16') from an ellipse template, and draw an isometric ellipse so that it is tangent to each side of the square drawn in the top plane. (If you do not have one of these handy templates, sketch the ellipse in by hand). Note that the ellipse is tangent to the isometric axes at four points as indicated in Fig. 5. Move down to the second set of isometric axes which represent the bottom plane. Take the handy template and draw isometric ellipse again. This time draw only the front half of the ellipse, since that portion is all that is visible when we view the cylinder from above. The two ellipses are joined with vertical lines at their outside points to complete the cylinder. The construction lines in Fig. 5 remain, to help you understand the procedure used in this drawing. To finish the drawing, erase the construction lines and shade the cylinder to enhance the visual effect of a three-dimensional object. Shading and shadows will be discussed later in this chapter when we refer to this isometric drawing of a cylinder again.

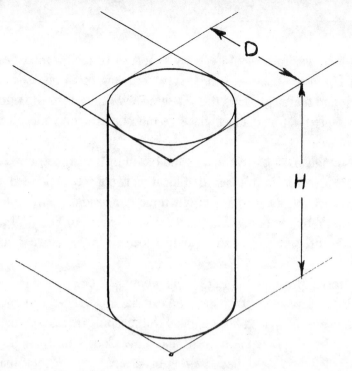

Fig. 5 Isometric pictorial of a right circular cylinder of diameter D and height H.

OBLIQUE DRAWINGS

Oblique pictorials and isometric pictorials are similar because both use parallel lines in constructing the three views. The difference between isometric and oblique pictorials is in the definition of the axes. In oblique drawings, we use a x,y, z coordinate system as shown in Fig. 6. The three coordinate axes divide the sheet into three regions for drawing the front, top and right-side views. With the axes defined as shown in Fig. 6, we are viewing the object from above looking from right to left. The z axis, which is the receding axis in Fig. 6, is drawn with a 45° angle relative to the x axis; however, other angles such as 30° or 60° are often employed.

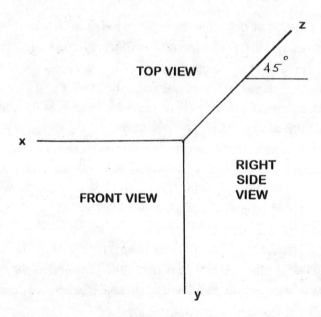

Fig. 6 Oblique axes use the x, y, z coordinate system.

Three different types of oblique drawings are frequently used for pictorials. They are illustrated in Fig. 7:

1. Cavalier oblique can be drawn with the receding axis at any angle from 0 to 90°, but the measurements along all three axes are the same scale.
2. Cabinet oblique can be drawn with the receding axis at any angle from 0 to 90°, but the measurements along this axis are half scale.
3. General oblique can be drawn with the receding axis at any angle from 0 to 90°, but the measurements along all this axis varies from half to full scale.

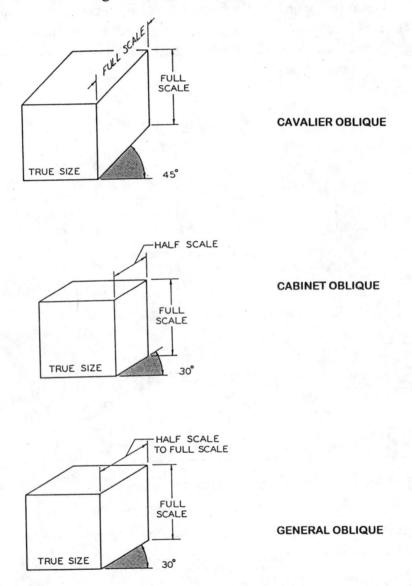

Fig. 7 A pictorial of a cube drawn with cavalier, cabinet and general oblique techniques.

If we examine the cube represented by the three types of oblique drawings in Fig. 7, it is evident that full scale (true length) measurements are used in the front view in all three types of oblique pictorials. The difference among them is the scale used along the receding axis. In cavalier oblique, the full scale measurements are made along the receding axis. This scale produces a drawing that looks out of proportion. The cube does not look like a cube.

In cabinet oblique half scale measurements are made along the receding axis. The resulting drawing is in better proportion than the cavalier oblique, but sometimes it appears that the depth dimension along the receding axis is too short. We prefer the general oblique where the measurement on the receding axis can be varied from ½ scale to full scale to give what appears to the eye to be the correct proportions.

To illustrate the procedure followed in drawing an oblique pictorial, examine the three-view drawing of a pair of connected rectangular blocks as shown in Fig. 8. To begin our oblique pictorial, draw the x, y, z axes using the lower left hand corner as the origin (point O). First draw the part of the front view (plane 1) that corresponds to the front of the large block. Then draw part of the top view (plane 2) of the large block. In drawing the front view, use true lengths (on plane 1) to measure the width and height of the large block. On the top view, establish the depth of the large block, maintaining perspective by using a scaling factor of ¾.

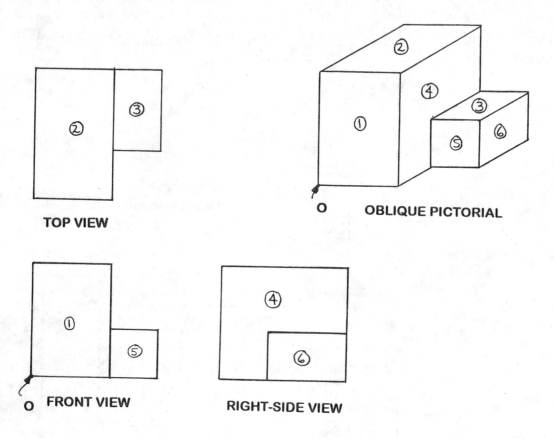

Fig. 8 Three-view drawing of a connected pair of rectangular blocks together with an oblique pictorial of the same object.

Note that the right-side view of the large block is obstructed by the smaller block. We handle this obstruction by drawing the small block. Using the information in the three-view drawing we can place plane 3 on the oblique pictorial. Plane 3 is located by working from the back edge in the top view. The width and height measurements required to locate and size plane 3 in the oblique pictorial are true lengths, but the depth is again scaled by ¾.

After we have drawn plane 3 locating the small block in the pictorial, it is easy to draw plane 4 by referring to the right-side view in the three-view drawing. Complete the drawing of the small block by dropping vertical lines down from three corners in plane 3 and closing the sides that form planes 5 and 6.

The resulting oblique pictorial clearly captures the relative proportions, and the positioning of the two blocks. A comparison of the pictorial with the three-view drawing demonstrates the advantages of the pictorial in visualizing the object. The pictorial is much more effective in visualizing the component. The three-view drawing is better for more precise definition of size and location. We dimension the three-view drawing and use it in the shop for manufacturing. Usually the pictorial is not dimensioned, and it is not use it as a detail drawing intended for the person manufacturing the component.

As final examples of preparing oblique pictorials, consider the drawing of a cylinder or a block with a circular hole as shown in Fig. 9. It is easy to draw a cylinder or a circle on an oblique pictorial, providing the required circles are placed on either the front plane or any plane parallel to the front plane. Since both the width and height dimensions are true length, and the x, y axes are orthogonal on the front view, the circle is not distorted into an ellipse. We can drawn the circle with a compass or a circle template. This is a significant advantage.

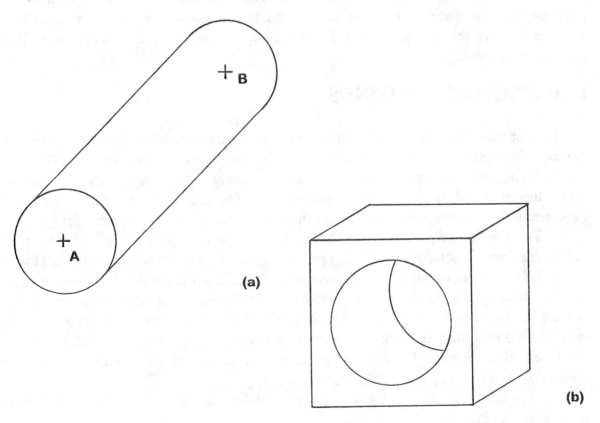

(a)

(b)

Fig. 9 Examples of oblique pictorials.
(a) Right circular cylinder. (b) Rectangular block with circular hole.

To represent the cylinder shown in Fig. 9a, we draw two circles centered at points A and B. Points A and B lie on the z (receding) axis which has been drawn at a 45° angle to the x axis. The circles are drawn in with construction lines, and the spacing of the two circles is scaled at a factor of ¾ relative to the length of the cylinder. The diameter of the circles are full scale. Place two lines parallel to the z axis are drawn tangent to the circles to give the sides of the cylinder. Draw the front circle with a heavy line, and darken the top right portion of the back circle. The resulting pictorial shows a somewhat distorted cylinder, The distortion is due to the fact that we have drawn the back circle with the same diameter as the front circle. Looking at a real right circular cylinder, the back circle would "appear" to be smaller than the front circle. We will address the apparent difference in the size of features located in different planes in the next section on perspective drawing.

The final example of oblique pictorial, presented in Fig. 9b, shows a simple rectangular block with a central hole. We have drawn the block so that the circle of the hole is see in its entirety in the front view.. The question is how to draw the circle in the back plane. Draw the square locating the back plane in the pictorial using light construction lines. Then find the center of this square and draw the circle with the same diameter as the front circle. Note that the two circles intersect, and that only a portion of the circle on the back plane is visible. Darken the construction lines on the back circle, over that part of the arc that is visible through the hole. The resulting pictorial is an effective three-dimensional drawing showing the appearance of a hole through a rectangular block. Our artist friends would critique us, because it too distorts the visual image of the object. The scale in the back plane and the front plane are the same and that produces distortion of the image. Usually this distortion is acceptable in engineering drawing, but we can improve our pictorials by preparing perspective drawings.

PERSPECTIVE DRAWINGS

Perspective drawings are pictorials that represent what we see either with our eyes or with a camera. This method of illustration is critical to the success of an artist making either sketches or paintings. Engineers also use perspective drawings, particularly when preparing visuals for folks who are not trained to read our more conventional three-view drawings. The tools used are the same as those described previously, except for adding a thumb tack and a piece of string to the list.

There is one very significant difference between perspective drawings and isometric or oblique drawings. In isometric or oblique drawings, the lines defining the edges (say top and bottom) are parallel to the axes; however, in perspective drawing, the lines defining some of the edges may not parallel. Drawing parallel often lines distorts the drawing, because we see parallel lines appearing to converge as they recede in space. The best illustration of this fact is a pair of railroad tracks, shown in Fig. 10. Looking down the tracks, you see that the tracks converge to a point and that the railroad ties appear to get shorter. The poles lining the track also appear to get shorter. Now everyone knows that the tracks are parallel. What's going on?

To understand what is going on, we need to define four terms that are used in describing prospective drawing.

1. Picture plane is the surface (i. e. the sheet of paper) of the pictorial. The edges of the paper represent the window through which you "see" a three-dimensional object that is the subject of the perspective drawing.

2. Horizon line divides the sky and the land or the sea if you are outdoors. In Fig. 10 this line is just below the level of the bridge that crosses the railroad tracks. The horizon line is at the level of your eyes and will change with your elevation. In a room where you cannot locate the true horizon because the walls of the block our view, we assume a horizon line at about the elevation of your eyes.

3. Viewing point and direction of view depends on the location of eyes relative to the object. You can look directly at the object, from left to right , right to left, downward, upward, etc. What you see changes markedly depending on these parameters. Look out a window at an object, and change where you stand and the direction of your view. Does the view of the object change?

4. Vanishing point is where parallel lines converge to a point as they recede into the distance. You can clearly identify the vanishing point in Fig. 10 where the tracks appear to meet.

Fig. 10 Photograph of railroad tracks showing the visual effect of convergence of parallel lines and foreshortening of objects in a distance.

ONE-POINT PERSPECTIVE

Depending on the view, an object can be represented by using one, two, or three-point perspective. Let's start with the simple one-point perspective and illustrate the approach by drawing a rectangular block. In one-point perspective, you place the front view of the block in the picture plane and show its true width and height as illustrated in Fig. 11. Then you draw a construction line to represent the horizon. The location of this line depends on the viewing point and the viewing direction. In Fig. 11, you are viewing the block straight on (not from the right or the left); however, you are above the block. Your eyes look downward and see the top surface of the block. The elevation is taken into account by raising the horizon line. Next, locate the vanishing point on the horizon line at the center point behind the front view, because we are looking straight on at the block. Draw construction lines from the vanishing point to the top corners of the block in the front view as shown in Fig. 11. The back edge on the top view is drawn parallel to the front top edge to establish the depth of the block. Note that the back edge is much shorter in length than the front edge. The shortening of the lines on the recessed planes give the illusion of the third dimension. The edges at the side are darkened to complete the pictorial. These edges are converging, again to give the illusion of depth on the two dimensional sheet of drawing paper.

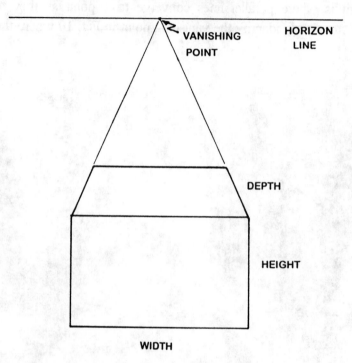

Fig. 11 Example of pictorial drawing with one-point perspective.

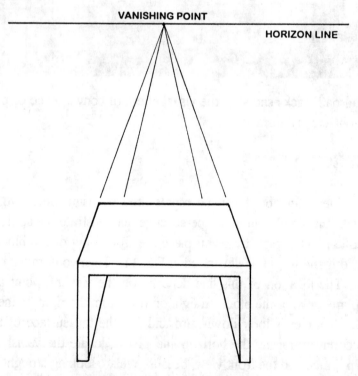

Fig. 12 One-point perspective drawing of a coffee table.

Another example of one-point perspective is the drawing of a coffee table presented in Fig. 12. The front of the table is drawn to scale in the picture plane as in Fig. 11. The width and height dimensions are true length. Coffee tables are low, so we stand looking downward, but straight on toward the top surface of the table. Draw the horizon line at eye level aligning the vanishing point with the center of the table. Next, draw light construction lines from the top outside corners or the table to the vanishing point. We also draw construction lines from the lower inside corners of the table legs as shown in Fig. 12. These construction lines define two triangles. We use the larger of these two triangles to draw the back top edge of the table, and then its edges to complete our perspective rendering of the top view. We use the smaller of the two triangles to draw in the bottom edge of legs that are visible under the table.

In a normal perspective drawing we would erase the construction lines, the vanishing point and the horizon line. We have not erased them in Fig. 12, because they were used to illustrate the procedure followed in making a one-point perspective drawing.

TWO-POINT PERSPECTIVE

The one-point perspective is useful when we view an object straight on so that its front view lies in the picture plane. However, if the object is rotated so that neither the front or side view is in the picture plane, as illustrated in Fig. 13, a two-point perspective is required. Again consider a rectangular block to illustrate the approach followed in drawing a two-point perspective pictorial. From Fig. 13, it is evident that only one edge of the block lies in the picture plane, so we will start with that fact ands proceed step by step to draw the block.

- Draw a vertical line, as indicated in Fig. 14, to establish the edge between left-side view and the front view.
- Draw the horizon line to reflect the fact that we are standing above the block, looking downward at an angle onto the top surface.
- Place two vanishing points on the horizon line. We have spaced the vanishing point located to the right of the vertical line (VP - R) farther from the vertical line than the vanishing point on the left-side (VP - L), because we are viewing the block at a slight angle from the right toward the left.
- Measure the true length of the vertical line (line 1) and label its ends with the letters A and B.
- Draw four construction lines connecting points A and B with VP - R and VP - L.
- Draw vertical lines (2 and 3) to locate the back edges of the left-side and the front of the block. The ends of these lines are located by the construction lines have been labeled (C, D) and (E, F).
- Draw top edge lines (4, 5) by connecting points F, A and D.
- Draw bottom edge lines (6, 7) by connecting points E, B and C.
- Draw construction lines from point F to VP - R, and from point D to VP - L, and locate point G at the intersection of these two lines.
- Draw sides (8, 9) by connecting points F, G, and D.

The two-point perspective drawing of the rectangular block is complete as shown in Fig. 14. The procedure may seem long when we outline it in steps, but it is not difficult. Once you understand the basic concepts of establishing the true length of the vertical line in the picture plane and the vanishing points on the horizon line, all of the other steps are routine.

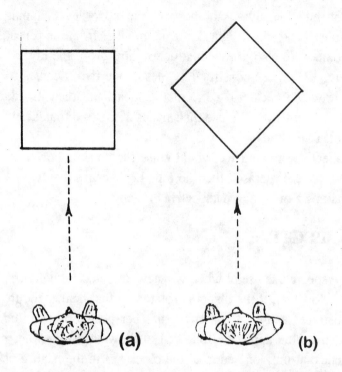

Fig. 13 Directions of viewing control the number of points used in perspective drawing.
    a.   Straight on --- one-point perspective.
    b.   Angle view --- two-point perspective.

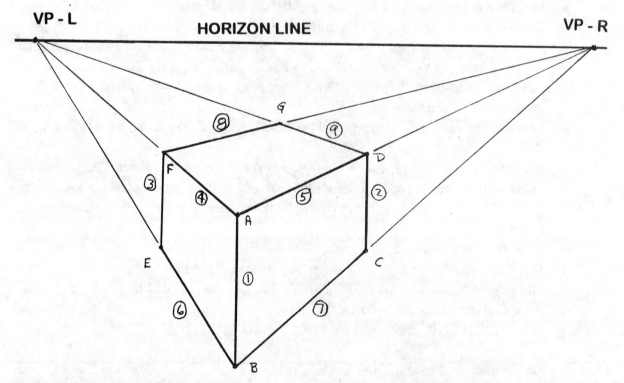

Fig. 14 Rectangular block represented with a two-point perspective drawing.

To complete the discussion of the two-point perspective, return to the example of the coffee table. This time, you will use two-point perspective to draw the table. Begin by placing the leg of the table in the picture plane as indicated in Fig. 15. The length of the leg is shown in true length on the vertical line that represents the edge of the leg. Establish the horizon line and the vanishing points to give the direction of the view that you want to represent. Draw construction lines from the ends of the vertical line to the vanishing points. Vertical lines are drawn to give the width and the depth of the table. The position of these vertical lines is not established by measurement, because the both the width and depth dimensions are not true length in the two-point perspective. Place the vertical lines in a position that maintains, "to the eye," the correct proportions of the table.

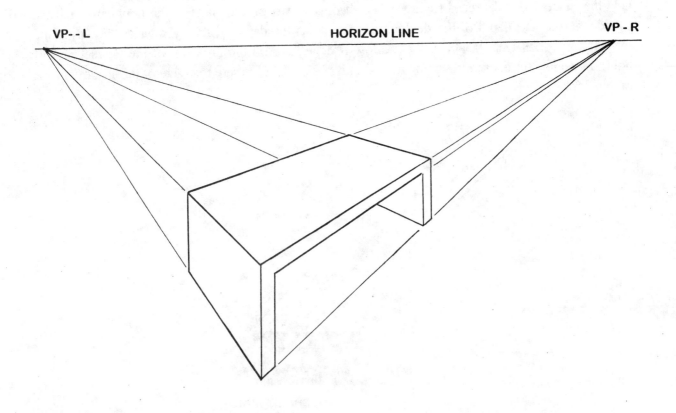

Fig. 15 Two-point perspective drawing of a coffee table.

Once the vertical lines that bound the left-side view and the front view are drawn, the remainder of the drawing is easy complete. Construction lines are drawn from the ends of the vertical lines to the vanishing points. The construction lines outline the table. It is necessary to darken only those segments of the construction lines that define the visible edges of the table.

Three-point perspectives are used when the object is very tall. Architects drawing a city view with skyscrapers would use three-point perspectives and taper the building as it extends into the sky. Engineers usually deal with smaller objects that usually can be represented in pictorials with either one or two-point perspective drawings. For this reason, we will not describe the methods used in preparing three-point perspectives. However, if you are interested in learning much more about perspective drawing, we recommend the excellent book by Powell [2].

SHADING AND SHADOWS

Add shading and shadows to our pictorial drawings makes them appear more realistic. Let's first distinguish between shading and shadow. If you light the object, some surfaces are exposed to this light and other surfaces are in the shade. There is a difference in the intensity of light reflected from these surfaces. Those in the shade are darkened slightly. A shadow is produce when an opaque object blocks the light. The shadow occurring in this region is shown as a dark area in the drawing. We show an example of shading and shadowing of a right circular cylinder in Fig 16. In this example, parallel light rays are illuminating the cylinder from the upper left. (They are included on the drawing only to show the logic for determining the shade and shadow regions). The right half of the cylinder is in the shade and is darkened slightly. The shadow is cast on the plane of the floor on which the cylinder rests. The depth of the shadow is the same as the depth of the cylinder as shown on the isometric pictorial. The length of the shadow is dependent on the direction of the parallel rays of light. Extending the lines representing the light rays helps define where the edge of the shadow forms on the floor.

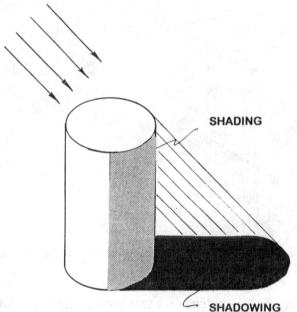

Fig. 16 Shading and shadowing of an isometric pictorial of a right circular cylinder.

Let's consider another example of shading and shadowing in a two-point perspective drawing of a cube. The drawing, presented in Fig. 17, is similar to that shown in Fig. 14, so we will assume that you can draw the cube in the two-point perspective pictorial.. In Fig. 17, we show the construction technique for determining the exact size of the shadow when the light is coming from a point source.

Suppose you have completed the two-point perspective drawing of the cube and are ready to shade and shadow the drawing. The process is first to select the location of the light source. It is your choice, but you will place the light source well above the horizon either to the left or the right of the cube. Second, you must select the vanishing point for the shadow. Again it is your choice as long as it is directly below the light source and in the ground (floor) plane. The reasons for these two constraints are evident. The shadow must vanish when the light source is directly overhead, and the shadow must always lie on the ground (floor) plane. You selected the vanishing point of the shadow on the horizon line; but it could have been placed anywhere on the ground plane under the light source.

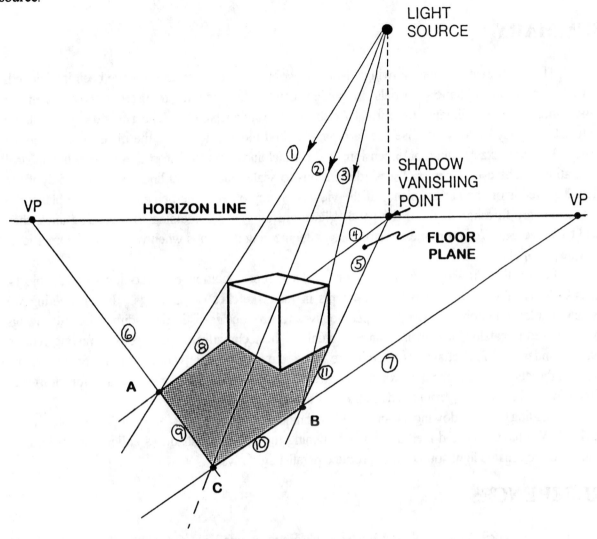

Fig. 17 Technique for determining shadow outline on a two-point perspective drawing.

Let's go though a step by step procedure for locating the shadow outline shown in Fig. 17:

1. Draw the light rays (1, 2, and 3) from the source through three of the top corners of the cube.
2. Draw two construction lines (4 and 5) from the vanishing point of the shadow through the two bottom external corners.
3. Locate points A and B at the intersections of the lines (1 and 6) and (3 and 7).
4. Draw lines 6 and 7 from the left and right vanishing points through points A and B.
5. The intersection of lines (6 and 7) locates point C.
6. Darken segments of the construction lines (8, 9, 10 and 11) to give the shadow outline.
7. Erase all of the construction lines and darken the shadow region within the outline.

We have completed the shadow, but the front and left-side views of the cube are in the shade. To complete the pictorial these two sides should be darkened slightly to indicate that they are in a shadow.

SUMMARY

Three different pictorial drawings, isometric, oblique and perspective, have been introduced. Both isometric and oblique pictorials are widely used in engineering to present three-dimension illustrations on a two-dimensional format (a sheet of drawing paper). The isometric and oblique pictorials are easy to draw because lines are parallel, and most if not all of the measurements are to scale. However, exact scaling of dimensions and parallel lines tend to distort these three-dimensional illustrations. The eye or camera "sees" dimensions to scale, and parallel lines as parallel only if the view depicted is on the picture plane. If the view is on any other plane behind the picture plane, the dimensions are foreshortened. If the parallel lines are on views containing the receding axes, they tend to converge. We ignore these distortions in engineering drawings when we prepare isometric or oblique pictorials.

Perspective drawings provide a more realistic representation of three-dimensional objects. Artists almost always use perspective concepts in their sketches or paintings. In engineering we usually employ one-point or two-point perspectives to produce realistic three-dimensional drawings that are used in catalogs and maintenance manuals. The techniques of perspective drawing require that we understand four quantities: the picture plane, the horizon line, vanishing point or points, and viewing direction. The techniques for drawing perspectives with converging lines and foreshortened dimensions follow directly from the definitions of these four quantities.

Shading and shadowing is an added technique for making pictorial drawings even more realistic. We have indicated methods for determining the outline of shadows either from point light sources or collimated light sources that produce parallel light rays.

REFERENCES

1. Franks, G. Pencil Drawing, Walter Foster Publishing, Laguna Hills, CA 1988.
2. Powell, W. F. Perspective, Walter Foster Publishing, Laguna Hills, CA 1989.
3. Earle, J. H., Engineering Design Graphics, Addison Wesley, Reading, MA 1983.

EXERCISES

1. What are the three types of pictorial drawings? Why do we use pictorial drawings in engineering applications?
2. Prepare an isometric pictorial drawing of a rectangular block 10 mm wide by 15 mm deep by 50 mm high. What scale should you use in preparing this drawing? Why?
3. Prepare an isometric pictorial drawing of the weighing machine that your team is developing.
4. Prepare an oblique pictorial drawing of the weighing machine that your team is developing.
5. Prepare a prospective pictorial drawing of the J. M. Patterson building. View the building from the intersection of the two frontage streets.
6. Draw a pictorial of a sphere with shading and shadowing.

CHAPTER 7

TABLES AND GRAPHS

INTRODUCTION

On many occasions we find ourselves with a collection of numerical data that we must present at a meeting, or in an engineering report. At other times, we may have results from a mathematical relationship that must be presented or reported. We have two choices in presenting numerical data. We can show the numbers in tabular form, or in a suitable graph. Both methods of presentation, the table or the graph, have advantages and disadvantages. The method that you choose will depend on your audience, the message that you are reporting, and the purpose of your presentation.

In this chapter, we will describe how to list data in a table using Microsoft's Word 7 as the word processing program. In a different chapter, we describe data entry into a spreadsheet which is also suitable method for preparing tables. We then introduce five common methods for representing data in the form of graphs which include:

1. Pie charts
2. Bar charts
3. X - Y graphs.
4. Semi-log graphs
5. Log - log graphs

Each type of graph is used for a different purpose and selection of the proper type of graph is essential in communicating effectively. For instance, the pie chart is used to show distributions (who gets the most and the least). The bar chart is used to compare one set of data with another set. X - Y graphs, the most frequently used type of graphing in engineering, show how Y varies with X.

When Y varies by very large amounts with small changes in X, we often use semi-log graphs which are capable of covering a very large range of Y values. When both X and Y vary over very large ranges, we display the results using a log-log graph representation.

TABLES

In Chapter 1 we described the responsibilities of Mechanical Engineers in their first position and presented statistical data in a table. Let's reproduce that table data, to show you how it is easy to use Word to prepare a table. We prepare the table heading, by selecting the center alignment, then identify the table number, and type the title. We often use a bold font for the table heading, to attract the reader's attention.

TABLE 1
Responsibilities of Mechanical Engineers in Their First Position

OK, we are now ready for the body of the table. Point the mouse to the menu row and select "Table". We have a choice of whether or not we want to show grid lines in the table. Let's click on the "Gridlines" label to use the grid lines to guide us as we enter the data. If we don't like them they can easily be removed at any time by clicking again on this label. Next, click on "Insert Table". The new menu that appears on the screen, requires some information about our table. We need to know beforehand the arrangement of the table (how many columns and rows). Let's say we want three columns and 15 rows. We can always add or delete either columns or rows later if we find that we were not correct in specifying the initial configuration. Finally, we select auto-format, and let Word initially control the width of the columns. Click back on the text, and the outline of the table appears. If the grid lines are dotted and light when the table first appears on the screen, they will not print as they are intended only to guide your typing. If you want the grid lines to show when the table is printed, click on Format/Borders and Shading. A new menu appears on the screen, and you click on the "Grid" button to choose the thickness of the grid lines that you want to outline each cell. The result of our first trial after we have typed the data into the correct cells is shown on the next page.

The result of our first attempt is fair, but it could be improved. The auto-format selection divided the width of the page into thirds for the width of the columns. This partitioning is not the best choice, considering the wasted space associated with the numerical entries and the double rows needed to accommodate the assignments. The first column titled ASSIGNMENT needs more width and the 2nd and 3rd columns are too wide. Let's modify the table by changing the column widths. We accomplish this by clicking on Table/Cell Height and Width. Select the column tab and adjust the width of the first column (3 in.), and click the Next Column button and adjust the width of both column 2 and 3 to 1 inch. The Table looks much better, and requires less space on the page, but it is not centered. To center it, click Table/Cell Height and Width and select the Rows tab. From the Rows menu select center alignment. The improved table is shown on the lower half of the next page. As a result of our modifications, it looks very professional.

ASSIGNMENT	% TIME	% TIME
Design Engineering		40
Product Design	24	
Systems Design	9	
Equipment Design	7	
Plant Engineering/ Operations/ Maintenance		13
Quality Control/ Reliability/ Standards		12
Production Engineering		12
Sales Engineering		5
Management		4
Engineering	3	
Corporate	1	
Computer Applications/ Systems Analysis		4
Basic Research and Development		3
Other Activities		7

TABLE 1
Responsibilities of Mechanical Engineers in Their First Position

ASSIGNMENT	% TIME	% TIME
Design Engineering		40
Product Design	24	
Systems Design	9	
Equipment Design	7	
Plant Engineering/ Operations/ Maintenance		13
Quality Control/ Reliability/ Standards		12
Production Engineering		12
Sales Engineering		5
Management		4
Engineering	3	
Corporate	1	
Computer Applications/ Systems Analysis		4
Basic Research and Development		3
Other Activities		7

Tables are effective in presenting precise numerical data. We do not need to estimate crudely a number by reading from a curve on an X-Y graph. If we have a reason to show our results with six significant figures, it is very easy to do so. A government agency that probably produces the most widely read tables in the United States is the Internal Revenue Service. They prepare tax tables each year that precisely show our tax obligations. We will present another example later that shows the advantage of a table in conveying numerical data with many significant figures. We will also show how spreadsheets can be employed to produce both graphs and tables in another chapter.

GRAPHS

We use graphs to visualize the data in both reports and presentations. The graphs are not intended to give precise results since we use tables for that purpose. The graphs show our audience a trend, a comparison, or a distribution quickly and effectively. There is no need for the audience to study the data to develop an understanding. Your graph shows the bottom line, and eliminates the time and effort required for the reader to analyze the data.

There are many different ways of presenting numerical data in a graphical format. Distribution of goodies are usually shown with pie charts, because we all quickly relate to getting a good size share of the pie. Comparisons are made with bar charts, with the height of each bar indicative of the magnitude of the quantity being compared. Linear graphs, where we plot a dependent parameter Y as a some function of an independent parameter X, is frequently employed. The data plotted can be generated with mathematical functions when we know the function Y(X). However, if the mathematical relation is not known, we conduct experiments and measure Y as we systematically change X. In both cases, the X-Y graphs represent the numerical data, and show a trend (increasing, neutral or flat, decreasing, oscillating, etc.). In some instances, the variations in X, Y or both of these quantities is extremely large, and it is not possible to clearly show the trends over the entire range of either X or Y on a graph with linear scaling. In these cases, we use a non-linear format, namely semi-log or log-log to present the widely ranging data. Let's consider five types of graphs and learn how to prepare them manual techniques. In a later chapter, we will show how to prepare these same graphs by using a spreadsheet.

PIE CHART

A pie chart is used to show how some quantity is distributed. In Table 1, we showed the various assignments for Mechanical Engineers starting there careers. Let's represent this same data in a pie chart as shown in Fig. 1. We begin by drawing a circle to represent our pie, and then we decide what fraction of the pie corresponds to each assignment. This determination is easy, if we remember that the whole pie contains an included angle of 360°. The slice of pie representing design engineering is a fraction of the whole pie, i. e. (360°)(0.40) = 144°. The size of the pie slices in terms of degrees for all of the engineering assignments are listed below:

1. Design engineering .... $(360°)(0.40) = 144°$.
2. Plant engineering, operations and maintenance .... $(360°)(0.13) = 47°$.
3. Quality Control, reliability and standards .... $(360)(0.121) = 43°$.
4. Production engineering .... $(360)(0.121) = 43°$.
5. Sales engineering .... $(360)(0.05) = 18°$.
6. Management .... $(360)(0.04) = 14°$.
7. Computer applications and systems analysis .... $(360)(0.04) = 14°$.
8. Basic research and development .... $(360)(0.03) = 11°$.
9. Other activities .... $(360)(0.07) = 25°$.

   We construct the pie chart by using a protractor to lay out these angles on the circle. The pie slices are then labeled as shown in Fig. 1. When we examine the pie chart, we can visualize the importance of design because the design slice is huge. We can also see the importance of producing product since these activities plant operations, quality control, and production combine to produce still another huge slice. The remainder of the pie, which is less that a quarter, is divided between, sales, management, computers and systems, research and development and other activities. These activities are important, but they represent opportunities for a much smaller fraction of the engineers beginning their careers. Compare the effectiveness of the data as presented in Table 1 and the pie chart of Fig. 1. The pie chart quickly leads the reader to the conclusion that design and production are where most of the opportunities exist for entry level positions in Mechanical Engineering.

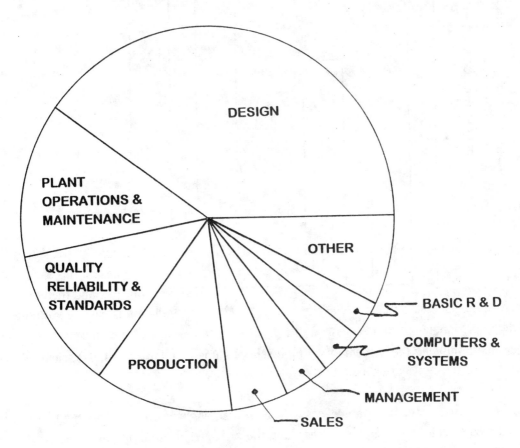

Fig. 1 A pie chart showing the type of initial assignments in industry for beginning engineers.

BAR CHART

Bar charts can also be used to show distribution, but they are better suited for illustrating comparisons. As an example of a bar chart, consider the results of a study by the Council of Chief State School Officers on the percentage of high school students graduating with three or more years of secondary mathematics and science in 1982 and 1994. We have prepared a bar chart showing the comparison in Fig. 2. The visual effect enhances the comparison. At a glance, we can see that a lot more students were taking mathematics and science in 1994 than in 1982. The second glance indicates that only about a third of the students were in math and science in 1982, while more than half of them were taking math and science in 1994. Finally, we can note that math seems a bit more popular than science in both 1982 and 1994. Three quick glances and the reader understands the data, because you have presented it in a visual format that effectively carries the message.

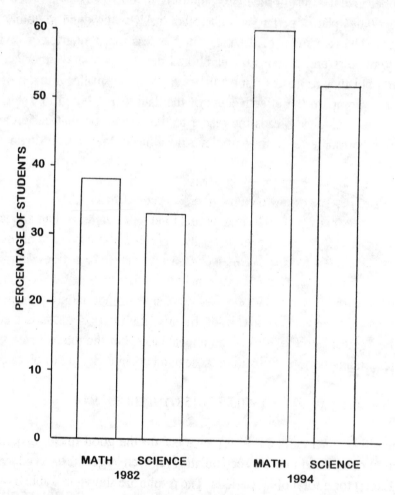

Fig. 2 A bar chart showing the percentage of high school students taking math and science in 1982 and 1994.

The preparation of the bar graph is easy. We used an engineering paper with a barely visible grid that facilitated establishing a suitable scale along the ordinate. A straight edge was employed in drawing the bars, and the grid lines visible on the engineering paper were used to scale each bar. The time to prepare the bar chart was about 10 minutes.

LINEAR X-Y GRAPHS

X-Y graphs are probably the most frequently employed graph used in engineering because it illustrates trends. The curves are produced by plotting Y the dependent variable along the ordinate as a function of the independent variable X which is displayed along the abscissa. Connecting the points with a curve or line segments indicates the trend in Y with changes in X. Graphing X-Y can be accomplished using linear scales for both X-Y, a linear scale for X and a log scale for Y, or log scales for both X and Y. We will cover all three of these methods of producing X-Y graphs.

We are going to need numerical results for our examples in discussing X-Y graphs. We will generate our numerical data using a very important equation that relates to money, and how you may choose to accumulate it, or spend it. Let's begin by assuming that you are going to accumulate some money. Sounds good? Suppose your rich uncle purchased a mutual fund for you at your birth. Your uncle plans to give you the proceeds of the fund on your 21st birthday. How much will you have accumulated when the big day arrives. The sum S accumulated is given by the following relation:

$$S = P(1 + i)^n \qquad (1)$$

where

   S is the sum accumulated.
   P is the amount of the initial investment.
   i is the interest rate for the compounding period.
   n is the number of periods over which the interest accumulates.

Suppose your rich uncle invested P = $5000 in a mutual fund that guaranteed an annual interest rate I = 10 %, and the interest is compounded semi-annually. We will also assume that uncle avoided the tax collectors (Federal, State and Local). Avoidance is not possible, but it does simplify the analysis. Let's calculate exactly how much money you can expect to collect on your 21st birthday. We need to compute S in Eq. (1) knowing that P = $5000. The number of compounding periods that will occur from your birth until you are 21 years of age is 42, because the fund compounds the interest earned twice each year. The interest rate I = 10% on an annual basis, but the interest rate i is only 5% per the semi-annual compounding period. Substitute these numbers into Eq. (1) to obtain:

$$S = \$5000(1 + 0.05)^{42} = \$5000(7.761587) = \$38,807.94$$

Wow! This is wonderful. You can buy a very nice car on the good uncle, or you can defer this option, and continue to compound interest accumulating an even larger sum. We have evaluated Eq. 1 in a spreadsheet (Excel) for a total of 80 periods. The results are shown in Table 2 on the next page.

In examining Table 2, we find that the uncle's $5,000 has grown over the years. The table gives accurate values at the end of each compounding period. For example, at the end of the 20th period, your tenth birthday, the principal in the mutual fund was $13,266.49. Note that the table conveys the exact numerical value with as many significant figures as required. If you maintain the fund until you are 40 years old, accumulating interest for 80 periods, you would have a total of $247, 807.21. Quite a growth from the $5,000 seed money planted by uncle.

Table 2

**Accumuated Sum from an Investment of $5000.00
at an Interest Rate of 10 %
Interested Compounded Semi-annually
No Taxes Paid on Interest Earned**

Periods	Multiplier	Sum	Periods	Multiplier	Sum
0	1.0000	$ 5,000.00	40	7.0400	$ 35,199.94
1	1.0500	$ 5,250.00	41	7.3920	$ 36,959.94
2	1.1025	$ 5,512.50	42	7.7616	$ 38,807.94
3	1.1576	$ 5,788.13	43	8.1497	$ 40,748.33
4	1.2155	$ 6,077.53	44	8.5572	$ 42,785.75
5	1.2763	$ 6,381.41	45	8.9850	$ 44,925.04
6	1.3401	$ 6,700.48	46	9.4343	$ 47,171.29
7	1.4071	$ 7,035.50	47	9.9060	$ 49,529.86
8	1.4775	$ 7,387.28	48	10.4013	$ 52,006.35
9	1.5513	$ 7,756.64	49	10.9213	$ 54,606.67
10	1.6289	$ 8,144.47	50	11.4674	$ 57,337.00
11	1.7103	$ 8,551.70	51	12.0408	$ 60,203.85
12	1.7959	$ 8,979.28	52	12.6428	$ 63,214.04
13	1.8856	$ 9,428.25	53	13.2749	$ 66,374.74
14	1.9799	$ 9,899.66	54	13.9387	$ 69,693.48
15	2.0789	$ 10,394.64	55	14.6356	$ 73,178.15
16	2.1829	$ 10,914.37	56	15.3674	$ 76,837.06
17	2.2920	$ 11,460.09	57	16.1358	$ 80,678.92
18	2.4066	$ 12,033.10	58	16.9426	$ 84,712.86
19	2.5270	$ 12,634.75	59	17.7897	$ 88,948.50
20	2.6533	$ 13,266.49	60	18.6792	$ 93,395.93
21	2.7860	$ 13,929.81	61	19.6131	$ 98,065.73
22	2.9253	$ 14,626.30	62	20.5938	$ 102,969.01
23	3.0715	$ 15,357.62	63	21.6235	$ 108,117.46
24	3.2251	$ 16,125.50	64	22.7047	$ 113,523.34
25	3.3864	$ 16,931.77	65	23.8399	$ 119,199.50
26	3.5557	$ 17,778.36	66	25.0319	$ 125,159.48
27	3.7335	$ 18,667.28	67	26.2835	$ 131,417.45
28	3.9201	$ 19,600.65	68	27.5977	$ 137,988.32
29	4.1161	$ 20,580.68	69	28.9775	$ 144,887.74
30	4.3219	$ 21,609.71	70	30.4264	$ 152,132.13
31	4.5380	$ 22,690.20	71	31.9477	$ 159,738.73
32	4.7649	$ 23,824.71	72	33.5451	$ 167,725.67
33	5.0032	$ 25,015.94	73	35.2224	$ 176,111.95
34	5.2533	$ 26,266.74	74	36.9835	$ 184,917.55
35	5.5160	$ 27,580.08	75	38.8327	$ 194,163.43
36	5.7918	$ 28,959.08	76	40.7743	$ 203,871.60
37	6.0814	$ 30,407.03	77	42.8130	$ 214,065.18
38	6.3855	$ 31,927.39	78	44.9537	$ 224,768.44
39	6.7048	$ 33,523.76	79	47.2014	$ 236,006.86
40	7.0400	$ 35,199.94	80	49.5614	$ 247,807.21

Let's represent the data shown in Table 2, as an X-Y graph. We first decide that the sum S is the dependent variable to be represented along the ordinate (the Y axis), and that your age is to be the independent parameter displayed along the abscissa (the X axis). We draw the X and Y axes on our engineering paper (which has a the faint grid), and apply a scale to each axis, as shown in Fig. 3. Scaling is very important because it determines the size of our graph. If the scale is too small, the graph looks like a postage stamp on your paper, and you have wasted an opportunity to show the graph to full advantage. However, if the scale is too large, we can not fit all of the data on the paper. To scale properly look at the range of the data to be covered for both the X and Y variables. In this case, the range for X varies from 0 to 21, and Y varies from $5,000 to $38, 807.94. We selected a scale of 1 inch equal to 5 years along the X axis, and a scale of 1 inch equals $5000 along the Y axis. The range displayed for X varied from 0 to 25, and for Y from 0 to $40,000. We add numbers to each axis, beginning with zero at the origin and incrementing in steps of 5 along both axes. Captions to both axes are added to remind the reader of the definitions of X and Y. These choices give the graph shown in Fig. 3, that is approximately 5 by 8 inches in size, which fits nicely on a single page with sufficient room for the figure caption and the margins.

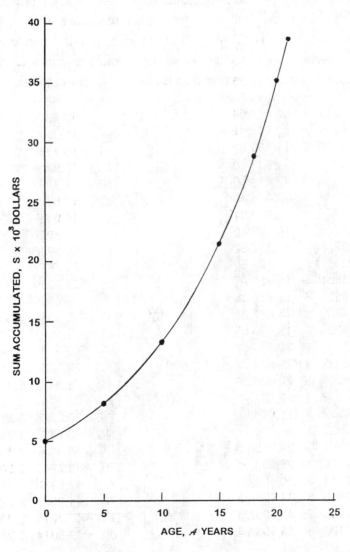

Fig. 3 A X-Y graph illustrating the sum accumulated with time from an initial investment of $5,000.

The data consisting of X and Y coordinates, shown in Table 2, locate the points that are plotted on the graph. We have used a drop compass to draw the small points, but, if you don't have this instrument, the points can be place by free hand. We have only plotted seven points to establish the curve. The number of points that are plotted depends on the function being graphed. In this case Y varies monotonically with respect to X. We can draw the curve accurately with only six to eight points when the function is relatively smooth and does not exhibit peaks or valleys.

Inspecting Fig. 3 clearly shows the trend of the accumulation sum S with respect to time (age). The sum increases continuously with time, and the rate of the increase (slope of the curve) is also increasing. The visual effect of the X-Y graph is dramatic. You know at a glance that you are getting rich. The only question is how rich? If we look at the ordinate and abscissa, we can estimate the sum accumulated for a given age, but the estimate may be in some error. The faint grid lines that we used in plotting the data do not show in the copy presented in Fig. 3. We lose accuracy in reading our graph without grid lines.

If we seek to prepare an X-Y graph that can be used to show both trends, while retaining some accuracy in reading the scales, a good quality graph paper with a fine grain grid should be employed. An example of this graph, plotted to the same scale on a graph paper with grid 20 lines to inch, is shown in Fig. 4. With these fine pitch grid lines, we can read the scale to $\pm (1/20)(5) = \pm 0.4$ years of age or $\pm \$400$ for the sum S. The resolution of the scale, that is possible with good quality graph paper, is a big improvement over the graph without visible gridlines as shown in Fig. 3. However, the best accuracy in reporting numerical data is obtained by using tables.

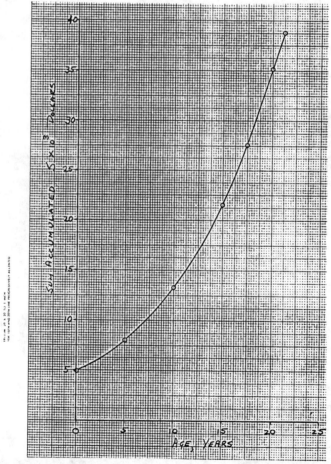

Fig. 4 A X-Y graph with fine pitch gridlines used for showing trends and reporting numerical values.

X-Y GRAPHS WITH SEMI-LOG SCALES

The results presented in Table 2, indicate that you can accumulate some very serious money from a relatively small initial investment if it is invested for a long time. If we try to plot the larger numbers associated with the sum S in Table 2 on the linear graph of Fig. 4, we would go off-scale. We could re-scale, but we would lose resolution particularly along the ordinate where the range in S is very large. A better approach is to use semi-log paper in preparing the graph. Semi-log paper, shown in Fig. 5, has a linear scale (in this example along the X axis), and a log scale (along the Y axis). Looking closer at the Y axis indicates that the graph paper is scaled with three log cycles over the 10 inches of the paper which are ruled. We have selected three log cycles, because it gives the range that we need to cover the range of values for the data given in Fig. 2. Of course, semi-log graph paper is commercially available with several different numbers of cycles on the log scale.

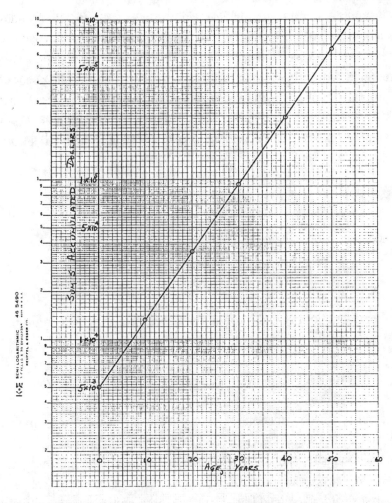

Fig. 5 A semi-log representation of the sum S accumulated with time.

Let's scale both axes on our semi-log graph paper. The X axis is scaled with one inch equal to 10 years, covering the range from 0 to 50 years. The Y axis scale is imposed upon us by the selection of the paper. (Remember that we selected three cycle paper, so we can cover a range of three decades). We label the first decade in thousands, the second in tens of thousands and the third in hundreds of thousands. Zero can not exist on a log scale, because the $\log_{10}(0)$ is not defined. In our graph in Fig. 5, we have placed the origin at $Y = \$2000$ and have provided for a margin at the bottom of the semi-log paper. It follows that the maximum value of Y which we can show on our three cycle paper is $\$1,000,000$.

We have labeled the Y scale at locations of 1 and 5 as designated on the log scale of the preprinted graph paper. The number is multiplied by 10^3 or 10^4 or 10^5 depending on the cycle involved. Next we place points on the graph at the coordinate locations defined in Table 2. In this example, we can draw a straight line through the data points. The reason for the linearity of the relation for the sum S on a semi-log graph paper is evident, if we take the log of both sides of Eq. (1) to give.

$$\log_{10} (S) = \log_{10} (5000)(1 + 0.05)^n = \log_{10} (5000) + \log_{10} (1 + 0.05) (n) \qquad (2)$$

Examination of the right hand side of Eq. (2) shows that $\log_{10} (S)$ is linear in the number of periods of accumulation (n). The intercept of the straight line with the Y axis (when $n = 0$) is $\$5000$. The slope of the straight line is $\log_{10}(1 + 0.05)$, which is the coefficient of n.

The use of semi-log paper has two advantages. First, it permits us to cover a very large range in the quantity represented along the Y axis. We have used a three cycle paper which covered a range from 2000 to 1,000,000. If the need existed, we could cover even a larger range by selecting a semi-log paper with more log cycles. The second advantage of semi-log representation is that it converts certain non-linear (exponential) functions, like Eq. (1), into linear relations that are much easier to interpret and to extrapolate.

X-Y GRAPHS WITH LOG-LOG SCALES

When you encounter power functions of the form:

$$Y = AX^k \qquad (3)$$

it is advantageous to represent both X and Y on log scales in preparing a X-Y graph. If we take the log of both sides of this power function we obtain:

$$\log_{10} (Y) = \log_{10} A + \log_{10} X^k = \log_{10} A + n \log_{10} X \qquad (4)$$

With the logarithmic form of the function $Y = AX^k$, the use of log-log graph paper is ideal. Representing both X and Y on a log scale gives us the opportunity to visualize the constants A and k in the power function. To illustrate this fact, we have taken as an example the function $Y = 2X^{1.4}$, and plotted it in Fig. 6. We cover values of X from 0.1 to 100 and values of Y from 1 to 1000. We have also shown the scales for $\log_{10} Y$ and $\log_{10} X$ in Fig. 6, to emphasize the difference between the two quantities.

There are three advantages for using log relations and log-log scales on graphs:

1. The linearization of non-linear power functions that often arise in engineering.
2. The ability to plot wide ranging numerical data.
3. Plotting a power relation on a log-log graph gives a straight line that is easy to interpret.

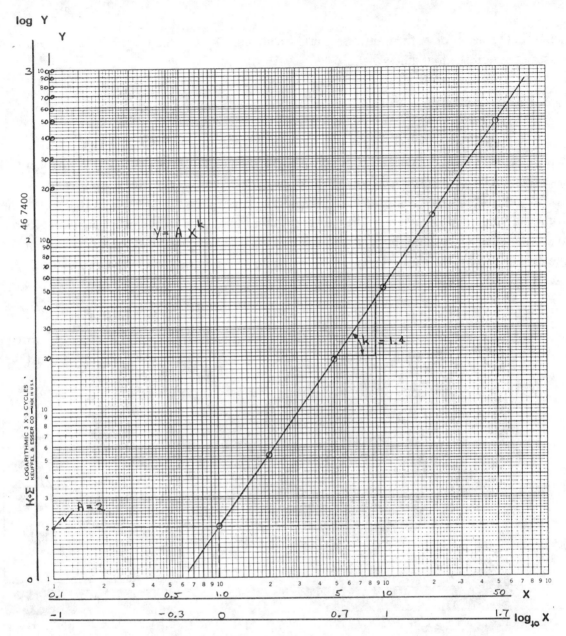

Fig. 6 A log-log graph of a power function $Y = AX^k$.

If the data is from experimental studies, it is often necessary to establish the constants A and k in a power relation like that shown in Eq. (3). The graphical procedure for determining these constants is given in Fig. 6. We find the constant A by setting $X = 1$, and reading the corresponding Y value as indicated with the dashed lines in Fig. 6. We find the constant k by determining the slope of the line. Remember in finding this slope, that it is necessary to work with the small numbers taken from the log scales. Accordingly, the slope k is given by:

$$k = (\log_{10} Y_2 - \log_{10} Y_1) / (\log_{10} X_2 - \log_{10} X_1) = (2.68 - 0.30)/(1.7 - 0) = 1.40. \qquad (5)$$

The log or log-log graphs are very important because many processes, that are widely used in engineering, exhibit non linear behavior which can be modeled with either exponential or power functions. The behavior of these processes are usually best represented by graphs with either one or both scales in terms of logarithms.

SPECIAL GRAPHS

There are many special graphs that have been developed to help the reader quickly absorb and understand numerical data. This brief chapter on Tables and Graphs does not permit a complete description of the many special and clever graphs that have been devised. However, we will introduce one of these, namely the geographical graph, which is based on a map of the world, the U. S., or some state. Superimposed on the map are contour lines dividing the entire geographical region into sub regions. The contour lines or sub regions are labeled to give the geographical distribution of some quantity across the mapped region. An example of a geographical graph, presented in Fig. 7, shows the weather prediction for the U. S. for April 22, 1997.

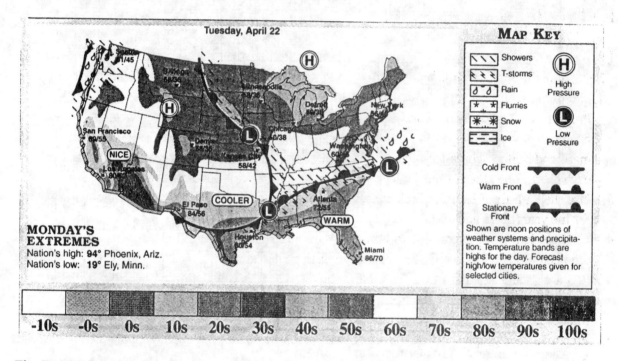

Fig. 7 A geographical graph showing the weather prediction for the U. S. for April 22, 1997.

SUMMARY

We have introduced tables as an approach for reporting data. They have the advantage in presenting precise numerical data. If you need six figure accuracy, it is easy to obtain in a tabular representation of the data. We also outlined the procedure for producing very tables presentable tables using Word 7.

Many different types of graphs are used to help in the visualization of the data. Graphs are much less precise than tables, but they more effectively show trends, comparisons, and distributions. We first introduced the pie chart which, as the name implies, is a circular area that is divided into pie like segments to indicate distribution of some quantity. Techniques for determining the size of the slices of pie were described.

We then introduced the bar chart that is used to compare the magnitudes of two or more quantities. The bar chart was illustrated, and the methods used to construct it were discussed.

Trends are best illustrated using line graphs including the linear, X-Y linear graphs, X-Y graphs with semi-log scales, and X-Y graphs with log-log scales. Linear X-Y graphs are used to show the trend of one variable with respect to another when the range of both variables is not too great. X-Y graphs are presented on semi-log graph paper when one variable has a very large range while the other variable has a more limited range. Log-log graph paper is employed to represent data fields where both variables change over very large ranges. We have provided examples for these three different types of graphs, and have described techniques used in their construction.

Finally, we have discussed special graphs that make use of maps to convey the concept of geography in the presentation of data. The map presented to demonstrate this technique is the weather map. It shows the use of contour lines dividing the U. S. into temperature zones.

EXERCISES

1. Write a memo for a group of young engineers starting their careers with your company that explains the policy of the engineering department relative to reporting data with tables or with graphs.
2. Prepare a table using a word processing program that shows the SAT scores for entering freshman at your College of Engineering for the time period from 1980 to 1996. Report both the math and verbal scores for women and men. Also tabulate the number of women and men in each class. The Dean of the College should be able to provide you with the data that is needed.
3. Use the data from exercise 2 to construct a bar chart comparing the scores of women and men for the years 1985 and 1995.
4. Use the data from exercise 2 to construct two pie charts showing the percentage of men and women enrolled as freshmen in the College of Engineering in 1985 and 1995.
5. Prepare a table like that shown in Table 2 showing the sum S with time. Use P = $1000, I = 8%, with interest compounded quarterly. Determine the Sum after each compounding period for a total of 20 years.
6. Prepare a X-Y graph showing the sum S with respect to time using the data generated in exercise 5.
7. Prepare a semi-log graph showing the sum S determined in exercise 5 on the log scale and the time on the linear scale.
8. Evaluate the power function $Y = 1.8 \, X^{1.75}$, and plot the result on a log-log graph. Let X vary from 1 to 100. Try to find log-log paper with a suitable number of cycles for both axes. Confirm that A = 1.8 and k = 1.75 from your graph.

PART III

SOFTWARE

APPLICATIONS

CHAPTER 8

KEY CAD Complete – for WINDOWS

INTRODUCTION

KEY CAD Complete is a Computer Aided Drawing (CAD) program that has been selected for your use in ENES 100. It is an entry level program, that is available to the student at a modest cost ($ 29.95 plus shipping, handling and taxes). The program is also relatively easy to learn to use, and a beginning student can produce high quality three-view and isometric drawings with a modest investment of time. All CAD programs have the advantage that they remove the requirement of manual dexterity from the drawing process. If you never learned to write or print clearly, you can breath a sigh of relief because you type printed matter in KEY CAD as text. If you can point with the mouse, hold it steady, click, and learn to exercise the menu and icon commands, you will be able to produce first class drawings in KEY CAD.

We will introduce you to the KEY CAD screen, the menu, the icon buttons, and the drawing pad. This introduction will be brief, because it is easier to learn by clicking than by reading. We also describe many of the more important menu items, and the lists of features available to you under each item. To illustrate the features of KEY CAD, we consider some simple examples, and show you how to produce engineering drawings. The examples will demonstrate how to prepare:

- A drawing block.
- A three-view drawing of a notched rectangular block.
- Dimensioning a three-view drawing.
- A three-view drawing of a more complex object with dimensions.
- An isometric drawing

THE KEY CAD SCREEN

When KEY CAD is loaded, the screen that is displayed is shown in Fig. 1. It is relatively simple. A line of menu items is located along the top edge of the screen, icon button pads are located on the left and right sides of the screen and along the top of the drawing pad just below the menu line. Finally, data showing the position of the cursor is shown below the drawing pad. The main portion of the screen is dedicated to the drawing pad. The file name of the drawing (in this case SCREEN1.KEY) appears along the line between the toolbar and the menu line after you save the file. The two strings of icon buttons may not be displayed on right side of your screen. We will show you how to display these strings later in the chapter.

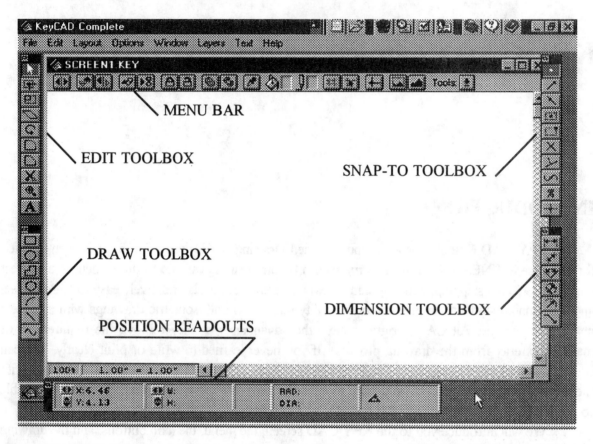

Fig. 1 The KEY CAD screen displayed on your monitor.

We move the cursor around the screen with the mouse and click either on one of the menu items or one of the icons in a tool bar. The keyboard can be used for some data entries with some of the tools, but we will focus on pointing, clicking and dragging with the mouse to make our drawings. Let's first examine the top bar on the screen which lists the menu.

The menu line contains the following headings:

- File…for handling the files, saving, printing output, and quitting.
- Edit…for editing with features such as cut, copy, paste, hide, duplicate, etc.
- Layout…for changing the screen, grid, dimensioning, and line styles.
- Option…for selecting fill patterns, lock state, position of origin and size.

- Window…for adjusting scrolling speed and displaying tool boxes.
- Layers…for locating and positioning layers in a multi-layer drawing.
- Text…to select the size and style of the font, and its alignment.
- Help…a complete indexed description of the program is available on screen.

Each menu heading has a pull down list. Use the mouse, point to a menu item, click the left mouse button (LMB), and a pull down list of selections is displayed on the screen. Move the cursor off to the side, and click the LMB to make the list disappear. The menu bar in KEY CAD is very similar to that found in your word processing program. Explore the menu, clicking on the various items until you begin to understand the features and options available to you in KEY CAD. Recognize that the LBM acts the same as the Enter key on the keyboard. We will start learning to draw with KEY CAD by using the mouse as our pencil. The features and options available in the menu and the icon button strings (called toolboxes) make it easy for us to draw by removing the need for manual dexterity.

THE TOOLBOXES AND THE TOOLS

As the menu and the icon strings suggest, KEY CAD has many capabilities that help us prepare high quality engineering drawings. The icon strings located on the left and right side of the screen are called toolboxes. The top-left string of icons is the edit tool box and the lower-left string is the draw tool box. We will introduce each icon in these two toolboxes, and then illustrate their use in preparing typical drawings. The treatment will be brief and introductory. No attempt will be made to cover all of the many capabilities and/or features of KEY CAD. For a more complete treatment, we refer you to the instruction manual [1].

THE DRAW TOOLBOX

The draw toolbox contains seven icons to identify the drawing tools available to you in KEY CAD as shown below:

RECTANGLE/SQUARE TOOL
OVAL /CIRCLE TOOL
POLYGON TOOL
MULIGON TOOL
ARC TOOL
LINE TOOL
FREE-FORMED SPLINE TOOL

RECTANGLE/SQUARE TOOL

Select the top icon, which is the rectangle/square tool, by clicking on it with the LMB. Then move the cursor to the drawing pad noting that the arrow of the cursor changes to a + when you locate the cursor over the drawing pad. The + permits you to more accurately position the cursor at the point on the drawing pad where you want to start constructing a rectangle. When you have positioned the cursor at the location for the upper left corner of the rectangle, depress the LMB holding it down while dragging the mouse to the location of the lower right corner. As you drag the mouse you will observe the rectangle expanding with the movement of the mouse. When you lift the LMB, the drawing of the rectangle is completed. Note the small square dots which are located on each corner. These little squares (vertex points) are drag points. You can click and drag on these points to resize or move the rectangle. When you click on a different icon in either the draw or edit toolbox, the drag points disappear. Try your hand at drawing a rectangle.

OVAL/CIRCLE TOOL

Click on the oval/circle icon which is the second button down in the string of drawing tools. When you move the cursor to the drawing pad the action is very similar to that described above for the rectangle. The cursor changes to a $+$ that permits you to locate a position on the drawing pad with more precision. Depress the LMB and hold it down as you develop either a circle or an oval (ellipse). When you lift the LMB the circle or oval is complete. Again you will note the small squares (drag points) at the corners of a rectangle surrounding the circle or oval.

If you want to draw circles or ovals from the center, click on the circle icon, and then click on Layout on the menu bar. On the Layout pull down menu select Draw from Center. If you want to draw a perfect circle instead of an oval that looks like a circle, hold down the Shift key as you drag with the mouse. Try to develop your technique in drawing both circles and ellipses.

POLYGON TOOL

The polygon tool aids you in drawing either open or closed polygons that may contain many different sides. This is a very useful tool in preparing three-view drawings. Select the polygon icon, and move the cursor to the point on the drawing pad where you will begin to draw a multi-sided figure. Click the LMB once at the start point and move the cursor to the next point on the polygon. You can watch the line that is formed in this process. When you are at the end-point of the first line click the LMB again to complete the first line. Repeat this click and drag process until all of the lines in the polygon are completed. At the last point in the polygon double click the mouse to end the process.

If your polygon consists of horizontal and vertical lines, hold down the Shift key while dragging the mouse.

MULTIGON TOOL

The multigon tool is like the polygon tool in that it is used to draw closed polygons. The difference is that the multigon tool is used when all sides of the polygon are equal. The multigon tool saves a lot of time in drawing hexagons, pentagons, octagons, etc. Select the multigon icon button and move the cursor to the drawing pad. Depress the LMB and hold the button down as you draw a multigon. When it is sufficiently large, release the LMB to complete the multigon. How many sides does your multigon have? Chances are high that you want some other number of sides. To control the number of sides for the multigon, select Edit from the menu while the multigon icon button is active. On the Edit pull down menu select Preferences. On the Preference Dialog Box, shown in

Fig. 2, type in the number of sides, say 6, to define the number of sides in the multigon. Now draw another multigon and confirm the fact that it is a hexagon.

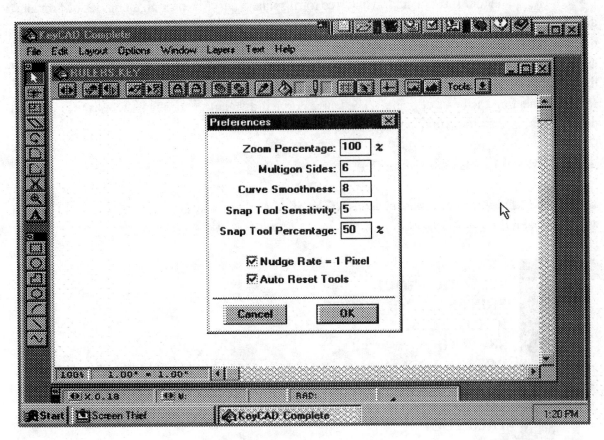

Fig. 2 The Preferences Dialog Box found under the Edit menu item.

ARC TOOL

The arc tool is used to draw either elliptical or circular 90° arcs. The procedure is similar to drawing a circle or an oval. Select the arc icon and move the cursor to the drawing pad. At the start point, depress the LMB and hold it down while dragging the cursor to the finish point. When you lift the LMB the arc is complete. Note the square drag points at the corners of the rectangle surrounding the arc. If you want to modify the arc, simply drag on one or more of the corners to resize, reshape or reposition the arc.

If your arc is to be circular in shape hold the Shift key down while dragging the mouse.

LINE TOOL

We use the line tool to draw lines of any length and at any angle. Select the line icon near the bottom of the icon string. Move the cursor to the drawing pad and locate the starting point for the line. Depress the LMB, drag the cursor to the end point for the line, and lift the LMB to complete the operation. Two small squares remain at the end points of the line. These are used as drag points to lengthen, shorten or reorient the line. If your lines are either horizontal or vertical, hold down the Shift key while dragging the mouse.

FREE-FORMED SPLINE TOOL

The spline tool is useful in drawing free form splines (curves) through a number of previously define points. To start drawing a free formed spline, select the spline icon and position the cursor on the first (end) point of the spline. Click the LMB and move to the next point and click the LMB again. Repeat this process until all of the points defining the position of the curve have been entered; then double click the LMB. As you locate all of the defining points, straight lines appear on the screen connecting each pair of points. The spline curve drawn automatically by KEY CAD, after the double click, is a smooth curve through all of the defining points.

THE EDIT TOOLBOX

The edit toolbox contains ten icons to identify the editing tools available to you in KEY CAD as shown below:

SELECTION TOOL
POINT SELECTION TOOL
RESIZE TOOL
SKEW TOOL
ROTATE TOOL
FILLET TOOL
CHAMFER TOOL
TRIM TOOL
ZOOM TOOL
TEXT TOOL

SELECTION (MOVE) TOOL

The top most icon in the edit toolbox is the selection tool. We use this tool to select and move individual objects to another location on the drawing pad. Click on the selection icon, move the cursor to an object on the drawing pad and click on it. The vertex points appear on the screen. Click and drag on a line in the object to relocate it. When dragging the object ghost lines are evident to represent the boundary lines of the object being moved.

We can select one or more objects. If you want to select several objects to be moved together, hold down the Shift key while clicking on each object. To move the group, simply select one line on one object, and click and drag that line. You will drag the group of the objects that you selected to a new location on the drawing pad. Draw a few objects on the drawing pad. Move one of these objects to a new location. Next, select several objects and move them all together to a new location.

POINT SELECTION TOOL

The point selection tool is used to reposition one or more points on the drawing pad. You may select points in two different ways. First, with the icon active, click the LMB on any vertex point within an object. That point is highlighted (a small solid square appears). You can click and drag these points to change the shape of the object. Ghost (dash) lines appear in the dragging operation to aid you in modifying the size and shape of the object.

It is also possible to select a number of points by dragging a rectangle about a portion of the object. The vertex points contained in the rectangle are highlighted. These vertex points are locked together and are moved in unison by clicking and dragging on any one of the points.

RESIZE (SCALE) TOOL

The resize tool is used to resize or change the scale of any object or group of objects that you have drawn. Activate the resize icon, point to the object to be rescaled with the cursor, and click. The vertex points on the object are highlighted and the object is ready to be resized. Click and drag one of the corner points to resize the object. Ghost (dash) lines appear to aid you in visualizing the new size of the object.

If there are several objects to be rescaled, they can be selected by dragging a rectangle about all of them. All of the vertex points within the rectangle are highlighted. Select one of these points, click and drag to resize the group of objects. Again ghost lines appear on the screen during the resizing operation to aid you in this operation.

SKEW TOOL

The skew tool is useful if you need to change the shape of an object by skewing it. Select the object first so that the vertex points are highlighted. Then with the skew icon active, click and drag on one of the vertex points until the selected object takes on an angular shape. The ghost lines are helpful in showing the skewed shape of the object as the vertex point is being dragged. The skew angle can also be determine from the position readouts located at the bottom of the screen.

ROTATE TOOL

We use the rotate tool to rotate an object from 0 to 360° about its center as an anchor point. Select the object that you intend to rotate, activate the rotate icon, and then click and drag a vertex point in the counter clock wise direction. A ghost image of the object is evident showing its position during the rotation process. When the object is in the correct orientation release the LMB.

FILLET TOOL

Fillets are circular arcs used in finishing machined components. The reentrant corner of a component is usually finished with a fillet by using the arc tool previously described under the drawing toolbox. If the fillet is on an external corner, the fillet tool is the appropriate technique to use in drawing the arc that removes the external (sharp) corner. To illustrate this tool, draw a rectangle, and activate the fillet icon. Click on a corner vertex point, and drag the point in toward the center of the rectangle. Ghost lines show the arc that forms to eliminate the external corner. When the radius of the arc is sufficiently large release the LMB.

CHAMFER TOOL

The chamfer tool is identical to the fillet tool in that it is used to eliminate sharp external corners. The only difference is that the chamfer is a straight line that removes the sharp corner instead of an arc. Chamfers are more commonly used in engineering practice because the flat surface produced by a straight line is easier to machine than the curved surface produced by an arc. The procedure for drawing a chamfer is identical to that described for the fillet tool.

TRIM TOOL

The trim tool is used to cut away portions of lines or arcs. It is like an eraser, and its use saves significant effort in redrawing and editing. First, select the object or line that is to be trimmed, and then activate the trim icon. Second, position the cursor over the point where you wish to trim (cut) the line, and click the LMB. A new vertex point appears on the line at the trim point. Click on the drawing pad to remove the vertex points. Third, select the excess line with the selection tool, and then press the Delete key to eliminate the unwanted line segment.

It is possible to trim segments from interior portions of lines and objects. However, it is necessary to mark two trim points before identifying the segment to be Deleted.

ZOOM TOOL

The zoom tool is very much like a zoom lens because it permits us to move in for a very close view of the details or move out for an overview of the drawing layout. To zoom in on an object, activate the zoom icon and place the cursor, which is a magnifying glass, over the object. Click the LMB and the object is magnified. The amount of magnification is set at a default value of 100 %, but it can be changed by clicking on Edit --- Preferences and adjusting the zoom percentage in the Dialog Box shown in Fig. 2.

The zoom out procedure is exactly the same, but you must hold down the Shift key after the zoom icon has been activated. With the Shift key depressed, the sign in the magnifying glass becomes negative. Clicking the LMB when the cursor is over the object reduces the size of the field providing an overview.

TEXT TOOL

The text tool is used to add notes, drawing blocks, bill of material, etc. to the drawing. KEY CAD supports all of the fonts that were installed with WINDOWS. Before using the text tool, click on Text on the Menu line, and select Fonts from the pull down menu. When the Dialog Box appears, select the type of font, its style and size. After these choices are made, activate the text icon.

With the text icon activated, move the cursor to the drawing pad and click in the approximate area where you want to add a note. A text box appears with drag points at its corners.

Type the note as if you were using a word processing program. When the note is complete, click on any other icon, and the text box with its drag points disappear.

The text can be edited at any time. Activate the select icon, place the cursor over the text and click the LMB. The text box reappears with its drag points. You can place the cursor in the message and edit using normal word processing techniques. You can click and drag on the corners of the text box to move or reshape the note. The entire note can be Delete after it has been selected simply by pressing the Delete key.

THE SNAP TO TOOLBOX

The Snap-To tools help people, like the author, who are not rock steady with the mouse. The Snap-To tool will take the cursor to a specified point when we move the cursor close enough to that point. The Snap toolbox contains ten icons displayed a string along the upper-right side of the screen. If they are not present on your screen, select Window from the menu. From the pull down menu select Toolboxes as illustrated in Fig. 3. On the second pull down menu you have four toolboxes including: Edit, Draw, Snap, Dimension as well as the display of Position data. We have already discussed the Edit and Draw toolboxes. Click on Snap and a string of ten icon buttons appear on the upper-right side of the screen. Click on Dimension and another string of six icon buttons appears on the lower-right of the screen.

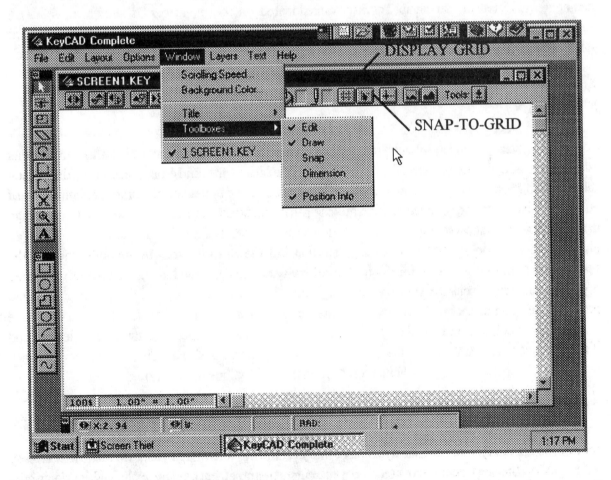

Fig. 3 The Window ---Toolboxes command to display the toolboxes.

THE SNAP TO ICONS

ANY POINT
END POINT
POINT ON LINE
CENTER POINT
CORNER POINT
INTERSECTION POINT
PERPENDICULAR
LINE CENTER
PERCENTAGE
ABSOLUTE POINT

ANY POINT

The any point tool is the default setting for KEY CAD. With the Any Point icon activated, you can choose any point in the field to set the initial point for some object such as a line or a rectangle. However, if the Snap-To-Grid icon (located on the menu bar as shown in Fig. 3) is engaged, the only points available in the field are the intersection points of the grid. We recommend that the Snap-To-Grid option be used only in very special cases where the grid geometry closely matches the geometry of the object being drawn.

END POINT

You will use the End-Point Snap tool to connect a line or some other object to an end point of an object already in the field of the drawing. When using the End-Point Snap-To feature, first select the object to which you plan to add a line. Small squares appear on the line to identify the end points to which you will connect. Activate both the End-Point icon and the Line icon. Begin to draw the new line, and you will note that it jumps to the closest end point on the object. If you do not observe the snap feature, increase the sensitivity of the Snap-To tool by going to the Edit --- Preferences in the menu. Adjustments to the sensitivity are made in the Dialog Box as shown in Fig. 2. We recommend a sensitivity of 5.

The End-Point icon will reset (deactivate) after each application. This is fine if you only have one end-point connection to make. However, if you have several lines to draw at a variety of end points then you will want to lock the End-Point icon in its active position. This lock is accomplished by double clicking on the icon. Clicking on any other Snap toolbox icon removes the lock.

POINT ON LINE

We use the Point-On-Line to connect the object being drawn to a point on a line that already exists on the drawing. In this snap-to mode, the cursor automatically snaps to the closest point on a line of the existing object. If the snap feature is not effective, you either have to position the cursor closer to the line or to increase the sensitivity of the Snap-To feature. Use the Edit---Preferences --- Dialog Box to increase the sensitivity.

In some cases we want to draw a line or some other object connecting two objects that already exist on our drawing. We can start the line with one snap feature, and end it with a different one. We start with the first snap feature, and press the Tab key to exercise the snap icons until the correct icon is activated prior to terminating our line. Try using the two different Snap-To commands because it is easier than the description implies.

CENTER POINT

We use the Center Point tool to connect objects being drawn to the center point of object that already exist on our drawing. The tool snaps the initial point to the closest center point within the closed area of an existing object. We use this feature to locate the center points of circles, ovals, multigons, etc.

CORNER POINT

We use the Corner Point tool to connect objects being drawn to the corner point of an existing object. The tool snaps the initial point to the nearest corner point of an existing object such as a rectangle or a multigon.

INTERSECTION POINT

We use the Intersection Point tool to connect objects being drawn to an intersection point that exists where two lines of an existing object cross. The tool snaps the initial point to this intersection point.

PERPENDICULAR

We use the Perpendicular tool when a line is to be drawn that is perpendicular to another line that already exists on the drawing. In using this feature we activate the line icon in the draw toolbox, and then select the Perpendicular Snap to tool. Next, the cursor is placed near the point of intersect of the existing line. Click and drag the line to the correct length and release the LMB.

LINE CENTER

We use the Line Center tool to connect a line or some other object to the center point of an existing line or arc. To draw a line to the midpoint of another object, activate the line icon, and then activate the Line Center icon. Move the cursor near the object center point, click and drag the mouse away from the object. The initial point of the line will snap to the center point of the object.

PERCENTAGE

The Percentage feature permits you to locate a point on a line or arc at a clearly defined point if you specify the location of that point with a percentage value along the length of the line or arc. The value is specified under Edit---Preferences----Preferences Dialog Box which is illustrated in Fig. 2. The procedure followed, after the percentage is specified, is identical to that covered under the Line Center tool.

ABSOLUTE POINT

The absolute point on a drawing is its origin. In KEY CAD the origin, where X = Y = 0, is in the lower left corner of the drawing pad. This is the reference point for all of the absolute readouts presented in the Position Box located at the bottom of the screen.

To draw a line from the absolute point, activate the line icon in the draw toolbox, and then activate the absolute icon. Place the cursor near the origin (absolute point) and click and drag to form a line. The starting point of the line is snapped to the origin.

STARTING A DRAWING

We still have more to learn about the icons and menus for KEY CAD, but we have covered enough to begin a drawing. We will introduce additional features of the software as we attempt to draw some simple features. Before beginning, we must get KEY CAD set-up to produce a drawing that is formatted correctly. Let's start with the paper by selecting File --- Paper Size and make certain that we are going to use paper which is 8 ½ by 11 in size. This size is correct for the laser printers available to you. Next, select File --- Page Setup and click on the Landscape dot. We usually prepare drawings with the paper in the landscape orientation.

OK! The paper is ready. Let's prepare the screen so that it is arranged to facilitate our drawing. Select Layout --- Rulers-Scale and check out the Dialog Box that appears (see Fig. 4). Let's use inches as our unit of measure in preparing the drawings. We will use 1 to 1 scale (screen to world), and type in 10 divisions to the inch. We will also click on the Layout/Rulers to display rulers along the top and left edges of the drawing pad as shown in Fig. 4. These rulers are useful in locating our position on the drawing paper.

In Fig. 4 you will note that only a small portion of the drawing pad is visible on the screen. The drawing pad is 8 ½ by 11 in size to correspond to the size of the paper, but we see only a portion of the upper left of the drawing pad. To work on the remaining portion of the pad it is necessary to use the scroll bars to position it on the screen. If we click on the Fit to Window icon on the menu bar at the location defined in Fig. 5, then the entire drawing pad is visible (48% size). We can now see the entire sheet of paper on the screen. Click to the Actual Size icon, also identified in Fig. 5, to the to revert back to actual size (100%). OK! We have the paper and the screen in control and can go from actual size to full view by clicking on one of two icon buttons.

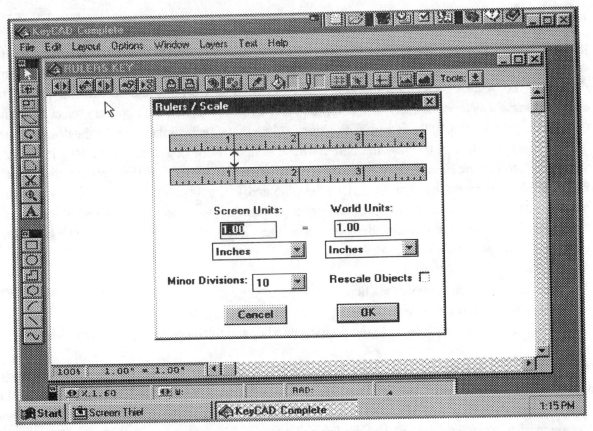

Fig. 4 The screen with rulers displayed.

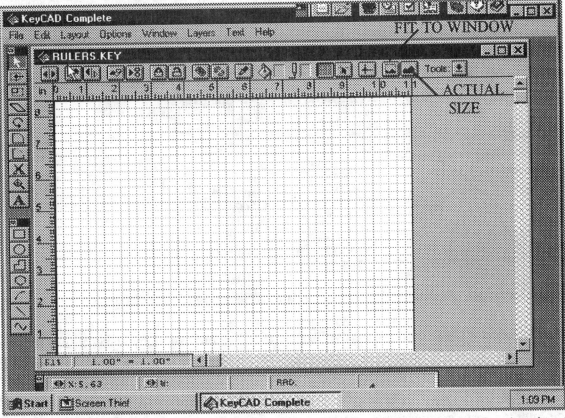

Fig. 5 Drawing pad reduced in size to fit the window (screen). The icon buttons for changing the size of the drawing are located on the Menu bar.

Let's start our first practice at drawing by preparing our own personalized drawing paper which includes a drawing block and a border. The idea here is to design this personalized paper, to name it, and save it to a file. We then have on file a sheet of paper suitable for any of our future drawings. To begin let's setup some grid lines to help us as we draw the border and set-up the drawing block. Click on Layout ---Grid Spacing, and when the Dialog Box appears select 0.25 and 0.25 for the grid spacing in the X and Y directions. Note that a rectangular grid fills the drawing pad.

Use the Line tool to draw the border which defines a drawing area 7.5 by 10.5 in. in size. We could also use the Rectangle or Polygon tools to draw the border, but these tools unfortunately eliminate the grid in the drawing area. Use the Line and Rectangle tools to construct the drawing block in the lower right hand corner of the paper. Before constructing the drawing block on our paper, let's decide beforehand something about its size. Try a block 0.75 inch high by about 2.00 inch wide, with three rows of information.

- On the top row, show the title.
- On the middle row, give the name of the team preparing the drawing.
- On the lower row, indicate the drawing number and the date.

This is an oversimplified drawing block, but it will serve our purpose for our initial drawings. Use the Text tool to add "Title of Drawing", "Team Name", "Date", and "Drawing No." We suggest using Arial Rounded MT Bold font (10 point) for lettering in the drawing block. The author found it easier to place the text in the drawing pad near the drawing block. Then in the edit mode, I dragged the text into the drawing block and positioned it on the proper line. An example of a drawing block is shown in Fig. 6. If you like the appearance of your drawing block, save it using the Save As command under File. We suggest a file name such as DWGBLOCK. Note that KEY CAD supports file names with eight or few alpha numeric symbols. You can start all of your drawings by loading this file. You only need to draw this block once. That is a very significant advantage of any CAD program --- no need to redraw the same object over and over again.

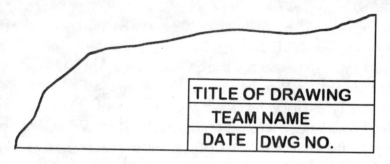

Fig. 6 An example of a simple drawing block that is sufficient for class drawings.

If you don't like the size of the font that is used, click on Text and you can change the size of the letters. If you notice an error in the text on your drawing, click on the Select icon in the Edit toolbox, mark the text that you want to change, and make the corrections in the typing box.

We hope you have recognized the second huge advantage of CAD. You don't need to print. We can drag text to any location on our drawings. For those of us that don't print well, breathe a sigh of relief.

THREE-VIEW DRAWINGS

To illustrate the use of KEY CAD in the preparation of three-view drawings, let's consider the notched block that was introduced in the chapter on three-view drawings (Figs. 1 and 2). This block is three inches wide, two inches high, one inch deep, with a one by one inch notch cut from the top. Open the file DWGBLOCK, and you are ready to start the drawing. The grid spacing that we established for the drawing block, 0.25 by 0.25 inch, is too small for this drawing since all dimensions are integer multiples of an inch. Change the grid pitch, by clicking on Layout --- Grid Spacing and make the X and Y pitch both equal to 1.000.

Start the drawing with the front view that is positioned in the lower left corner of the paper. Activate the Snap-To-Grid icon button that is on the menu bar. This will drive all of the vertex points on our drawing to the intersection of the grid lines. Since our object is to be constructed with line segments that are all multiples of the grid pitch, this method of snap-to is the most efficient. Select the Polygon tool from the drawing toolbox, and construct the front view. With the Polygon tool the front view is constructed by clicking the LMB at each corner (count eight corners). Remember to double click when you close the polygon to terminate the tool.

Let's draw the top view next. Select the Rectangle tool from the Drawing toolbox, and move the cursor to a location one inch above the front view. Start the lower-left corner in line with the left side of the front view. With the grid lines visible, it is easy to keep the views in proper alignment. Depress the LMB and drag the rectangle to the upper-right corner. Check to make sure that you have a rectangle one by three inches in size before you release the LMB. To complete the top view, we need to add two lines representing the edges of the notch. Select the Line tool, and draw the two vertical lines to complete the drawing. Note again the value of the grid lines in projecting the locations for lines between the front view and top and side views. To complete the drawing, let's construct the side view. Again select the Rectangle tool, move the cursor to a location one inch to the right of the front view, and start at the lower-left corner of the rectangle. Drag the cursor to the upper right corner, and check to ascertain that you have a rectangle one by two inch in size before releasing the LMB. We have one more line to draw to complete the side view. It is a dashed line to define the location of the bottom of the notch. But we are drawing only solid lines in KEY CAD. To change the type of line that is drawn in KEY CAD, click on Layout --- Lines --- Type, and a Dialog Box appears giving you a selection five different types of lines, as shown in Fig 7. Click on the dot for the dashed line and complete the drawing. If you need to draw solid lines again, it is necessary to return to the Line Types Dialog Box and change the line style.

Look over the drawing to check that it is complete and correct. If you are satisfied, use File --- Save As and provide a name for the file such as NOTCHBLK. Finally, print the drawing by using File --- Print. In checking your work, use the drawing shown in Fig. 8, to determine if you have positioned all of the lines correctly in each of the three views, and have arranged the views properly on the paper.

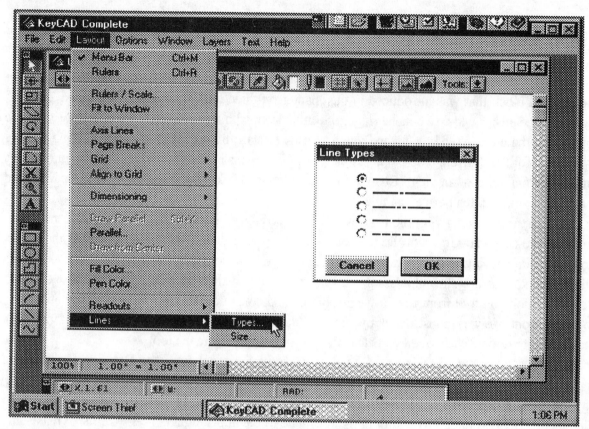

Fig. 7 Procedure for changing the type of lines constructed in KEY CAD.

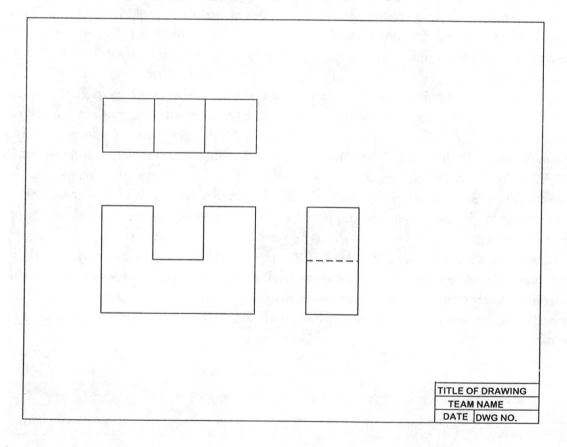

Fig. 8 A three-view drawing of the notched block prepared with KEY CAD.

DIMENSIONING WITH KEY CAD

The drawing shown in Fig. 8 looks good, except it does not have dimensions. Let's add them, using the automatic dimensioning capabilities of KEY CAD. On the right side of the screen, a Dimension toolbox with six icon buttons should be visible. If it is missing, display the Dimension toolbar by clicking on Windows --- Toolbox ---Dimension as illustrated in Fig. 3. Alternatively, you can click on the Tools icon on the menu bar, and add or delete toolboxes. The six icon buttons in the Dimension toolbox are defined below:

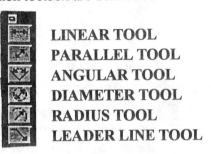

LINEAR TOOL
PARALLEL TOOL
ANGULAR TOOL
DIAMETER TOOL
RADIUS TOOL
LEADER LINE TOOL

KEY CAD dimensions the individual features on the drawing automatically. When you position the cursor at the appropriate vertex points, the program determines the dimension and automatically constructions the leader lines, the dimension line (with a selection of the termination), and shows the numerical value and units. To select the type of termination for the dimension line click on Layout ---- Dimensioning ---Arrowheads ---- and make your selection as shown in Fig. 9. You should also select the font size and style for the dimension numbers from the Text menu prior to selection of the appropriate Dimension tool.

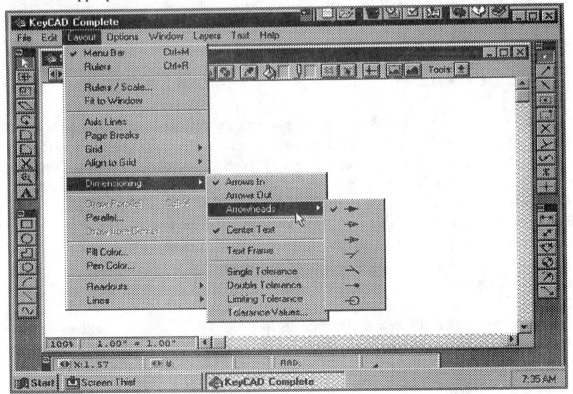

Fig. 9 Procedure for selecting the termination for the dimension line.

LINEAR TOOL

The Linear dimension tool automatically determines the horizontal distance between any two points (see Fig. 10a for an example). The procedure is to select the Linear tool with the appropriate Snap-To tool. Position the cursor at one of the vertex points of the horizontal line being dimensioned, depress the LMB, drag the cursor to the other vertex point, and release the LMB. Next, drag the cursor with the dimension line and text outward from the object. When the dimension line and text are in the proper position click the LMB to set the dimension in place.

PARALLEL TOOL

The Parallel dimension tool automatically determines the distance between any two points (see Fig. 10b for an example). The procedure is to select the Parallel tool with the appropriate Snap-To tool. Then position the cursor at one of the vertex points of the vertical or inclined line being dimensioned, and depress the LMB. Drag the cursor to the other vertex point and release the LMB. Next, drag the cursor with the dimension line and text outward from the object. When the dimension line and text are in the proper position click the LMB to set the dimension in place.

ANGULAR TOOL

The Angular tool automatically determines the angle (in degrees) between two lines as illustrated in Fig. 10c. The procedure is to select the Angle tool and the appropriate Snap-To tool for the first vertex point. Position the cursor near the first vertex point, depress the LMB and drag to the second point before releasing the LMB. Now drag the cursor away from the line in a circular motion to form the required angle. Try it a few times in practice because the Angle tool is a little tricky.

DIAMETER TOOL

As the name implies the Diameter tool automatically determines the diameter of any circle that we have constructed. This dimensioning tool does not require the use of a Snap-To tool since it automatically snaps to the boundaries of the circle. To dimension a circle, select the Diameter tool, place the cursor near the boundary of the circle, and depress the LMB. Drag the cursor to the desired text position and release the LMB. If you release the LMB in the center of the circle, the result is the dimension illustrated in Fig. 10d. However, if you drag the dimension text outside the circle, the result is as presented in Fig. 10e.

RADIUS TOOL

The Radius tool is identical to the Diameter tool except that it used to designate the radius of a curve or circle instead of the diameter. The radius dimension is shown in Fig. 10 f.

LEADER LINE TOOL

The leader line is used as a pointer. We point to an object and then usually make a note next to the leader as shown in Fig. 10g. To construct a leader, activate the Leader icon and place the cursor at the location where you want the arrowhead. Depress the LMB, drag the cursor to the location of the knee of the leader line, and release the LMB. Continue to drag the cursor to the end point of the leader and click again.

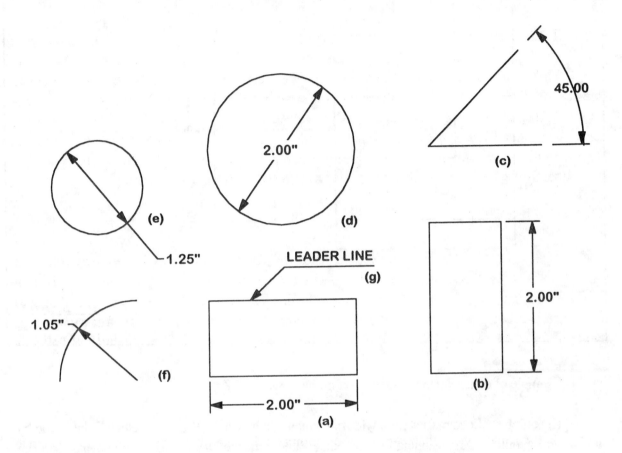

Fig. 10 Examples of the capabilities of the Dimension tools.

OK! Now that we understand the use of the Dimension tools, let's dimension the drawing illustrated in Fig 8. It is an easy drawing to dimension because all of the lines are either vertical or horizontal. We only need to use the Linear and Parallel dimensioning tools. Let's start with the Linear tool and dimension the 3", and the two 1" lines. Just follow the procedure described above and check your result with the dimensioned drawing shown in Fig. 11.

Next dimension the three vertical distances which are all 1" in length. Select the Parallel tool and proceed as described above. Again check your results with those shown in Fig. 11. If your results differ make sure that you have activated the Snap-To-Grid icon on the menu bar. Remember too that you can adjust the size and style of the font used in dimensioning with the Text command from the menu. Finally, the style of arrow heads and the position of the text on the dimension line are changed under Layout ---Dimensioning --- Arrowheads.

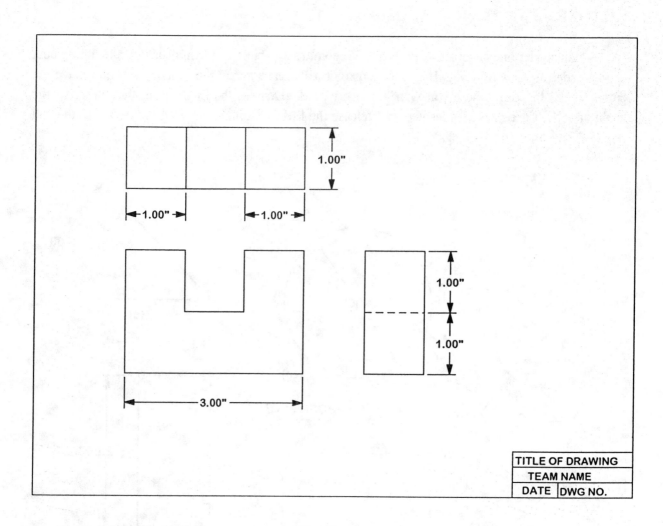

Fig. 11 Example of dimensioning a three-view drawing with KEY CAD.

The number of decimals shown with the dimension line is 2. We have selected this value by clicking on Layout --- Dimensioning --- Tolerance values. In the Dialog Box that appears, you can specify the number of decimals used in stating the dimensions. This is an important decision as the cost of manufacturing the part increases with the number of decimals specified. Suppose that you show a dimension of 3.000" for the width of the block. That is OK, if you want to fabricate the block with three decimal precision; however, be aware that the cost of manufacturing a part with three decimal precision is significant. If you need this degree of precision, use three decimals in writing the dimensions; however, if you are satisfied with one or two decimal accuracy, change the settings for the number of decimals and reduce the manufacturing cost for the component.

When using the dimensioning tools, KEY CAD places the units selected on each dimension. For example, the width of the block in Fig. 11 is given as 3.00". This is unfortunate because the units for the dimensions used in the drawing should be stated once, and only once, in either the drawing block, or a note placed near the drawing block. We can avoid this problem by manually drawing the leader and dimension lines, and then placing the dimensions on the drawing with the Text command. However, this approach is very time consuming. We will accept the placement of units on the dimensions in this course, but you should recognize that it is not an accepted drafting practice.

A THREE-VIEW DRAWING WITH A HOLE AND ARC

Our first illustration to show some of the neat features in KEY CAD was a simple three-view line drawing. Let's move on to a slightly more complex drawing involving features such as rectangles, circles, and an arc as shown in Fig. 12.

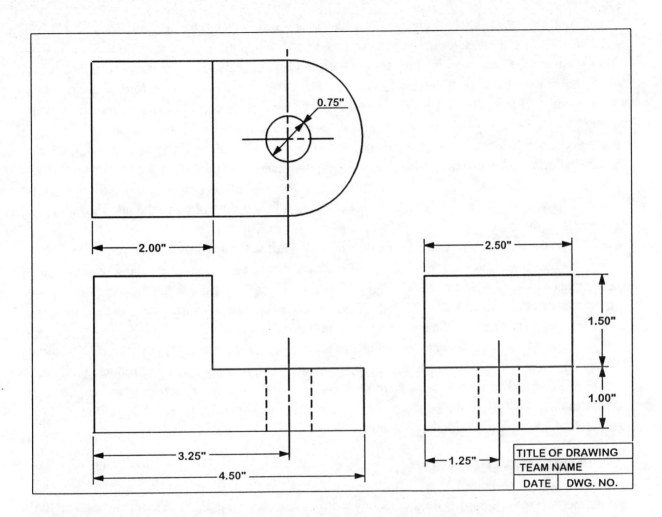

Fig. 12 A three-view drawing made with lines, rectangles, circles and an arc.

We open our DRWBLOCK file, but before beginning the drawing, we click on Layout --- Grid --- Grid Spacing and adjust the grid so that it is on 0.25 inch centers. This selection was made because all of the dimensions in the drawing are multiples of ¼ inch. Let's also click on the Snap-To-Grid Point icon in the Menu bar. Finally, select Layout --- Lines --- Size, and in the Dialog Box adjust the size (weight) of the lines to 1. This will provide us with a heavier line for the three-view drawing.

Let's begin drawing the outline of the front view by selecting the polygon tool. We start the outline by placing the cursor at the location X = Y = 1.00 and click at each of the six corners. Double click to disable the polygon tool at the starting point. Move up on the drawing pad an inch above the front view and begin to draw the top view. Use the Rectangle tool to draw a rectangle 2 by 2.5 inches in size to from the left side of the top view. Use the Line tool to draw horizontal lines

1.25 inch long extending to the right from the rectangle. Finally, complete the outline of the top view by using the Arc tool twice to sketch two 90° circular arcs to form the rounded end of the object.

Move to the right side of the drawing pad and draw the outline of the side view one inch to the right of the front view. Select the Rectangle tool, depress the Shift key and draw a square with a side of 2.5 inches. Then select the Line tool and draw the horizontal line that positions the step on the block.

Let's return to the top view and draw the circle. First, locate the center of the circle which is 3.25 inches from the left edge along the horizontal centerline in the top view. Click on the circle icon and then under the Layout menu click on Draw from Center. Now align the cursor with the center of the hole in the block, and depress the LMB. Drag the cursor outward to form a circle with a diameter of 0.75 inch. Next draw the center lines showing the location of the center of the hole. Under Layout --- Lines ----Types, select lines that are solid with a single dash. Activate the Line tool and draw centerlines on the top view. Continue to use this tool to draw the center lines through the hole in the front and side views. Next, under Layout --- Lines --- Types, select the dashed line. Then use the Line tool to draw the hidden lines representing the sides of the hole in both the front and side views. Check to make certain that the drawing you have produced looks like the one shown in Fig. 12 except for dimensions.

To complete the drawing add the dimensions. Start with the Linear dimension tool with the Snap-To-Grid point icon engaged. Use the Linear dimensioning tool to place the four dimensions defining horizontal distances of 2.00, 3.25, 4.50, and 1.25 inches. Next use the Parallel tool to place the two vertical dimensions --- 1.00 and 1.5 inches. Finally, use the Circle tool to dimension the circle making certain that the dimension is placed outside the circle.

Finally, check the drawing for completeness and accuracy. If it passes your inspection, print it. We hope this example helps you appreciate the power of a CAD program in producing relatively complex drawings that look good and are accurate. It may have taken longer than you expected to complete the drawing including dimensions, but remember the time required will decrease rapidly as you grow more proficient with the program.

ISOMETRIC DRAWINGS

KEY CAD can also be employed to prepare isometric drawings. The CAD procedure is almost identical to that used in previous sections to prepare three-view drawing. The only difference is that we construct lines in the three isometric directions instead of the X and Y directions. To provide the three directions required for isometric drawings, we use the on-line measurement of the angle of the line that is provided in the readouts at the bottom-right side of the screen. The lines in the isometric drawing are in the 30°, 150°, or 270° (vertical) orientations. As you draw with the cursor, keep an eye on the angle as you establish the length of the line. Only when both the length and the angle are correct, should you release the LMB. We have drawn the isometric pictorial, shown in Fig. 13, using the Line tool. Precise location of the end points of the lines is a bit more difficult because establishing the correct length and the correct orientation of the line simultaneously requires considerable patience with the mouse. The alignment problem is alleviated, to some degree, by the using the zoom command to enlarge the intersection points. We then select an offending line and drag the vertex point to give more precise junctions where several lines meet.

Try drawing an isometric with KEY CAD. We show an example in Fig. 13 for your guidance. If you prefer, prepare an isometric drawing of your choice.

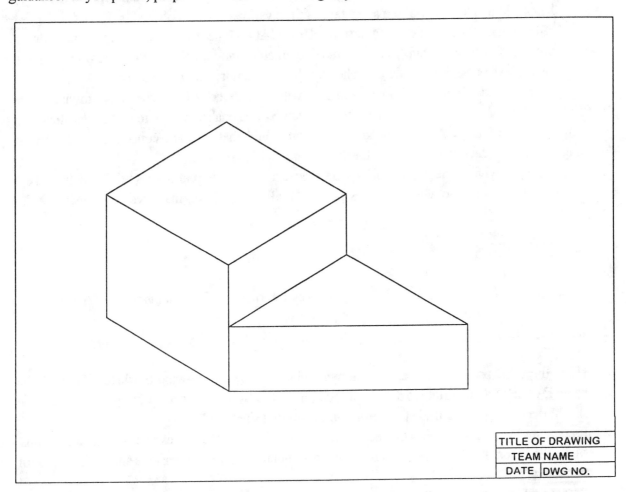

TITLE OF DRAWING	
TEAM NAME	
DATE	DWG NO.

Fig. 13 An example of an isometric drawing prepared with KEY CAD.

SUMMARY

KEY CAD Complete is a computer aided drawing program that has been selected for your use in this course. It is a low cost program that is available to the student at a modest price. The program is also relatively easy to learn to use, and a beginning student can produce high quality, three-view drawings after a modest investment of time in learning the menu and icon options.

We have introduced the KEY CAD screen, defining the menu line, the four toolboxes with their icon tools, and the working area for the drawing. For most applications, we recommend that you use the mouse to draw because it is faster than the keyboard options.

Many, but certainly not all, of the operations possible in KEY CAD have been described. We emphasized the more basic menu items in this coverage. If you become familiar with the choices in the four toolboxes --- Draw, Edit, Snap-To, and Dimension, you will be able to produce excellent engineering drawings. We suggest that you use the descriptions of the operations presented here as a reference source. Try to master the commands by attempting to duplicate the drawings shown in the chapter.

We first describe how to prepare a drawing block. It is a simple drawing block, but it will suffice for this class. We then lead you through a three-view drawing of a simple notched block. Try to draw this block, and learn about the advantages of the various features offered by KEY CAD.

Dimensioning is very easy and fast in KEY CAD. The Dimension toolbox provides a painless process for dimensioning lengths, radii, diameters, and angles. No measurements are necessary and the dimensioning is accomplished by clicking and dragging the cursor.

We introduced drawing of circles and arcs with a more complex three-view drawing. The description of the approach used to make the drawing is reasonably complete. We trust that you have mastered the KEY CAD commands by this time, and that you are concentrating on a neat drawing layout, and the completeness of the drawing and dimensioning.

Finally, we give a brief treatment on isometric drawing. The treatment is brief only because it is not difficult to draw isometrics with KEY CAD after you have mastered three-view (X-Y) drawings.[1]

REFERENCES

1. Anon, Key CAD Complete for Windows, SoftKey International Inc., Cambridge, MA 1994.

EXERCISES

1. Using the CAD program to prepare a drawing block. Save the drawing block to a file with the name DWGBLOCK. Print the drawing block on an 8 ½ by 11 inch sheet of paper and check the size of the drawing area inside the border on the printed sheet.

2. Prepare a CAD drawing of a notched block like that shown in Fig. 6, except for the dimensions. Make the block 4" wide by 3" high by 2" deep. Retain the same dimensions shown in Fig. 6 for the notch.

3. Prepare a CAD drawing of a support pad made of steel. The pad is a rectangle block one inch thick, by six inches wide, and four inches deep. The block has ½ inch diameter holes, that are located a distance of one inch from each edge near all four of the corners the block. Sharp corners are taboo, so either chamfer or round them. Your pick.

4. Dimension the drawing prepared for Exercise 2.

5. Dimension the drawing prepared for Exercise 3.

6. Prepare an isometric drawing of the J. M. Patterson Building. Show the view from the intersection of the two streets fronting the building.

---

[1] The DOS version of KEY CAD is superior to the Windows version for preparing isometric drawings because the three isometric directions are defined with a three line cursor in the DOS version. These three cursor lines are a constant reminder of the orientation of every line on each isometric plane. Except for the omission of the three line cursor in KEY CAD for Windows, the Window version is much superior to the DOS version of the program.

CHAPTER 9

MICROSOFT EXCEL

INTRODUCTION

Microsoft® EXCEL is a spreadsheet program that is an extremely important tool in engineering, because it is useful in so many different applications. We will cover three of these applications including, preparation of a parts list, performing calculations, and drawing graphs. There are several different spreadsheet programs on the market, namely EXCEL by Microsoft, QUATTRO PRO by Borland, and LOTUS 1 • 2 • 3 by Lotus. All of these programs have a similar format, and they all have essentially the same capabilities. If you learn to use one program, it is relatively easy to switch to a different program, with a modest investment in time. We have selected EXCEL because it has been adopted by many universities, and you will probably have access to it on the computer net during your tenure in college.

The purpose of this chapter is to show you how to navigate on the EXCEL spreadsheet, to make tables, perform calculations, plot several types of graphs, print the output, and save a file. You should recognize that these objectives are limited. We are concentrating on developing your entry level skills. Hopefully you will find the spreadsheet tool important enough to develop a higher skill level with independent study. More complete and detailed treatments are given in references [1, 2].

THE SCREEN

You load EXCEL from WINDOWS by clicking on the EXCEL icon, and the program opens by displaying a spreadsheet as shown in Fig. 1. Let's explore this display. At the top you will note five rows, then below in the center is a big table that is the spreadsheet proper, and finally at the bottom two more rows.

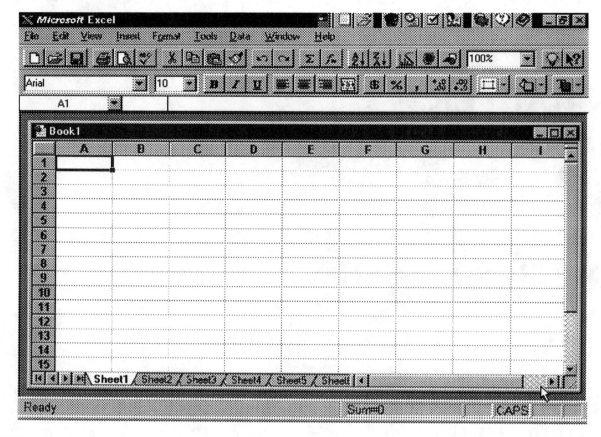

Fig. 1 The screen display in Microsoft Excel

THE FIVE TOP ROWS

The top most row, called the title bar, tells you that you are working in the program EXCEL, and the name of the file that is open (Book 1 until you save and name the file). The three buttons to the far right side of the title row are common to all Microsoft® programs. The left hand button with the dash symbol places the program aside, so that you can work in a different program. The middle button permits you to either expand your view of the screen or to shrink it. The right button with the X closes both EXCEL, and the file upon which you are working. Do not use this button if you intend to use EXCEL again during your current session. Try clicking on these three buttons until you understand their functions.

The second row from the top is the called the menu bar, which lists nine different features included in the program beginning with File and ending with Help. The menu bar is also common to all of the Microsoft® programs. Use your mouse, select one of the menu items, and click. A pull down menu is displayed. We will not list the items on each of these pull down menus, because you will learn faster by clicking than by reading. The purpose of each of the menu entries is listed below:

1. File is to open or close, save, or print a file or exit EXCEL.
2. Edit is to change entries that you have already made.
3. View is to change the appearance of the screen.
4. Insert is to add cells, rows and columns to the spreadsheet.
5. Format is to change the height of rows and the width of columns.

6. Tools are to check spelling, audit calculations and for developing macros.
7. Data is to provide a means for sorting, filtering and forming data entries.
8. Window is to enable you to open a new window.
9. Help is to provide answers to your questions as you attempt to learn the particular program that is loaded.

The third row down exhibits a toolbar with about twenty buttons. Again this is a standard toolbar (unless someone has customized it) that is common to all of the Microsoft® applications. The icons on each button give a good clue as to the purpose of each button, but if you are in doubt, use the mouse and point to a particular button. Don't click. Wait for a moment, and a small label appears to tell you the function of each button. Again point and click until you understand.

The fourth row down is the formatting tool bar, and again it is common to all of the Microsoft® applications. Icon labels on each button give us a hint of the purpose of each button, but if we are in doubt, point to the button and a label will appear defining the function. Try changing the font type and size.

The last row gives information about the spreadsheet. The block to the left defines the cell location on the spreadsheet that is active (A1 shows when you begin a new book or file). The longer block on the right, called the formula bar, is activated when you make an entry into a cell. The numbers or letters that you have entered into the cell appear in this block. If you wish to edit the entry, without completely retyping it, activate the cell, use the mouse to place the cursor in the formula block, and make the necessary changes.

OK, we have briefly defined the top five rows, but it is up to you to try to develop an understanding of the menu items and the buttons by pointing and clicking. Don't be afraid to make the computer beep; that is part of the learning process.

THE SPREADSHEET

Most of the screen is taken up by the spreadsheet, which is simply a big table with lots of columns and rows. The columns are labeled across the top with letters A, B, C, to I and beyond, and the rows are labeled down the left side with numbers 1, 2, 3, to 15 and many more. Each little block, called a cell, is identified with its coordinates (i. e. The cursor in Fig. 1 is pointed at cell D7). Note that we put the letter (column location) first, and the number (row location) second. The cells are where we enter either numbers, text or formulas on the spreadsheet.

We can move about on the spreadsheet using either the mouse or the key board. When the pointer is on the spreadsheet, the location is identified with a large plus sign. We can use the mouse to point to any cell on the screen, but if the cell of interest is off screen, the scroll buttons located to left or below the spreadsheet are used to bring the cells of interest into the field of view. When the big plus sign (see Fig. 1) is pointed at the correct cell --- click, that cell becomes active, and is ready to receive the data that we enter. We can also move the active cell by using the arrow keys, the tab key, the shift-tab keys, and the control-arrow keys. Try them and watch how the active cell moves about the spreadsheet.

The spreadsheet is much larger than it looks on the screen, which shows only columns A through I and rows 1 through 15. Explore down on the spreadsheet by holding the control and the arrow-down keys, and note that the number of the last row is 16,384. Depressing the control and the arrow-right keys moves the active cell to the far right of the spreadsheet, and indicates a column heading of IV which is the 256th column. We have a total of (256)(16,384) = 4,194,304 cells on the spreadsheet. Let's hope we never have to use all of them!

At the bottom of the spreadsheet, we have tabs that give the sheet numbers. When opening EXCEL and starting a file, we begin with a workbook for a given project. The program automatically establishes 16 sheets (pages) in our workbook, although only six are visible to us. We can develop several spreadsheets in this file by clicking on the tabs to move from one sheet to another. The buttons to the left of these tabs change the sheet numbers on the tabs, permitting us to gain access to sheets 7 though 16.

The row at the very bottom of the spreadsheet is called the status bar. It provides a brief statement about the function of any button that is being activated, and other information about the status of the program.

PREPARING A PARTS LIST

Let's begin to learn how to use EXCEL by preparing a parts list for the weighing machine (scale). Load the program from WINDOWS by clicking the Start button, select Programs, and then Microsoft Excel. The screen display should look like Fig. 1, indicating that we are ready to start. We recommend that you title the file even before beginning to develop the parts list. Click on File and select Save As. When the Save As page is displayed, you enter a suitable file name for the project (PARTS LIST), and click the save button. The file name is now displayed on the top row of the spreadsheet.

We start the parts list by typing its title in cell A1. Since the title should be obvious and easy to read, we use bold font and increase the font size to 16 point. We skip a row and position the cursor on cell A3 to begin organizing the section headings as shown below:

PARTS LIST FOR A PLATFORM SCALE				
NUMBER	NAME OF ITEM	DESCRIPTION OR DRAWING NO.	QUANTITY	PRICE
1	PLATFORM	WOODEN PLATE---- DWG. NO. 100-01	1	$ 1.20
2	BASE SUPPORT	WOODEN BOX -------DWG. NO. 100-02	1	$ 2.55
3	POST	PVC PIPE 2 IN. DIA. DWG. NO. 100-03	1	$ 3.50
4	LINKAGE	1/8 IN. DIA. WIRE CABLE --CUT TO LENGTH	1	$ 2.75
5	TOP LEVER	1/4 x 1 IN. AL. BAR --DWG. NO. 100-04	1	$ 3.00
6	MIDDLE LEVER	1/4 x 1 IN. AL. BAR --DWG. NO. 100-05	1	$ 5.00
7	LOWER LEVER	1/4 x 1 IN. AL. BAR --DWG. NO. 100-05	1	$ 3.00
8	BOLTS & NUTS	1/4- 20, 3/4 IN. LONG WITH LOCK NUTS	9	$ 2.70
9	COUNTER WEIGHT	AL BAR 1 IN. DIA. ---DWG. NO. 100-06	1	$ 2.75
10	TARE WEIGHT	AL BAR 1 IN. DIA. ---DWG. NO. 100-07	1	$ 2.75
11	TARE SCREW	NO. 8-32 BRASS THUMB SCREW 1/2 IN. LONG	1	$ 0.85
12	SUPPORT BARS	1/8 x 3/4 IN. STEEL BAR --- DWG. 100-08	4	$ 4.40

We have five section headings (columns) in our parts list:

    A. A number which refers to the part number. Remember that every unique part has its very own part number.
    B. The name of the item.
    C. A description of the item or a drawing number defining the part.
    D. The quantity of parts that will be needed to build a single prototype.
    E. The unit price for the item.

When we type these column headings in cells A3, B3, C3, D3, and E3, it is clear that the columns are not the correct width to accommodate the text. The column widths are adjusted by pointing to the line between A and B on the column headings. When the pointer is positioned correctly, you will see a short vertical line with arrows to the left and right. Depress the left mouse button, and drag the column boundary either to the left or the right to adjust the width of column A. Move to the position on the column headings between column B and C, and repeat the process to size the width of column B. Continue to adjust all of the column widths until they are all of the proper width to best represent the data included in the parts list..

In our example parts list, we have shown 12 part numbers. You may have more part numbers, because we have not made a serious attempt to be complete in this listing. We type the necessary information for the parts list into the table that is provided by the spread sheet. Our only concern in this process is that we have activated the correct cell before making an entry. If one of our entries is too long, we can change the column width at any time to accommodate it. However, keep the descriptions short because we want the parts list to fit the page.

When you have completed the entries, work on the spreadsheet to improve its appearance. We have centered the part numbers, the quantities and the prices. We center by marking all of the entries in a given column (point, drag with the mouse, and watch a black background color fill the column). When the black covers the column entries that you wish to center, move the pointer to the formatting bar, and click on the centering button. On the column for price, we initially made the entries with the default format; however, prices are in terms of dollars. We changed the format of the column by marking it, and then click on the $ button on the formatting toolbar to show the prices in a currency format.

We are ready to print the parts list, but we want to check to see if it fits the page, has the correct margins, and is readable. We click on File then Print Preview, to review the appearance of the spreadsheet that we will see when it is printed. To change its appearance click on the Setup button and then choose the sheet tab. From this page, point and click on gridlines, so that all of the cell boundaries are visible when the spreadsheet is printed. Printing the cell gridlines makes the parts list much easier to read. Next click on the page tab, and select either the portrait or landscape mode. The default setting is the portrait mode because it is more common, but if the spreadsheet is much wider than it is long, the landscape mode is preferred. You can also adjust the size and the quality (dots per inch) of the printing on this page. Finally, click on the margin tab, and change the margins so that the spreadsheet is positioned where you want it to be on the page.

When you have made all of the modifications needed, and the spreadsheet looks professional (that means very good), click on the print button, select the number of copies, and print the spreadsheet. Your parts list should compare favorably to the one shown above.

PERFORMING CALCULATIONS

One of the most important advantages of any spreadsheet program is its capability to perform calculations. EXCEL performs not just one calculation, but a whole sheet full of calculations, almost instantaneously. The results can be displayed in two different ways --- as an accurate table, or as a suitable graph. In this section, we will show you two simple examples illustrating the computing power of EXCEL, and illustrate the results in tabular format as produced by the spreadsheet program.

THE RICH UNCLE EXAMPLE

For the first example, consider your rich uncle that we introduced in a previous chapter on Tables and Graphs. Recall that we discussed the compound interest which you earned from his initial investment of $5000. We will now describe the results of this table, presented on the next page, which is an analysis of the growth of the investment. Let's recall the accumulation relation, and the input data from the previous chapter on Tables and Graphs where:

$$S = P(1 + i)^n = PM \qquad (1)$$

Uncle contributed P = $5,000.00, and the interest rate (i) for the six month compounding period was 5%. (The annual interest rate was 10%). We want to track the sum S accumulated for a total of 80 compounding periods (40years). Recall that n is the number of compounding periods, and M is a multiplier dependent on n.

Load EXCEL, and Save As --- compound interest --- to establish the file name. In the results shown on the next page, we titled the spreadsheet and entered the parameters controlling the results in the first five rows (A1 to A5). We skip a row, and define the headings for Periods, Multiplier and Sum in row 7. Initially, we define only three columns (A, B and C). We place the active cell at location A8, and begin to enter the number of the period starting with zero. We could type all 40 entries into the A column, but it is easier to fill in the entries automatically. To let EXCEL do the work, make sure the active cell is on the zero entry in cell A8, click on Edit, select Fill, and then Series. A Series page is displayed, and you select Column to indicate that you are filling entries into column A. You also select Linear, because the series that you are using to fill in column A is a linear series (i. e. 1, 2, 3, etc.). At the bottom of the page note two small boxes; one for Step and the other for Stop. We enter 1 for step and 40 for stop, then click on the OK button. The program fills all of the cells from A8 to A48 with numbers, starting with 0 increasing in steps of 1 until it stops at the number 40. The Fill command listed in the Edit menu item saves a lot of time.

Let's go next to column B, and activate cell B8. We want to calculate the multiplier in this column. The multiplier M is a number that increases with the number of the compounding periods, and is given by $M = (1 + i)^n = (1.05)^n$. With cell B8 active, click in the formula bar in the 5th row. We begin by notifying EXCEL that we intend to feed it a formula by typing =, then we type the formula. In this case our entry would look like =(1.05)^A8. We use the cell address A8 because it contains the value of n for the row of calculations being considered. We use the ^ symbol to indicate that we are raising (1.05) to a power. Just to the left of the formula bar, note the buttons labeled X, ✓, and f_x.

DETERMINING THE RUNNING SUM						
FOR AN INITIAL INVESTMENT						
ANNUAL INTEREST RATE OF 10%						
COMPOUND INTEREST SEMI-ANNUALLY						
NO TAXES PAID						
Periods	Multiplier	Sum		Periods	Multiplier	Sum
0	1.0000	$ 5,000.00		40	7.0400	$ 35,199.94
1	1.0500	$ 5,250.00		41	7.3920	$ 36,959.94
2	1.1025	$ 5,512.50		42	7.7616	$ 38,807.94
3	1.1576	$ 5,788.13		43	8.1497	$ 40,748.33
4	1.2155	$ 6,077.53		44	8.5572	$ 42,785.75
5	1.2763	$ 6,381.41		45	8.9850	$ 44,925.04
6	1.3401	$ 6,700.48		46	9.4343	$ 47,171.29
7	1.4071	$ 7,035.50		47	9.9060	$ 49,529.86
8	1.4775	$ 7,387.28		48	10.4013	$ 52,006.35
9	1.5513	$ 7,756.64		49	10.9213	$ 54,606.67
10	1.6289	$ 8,144.47		50	11.4674	$ 57,337.00
11	1.7103	$ 8,551.70		51	12.0408	$ 60,203.85
12	1.7959	$ 8,979.28		52	12.6428	$ 63,214.04
13	1.8856	$ 9,428.25		53	13.2749	$ 66,374.74
14	1.9799	$ 9,899.66		54	13.9387	$ 69,693.48
15	2.0789	$ 10,394.64		55	14.6356	$ 73,178.15
16	2.1829	$ 10,914.37		56	15.3674	$ 76,837.06
17	2.2920	$ 11,460.09		57	16.1358	$ 80,678.92
18	2.4066	$ 12,033.10		58	16.9426	$ 84,712.86
19	2.5270	$ 12,634.75		59	17.7897	$ 88,948.50
20	2.6533	$ 13,266.49		60	18.6792	$ 93,395.93
21	2.7860	$ 13,929.81		61	19.6131	$ 98,065.73
22	2.9253	$ 14,626.30		62	20.5938	$ 102,969.01
23	3.0715	$ 15,357.62		63	21.6235	$ 108,117.46
24	3.2251	$ 16,125.50		64	22.7047	$ 113,523.34
25	3.3864	$ 16,931.77		65	23.8399	$ 119,199.50
26	3.5557	$ 17,778.36		66	25.0319	$ 125,159.48
27	3.7335	$ 18,667.28		67	26.2835	$ 131,417.45
28	3.9201	$ 19,600.65		68	27.5977	$ 137,988.32
29	4.1161	$ 20,580.68		69	28.9775	$ 144,887.74
30	4.3219	$ 21,609.71		70	30.4264	$ 152,132.13
31	4.5380	$ 22,690.20		71	31.9477	$ 159,738.73
32	4.7649	$ 23,824.71		72	33.5451	$ 167,725.67
33	5.0032	$ 25,015.94		73	35.2224	$ 176,111.95
34	5.2533	$ 26,266.74		74	36.9835	$ 184,917.55
35	5.5160	$ 27,580.08		75	38.8327	$ 194,163.43
36	5.7918	$ 28,959.08		76	40.7743	$ 203,871.60
37	6.0814	$ 30,407.03		77	42.8130	$ 214,065.18
38	6.3855	$ 31,927.39		78	44.9537	$ 224,768.44
39	6.7048	$ 33,523.76		79	47.2014	$ 236,006.86
40	7.0400	$ 35,199.94		80	49.5614	$ 247,807.21

After you have entered the formula, click the ✔ button, and look at cell B8. It should read 1, because your entry $=(1.05)\wedge A8 = (1.05)^n = (1.05)^0 = 1$. OK, you have used EXCEL to compute the first multiplier M; let's determine the other 40 multipliers by copying the first one. Activate B8, and then click the Copy button on the toolbar. You have just stored the formula $=(1.05)\wedge A8$ in temporary memory, and it will be stored there until it is replaced the next time you use the copy command. Next mark the B column starting with B9 to B48 by dragging the mouse pointer down the column. Check that the region is black where you want to determine the multiplier term, then click on the Paste button. The cells B8 to B48 show the multiplier M superimposed on the black background color of the column. The black background is eliminated by clicking on any cell outside this region. In performing the calculations down the B column from the Copy, then Paste commands, EXCEL modified the formula $=(1.05)\wedge A8$. Activate cell B9, and note the formula has changed to $=(1.05)\wedge A9$. EXCEL modifies the formula to accommodate the changes in n, as we move down the column of calculations. In this instance, we needed EXCEL to make this modification. However, in the next example, this type of modification leads to errors, and we will show the technique necessary to control the way EXCEL modifies the formulas in the Copy and Paste commands.

Looking at the results for M in column B, we note that the number of decimals is not consistent from row to row. Select any cell in column B, click on either one of the two decimal buttons on the formatting toolbar, and watch the number of decimals change. Click on one or the other of these buttons until you display the multiplier with four digits after the decimal point.

To determine the sum S that is accumulated with time, we need to multiply the multiplier M by the initial investment of P = 5000. We select cell C8, and type =5000*B8. The symbol * is used to indicate multiplication. When we click on the check button, the entry in cell C8 changes from =5000*B8 to 5000. EXCEL noted that cell B8 contained the number 1, provided the calculation of (5000)(1), and displayed the results in cell C8. When we need to incorporate a cell address into a formula, we can type it, or we can point to the cell with the proper address, and click the left mouse button to enter it into the formula. In the point and click approach, one of the math symbols (+, -, * or /) must precede the pointing.

Look again at cell C8, and note that the result 5000 is not in the correct format. We need to show the results in a currency format, to indicate that the sum accumulated from Uncle's investment is in terms of dollars. To change the format:

1. Mark the entire column from C8 to C48, by dragging with the mouse.
2. Check that the black background covers only this region.
3. Point and click on the $ button on the formatting bar.
4. Clear the black background by clicking on any cell not in the black region.

When we converted the cell C8 into a currency format, another problem occurred. The result of 5000 changed into #######. This symbol indicates that the column is not wide enough to display the result. To widen column C, we point to the row for the column headings at a location between C and D. When the line with the double arrows shows in place of the pointer, depress the left mouse button, and drag to the right to widen column C. With a wider column the result in cell C8 reads $5000.00.

We have one result for the sum S in cell C8. Let's arrange for EXCEL to complete the calculations. Activate cell C8 because it has the required formula, and click on the copy button on the toolbar. Next mark the cells C9 through C48 where we want the formula to apply. The black background verifies that the region is marked correctly. Next, we click on the paste button, and

EXCEL applies the formula to all of the marked cells in the C column, and calculates the results in each cell. Scroll down the column and check to see that all of the results are displayed. If you encounter ####### the column needs to be widened. The result in cell C48 should read $35,199.94.

We have computed the accumulated sum S for 40 periods, but we originally stated that we were to determine S for 80 periods. Why did we stop at n = 40? We divided the calculation into two parts so that the output, when printed, would fit better on one page. It is easy to extend the calculation to n = 80 by using the copy command.

To complete our work mark the block with corners at A7 and C8, and click on the copy button. We now have our headings and the initial formula in temporary storage (memory). Move the cursor and activate cell E7. We skip column D to provide space between the two lists of results in our table. After activating cell E7, click on the Paste button, and the results from the A7 to C8 block are copied (with modifications for the shift in columns) into the E7 to G8 block. Look at cell E8 and note that the period shown is n = 0. Let's change that value to 40 by editing the numbers displayed on the formula bar. When cell E8 is changed to 40, the results in E9 and G9 check with previous results for the period n = 40. This fact implies that the program in EXCEL has adjusted the formulas to account for the fact that we have changed the columns in which the calculations are made. Activate cells F8 and G8 and compare the formulas in them with those in cells B8 and C8. Do you see the changes automatically made by EXCEL, as we copied formulas from one column to another.

To complete the table, activate cell E8 and fill in the entries for n = 40 to 80. We again use the automatic number generator in EXCEL. Click on Edit, select Fill, and then Series. On the pull-down series page, click on Column, Linear, and indicate a step of 1 with a stop at 80. Click on the OK button, and then scroll down the spreadsheet to check that you have filled in the correct entries in column E. Next, mark cells F8 and G8, and copy them. Mark the F8 - G48 block, and paste the formulas from F8 and G8 into this block. Clear the black background, and then scroll down the G column. You will probably note the symbol ####### for some of the results. Widen column G until all of the results are shown in the currency format. As a check, you should have the result $ 247,807.21 in cell G48.

We are now ready to print the table showing all of the results. Click on File, then click on Print Review. The screen shows the spreadsheet as it will appear when printed. We note that it could be improved with gridlines and a larger left hand margin. To show the gridlines, click the Set-up button. From the pull-down book which appears on the screen, select the sheet tab. Click on the gridlines square, and note that a check appears to indicate that the gridlines will be printed. Next click on the OK button to return to the preview of the spreadsheet. To adjust the margins, simply point, click and drag a margin line until the table is centered in the sheet that is displayed in the print review display. Now click on the Print button to print the spreadsheet.

We will return to this spreadsheet later when we graph the results.

STRAIN ON A SIMPLY SUPPORTED BEAM

In chapter 3 on Mechanical Aspects of Weighing Machines, we introduced an equation for determining the strain ε in a simply supported beam when it is subjected to a central load of W_u. The relationship for the strain is given below:

$$\varepsilon = 3 \, W_u \, S/(2bh^2 \, E) \qquad\qquad (2)$$

where:

W$_u$ is the load (weight) in pounds or Newtons.
S is the span of the beam in inches or meters.
b is the depth of the section of the beam in inches or meters.
h is the height of the section of the beam in inches or meters.
E is the elastic modulus in psi or Pascals.

Let's program this relation in EXCEL to determine the strain induced in the beam at a location under the load point. As we examine Eq. (2), note that the strain ε depends on five variables, namely S, b, h, E and W$_u$. For this example, let's fix three of these variable at reasonable values as shown below:

• S = 6 in.
• b = ¼ in.
• E = 10.6 x 10^6 psi. (An aluminum alloy).

We will treat the weight W$_u$, and the height h as variables, and explore solution space for the strain ε. Let's consider the load W$_u$ increasing from 0 to a maximum of 250 lb in steps of 25 lb. Also we will consider that the height h of the cross-section of the beam varies from 0.1 in. to 0.7 in. in steps of 0.1 in. To determine the strain, we arrange the spread sheet with one variable W$_u$ displayed in column A, and the other variable, the height, arranged along the 8th row. This arrangement of the spreadsheet is shown below:

DETERMINE STRAIN							
FOR A SIMPLY SUPPORTED BEAM							
SPAN S = 6 in., DEPTH b = 1/4 in., ELASTIC MODULUS E = 10.6 x 10^6							
WEIGHT Wu AND HEIGHT h ARE VARIABLES							
STRAIN = 3WuS/(2bh^2E)							
HEIGHT	0.1	0.2	0.3	0.4	0.5	0.6	0.7
Wu (lb.)	in.	in.	in.	in.	in.	in.	in.
0	0	0	0	0	0	0	0
25	0.008491	0.002123	0.000943	0.000531	0.000340	0.000236	0.000173
50	0.016981	0.004245	0.001887	0.001061	0.000679	0.000472	0.000347
75	0.025472	0.006368	0.002830	0.001592	0.001019	0.000708	0.000520
100	0.033962	0.008491	0.003774	0.002123	0.001358	0.000943	0.000693
125	0.042453	0.010613	0.004717	0.002653	0.001698	0.001179	0.000866
150	0.050943	0.012736	0.005660	0.003184	0.002038	0.001415	0.001040
175	0.059434	0.014858	0.006604	0.003715	0.002377	0.001651	0.001213
200	0.067925	0.016981	0.007547	0.004245	0.002717	0.001887	0.001386
225	0.076415	0.019104	0.008491	0.004776	0.003057	0.002123	0.001559
250	0.084906	0.021226	0.009434	0.005307	0.003396	0.002358	0.001733

We have selected the ranges for the load and the height to cover the region of interest. The weight in the specification given in Chapter 2 goes to a maximum of 250 lb, and the calibration will be conducted using 25 lb weights to verify the accuracy of the scale. The height of the beam is initially too small, and the resulting strains, even at small loads, are excessive. However, when the height is increased to the maximum value of 0.7 in., the strains are much lower.

In the spreadsheet illustrated above, we have entered the height data in row 8, columns B to H. In rows 1 to 4, we provide the title of the spreadsheet, and information related to the calculation performed in the spreadsheet. On row 6, we have shown the equation that has been evaluated in the spreadsheet. Both rows 5 and 7 are blank to provide space above and below the equation. In row 9, we have entered the W_u heading, its units, and the units for the height of the beam. The spreadsheet is now organized with an arrangement that displays the two variables, and gives their units of measure.

To begin the programming of the equation for the strain, let's fill column A with information about time. Type 0 in cell A9, and keep this cell active. Then click Edit on the menu bar, select Fill, and Series. On the series pull down page, select Column, Linear, and enter step = 25 and stop = 250. Finally, click the OK button, and then check to determine if the numbers filling the A column are correct.

You should activate cell B10 and type the formula for the strain, namely =3*A10*6/(1*(1/4)*10.6*10^6*(B8)^2). (Note that this entry is correct for cell B10, but it will give us problems later when we try to use the copy and paste commands to complete the spreadsheet). Let's proceed by copying the formula in cell B10 into cells B11 to B20. To copy the equation into the other cells in column B, position the cursor on cell B10 and click on Copy. Then mark the column from B11 to B20. Next click on Paste. When we examine our results, they are clearly in error. To trouble shoot the problem activate cell B11, and examine the formula displayed in the formula bar. You will see =3*A11*6/(1*(1/4)*10.6*10^6*(B9)^2). Unfortunately this relation is not correct. In the copy-paste operation, EXCEL modified our formula indexing the row locations on both the A and B column entries by 1. This indexing was OK for the column A entries, but we do not want to index the column B entries. Instead we want to use the data in cell B8 for all the calculations in column B. To accomplish this, we modify the formula in cell B10 to read =3*A10*6/(1*(1/4)*10.6*10^6*(B$8)^2). The $ symbol before the number 8 locks the value contained in the B cell in row 8 into the equation, as we perform the paste operation and EXCEL indexes down the B column in modifying the formula. We now copy cell B10, and paste down the B column from B11 to B20 to generate the correct results for strain corresponding to a beam with a height of 0.1 in.

The copy and paste operation worked well on column B after we fixed the B$8 entry. Let's try to copy the modified formula in cell B10 into the row of cells C10 to H10. If you copy, paste, and then examine the formulas in these cells, you will note that EXCEL has not given you the formulas that you need. Again you must modify the entry in cell B10 to properly perform the copy and paste operation. The equation in cell B10 reads =3*A10*6/(1*(1/4)*10.6*10^6*(B$8)^2). As we copy along a row to the right, the cell address A10 changes to B10, C10, D10 etc. We need to multiply by the weight, and do not want the A column designation to change in the copy-paste operation. To modify our entry in cell B10, we type =3*$A10*6/(1*(1/4)*10.6*10^6*(B$8)^2).. The $ symbol before the A locks column A into the formula, so that it remains fixed as we change columns in the copy-paste operation.

Now that we have corrected our entry into cell B10, we copy and paste the formula to cells C10 to H10 along row 10. Then copy this row (C10 to H10) and paste to the block C11 to H20. The program executes these calculations and displays the results as shown in the previous spreadsheet. For the maximum load of 250 lb, the strain varies from 0.084906 for a beam 0.1 in. high to 0.001733 for a beam 0.7 in. high. You should use your calculator to check a few values of the strain to insure that you did not make a mistake when programming the equation into the EXCEL spreadsheet. Finally the spreadsheet does not indicate the units for the strain. In this case, the absence of units does not present a problem because strain is a dimensionless quantity.

You may also want to change the number of decimals used in reporting the results. We usually use six decimal places to represent the strain. This is a very large number of decimals places to carry but strains are small and often two or three of the places are used to carry zeros. You can check the spreadsheet results for the strain to verify this fact. You may change the number of decimals, by marking the block containing the results (B10 to H20). Then click on the increasing or decreasing decimal keys in the formatting toolbar until the number of decimals shown is six.

Try these examples and learn the power of EXCEL in performing not one, but a whole sheet full of calculations. Use EXCEL to solve the homework problems assigned in Physics and Math courses. You can determine the single answer usually sought in an assigned problem, and then go on to explore solution space in EXCEL. Try exploring solution space; you will like the results.

GRAPHS WITH EXCEL

EXCEL can be employed to produce several types of graphs easily and quickly after you have learned the procedure. To show the method for producing graphs, we will use examples demonstrate the capabilities of EXCEL in preparing pie, bar, and X-Y graphs. We will begin by discussing the pie graph.

PIE CHARTS

In a previous chapter on Tables and Graphs, we demonstrated the manual technique for preparing a pie chart. Remember that this chart is used primarily to show the distribution of some quantity. The larger the piece of pie the larger the share of the distribution. To illustrate how to make a pie chart in EXCEL, we open a new book and title it --- PIE CHART --- using the As Save selection under the File menu item. We use the top rows to title the spreadsheet, and enter the tabular data, that describes the distribution of time in various assignments for Mechanical Engineers in their initial positions in industry in cells B9 to B17. These entries are shown in the spread sheet presented on the next page.

This pie chart looks very good. How did we generate it? After the numerical data on the distribution has been entered in cells B9 to B17, we mark them. Then click on the chart wizard button on the toolbar. When we depress the chart wizard button, the pointer becomes active. We move the active pointer to an area below the tabular data to create a box in which the pie chart will be located. We make the box large enough to permit us to view the chart when it is generated. Don't worry about the size of the box since we can change its size later to improve the appearance of our visual display.

The chart wizard presents five pull down pages, that lead us through the steps necessary to convert the data displayed on the spreadsheet into a graph of one type or another. The first page (Step 1) identifies the range of data. In our example, we have already marked the data in cells B9 to B17, so we confirm that these cells are correct in the dialog box. Clicking the Next button moves us to the next page (Step 2), where we find that EXCEL offers 15 options for our chart. Two of these options are for pie charts. Let's select the three dimension pie chart by clicking on its button, and then the Next button. The next page (Step 3), shows seven different options for displaying our three dimensional pie chart, and we select option seven.

The page for Step 4 previews our pie chart. Since it looks perfect, we do not alter the default settings in the row and column dialog boxes. Finally, the last page (Step5) is to assist us in titling the graph, and providing legends to identify the different slices of the pie. We click on the legend dot, and add our title. Except for sizing the chart and locating it on the spreadsheet, we have completed the pie chart. We click the finish button on the Step 5 page, and examine the results. If the chart is too small or too large, click on a corner of the chart and drag the corner to adjust the size.

When you are satisfied with the appearance on the screen, click on File and select Print Review. You can add gridlines and adjust margins if you wish on the Print Review display. Print the spreadsheet and the pie chart, and observe its professional appearance.

PIE CHART

RESPONSIBILITIES OF MECHANICAL ENGINEERS
FIRST ASSIGNMENT IN PERCENT TIME

ASSIGNMENT	PERCENT
1. DESIGN ENGINEERING	40
2. PLANT ENGINEERING, OPERATIONS, MAINTENANC	13
3. QUALITY CONTROL, RELIABILITY, STANDARDS	12
4. PRODUCTION ENGINEERING	12
5. SALES ENGINEERING	5
6. MANAGEMENT	4
7. COMPUTER APPLICATIONS, SYSTEMS ANALYSIS	4
8. BASIC RESEARCH AND DEVELOPMENT	3
9. OTHER ACTIVITIES	7
TOTAL	100

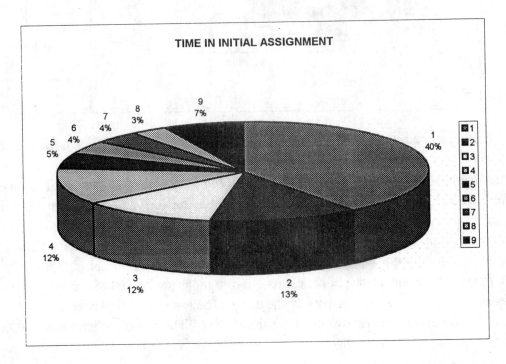

TIME IN INITIAL ASSIGNMENT

BAR CHARTS

Bar charts are most suited for showing comparisons. In the chapter on Tables and Graphs, we introduced an example of a bar chart that compared the percentage of high school students graduating with three or more years of secondary mathematics and science in 1982 and 1994. It is a very simple comparison, but it indicates the power of visual graphics to carry a message.

We begin by loading EXCEL, which opens to a new book. Again we title this book by clicking on File, Save As, and then we name the file ---BAR CHART. We enter the title of the new spreadsheet in the first five rows, and then enter the data for the graduating high school students in the block A7 to C9, as illustrated below:

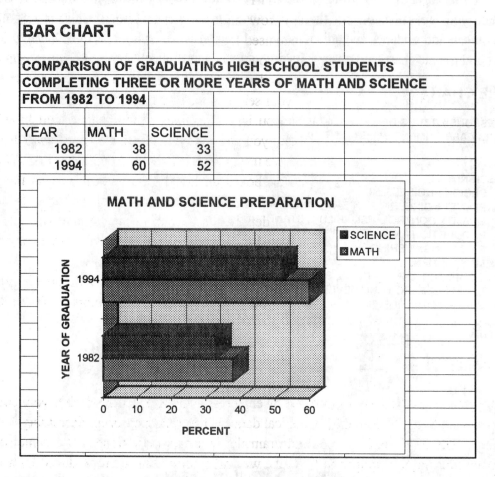

BAR CHART							
COMPARISON OF GRADUATING HIGH SCHOOL STUDENTS							
COMPLETING THREE OR MORE YEARS OF MATH AND SCIENCE							
FROM 1982 TO 1994							
YEAR	MATH	SCIENCE					
1982	38	33					
1994	60	52					

In the spreadsheet, we note that the column headings are in row 7 which is the first row of the data entries. Note also that years involved in the comparison are given in column A. We begin to create our bar chart by marking the region on the spreadsheet where the data exists, namely A7 to C9. We then click on the Chart Wizard button, and move the cursor to cell B10 to mark a box for the approximate location of our bar chart.

Step 1 from the Chart Wizard shows the range for the data, that we have already marked. Moving to the next page (Step 2) shows illustrations for 15 different graph options. Of these 15 options, four have the appearance of bar charts. Two options are labeled bar charts, and two others are labeled column charts. They are nearly the same. The bar charts display comparison categories along the Y axis and numerical values along the X axis. The column charts display comparison categories along the X axis and values along the Y axis. We consider both the column and the bar chart representations in EXCEL to be suitable for preparing bar charts. Take your pick; we have selected the three dimensional bar chart option in this example. The next page (Step 3), gives five options for the format of our bar chart. We selected option 4 which shows gridlines in the background of the chart.

The next page (Step 4) is most important in arranging the appearance of our bar chart. It shows a preview of the chart that changes as we modify the entries in the dialog boxes. We click on the column dot because the numerical data showing the comparisons is arranged in columns B and C. In the top dialog box for columns, we select 1 which corresponds to the 1st column in the marked data block. The numbers in column A (1982 and 1994) are used for the label along the vertical axis. In the lower dialog box, we select row 1 the 1st row in the marked data block, which is actually row 7 on the spreadsheet. The entries in this row are used for the legend text. This description is long and may be somewhat confusing. We suggest that you change the column and row settings, and observe the changes to the bar chart. You will soon understand that EXCEL can quickly image the data in the preview area, and you can easily correct your selections to give the correct display.

The final page (Step 5) is to assist you in adding titles to the chart. You have already provided the data for the legends in Step 4, but you need to add a title in the appropriate dialog box. The title must be short and to the point; we have used MATH AND SCIENCE PREPARATION for this example. We have also used the dialog boxes to caption the vertical axis as the year of graduation and the horizontal axis as percent. When you click on finish, the bar chart is displayed on the spread sheet. The appearance of the chart depends on the size of the box that was marked just after clicking on the Chart Wizard button. If the box is too small, click on the suitable corner or side marker, and drag to enlarge the view.

When you are satisfied with the appearance of the spreadsheet, click on File and Print Preview. Add gridlines, adjust the margins as you wish, and then print the output. Try to duplicate the spreadsheet and chart shown above.

X-Y GRAPHS

The graph most frequently employed by engineers is the X-Y graph which is very effective in visually conveying trends indicated by numerical data. Is the trend increasing, decreasing, level, or is it oscillating? The X-Y graph quickly and dramatically shows these trends. To demonstrate the method for producing X-Y graphs in EXCEL, we will consider the strain produced by a centrally loaded simply supported beam as predicted by Eq. (2).

We have previously used Eq. (2) to demonstrate how to perform calculations using EXCEL. Let's save some work by recalling the spreadsheet showing the results of evaluating Eq. (2). These calculations provide the numerical data necessary for preparing our X-Y graph. In examining this spreadsheet, we note that the units for the height of the beam shown in row 9 (cells B9 to H9) will cause difficulty in Step 4 of the Chart Wizard. The legends for our graph are supposed to be located in the first row of the data block used for the X-Y graph, and not the units for the height of the beam. To correct this difficulty with our spreadsheet, we mark cells B9 to H9 and delete the contents. Then we copy the height values from cells B8 to H8 into cells B9 to H9. Finally, we edit the entries of these cells as shown in the spreadsheet below:

DETERMINE STRAIN

FOR A SIMPLY SUPPORTED BEAM

SPAN S = 6 in., DEPTH b = 1/4 in., ELASTIC MODULUS E = 10.6 x 10^6

WEIGHT Wu AND HEIGHT h ARE VARIABLES

STRAIN = 3WuS/(2bh^2E)

HEIGHT	0.1	0.2	0.3	0.4	0.5	0.6	0.7	
Wu (lb.)	h = 0.1 in.	h = 0.2 in.	h = 0.3 in.	h = 0.4 in.	h = 0.5 in.	h = 0.6 in.	h = 0.7 in.	
0	0	0	0	0	0	0	0	
25	0.008491	0.002123	0.000943	0.000531	0.000340	0.000236	0.000173	
50	0.016981	0.004245	0.001887	0.001061	0.000679	0.000472	0.000347	
75	0.025472	0.006368	0.002830	0.001592	0.001019	0.000708	0.000520	
100	0.033962	0.008491	0.003774	0.002123	0.001358	0.000943	0.000693	
125	0.042453	0.010613	0.004717	0.002653	0.001698	0.001179	0.000866	
150	0.050943	0.012736	0.005660	0.003184	0.002038	0.001415	0.001040	
175	0.059434	0.014858	0.006604	0.003715	0.002377	0.001651	0.001213	
200	0.067925	0.016981	0.007547	0.004245	0.002717	0.001887	0.001386	
225	0.076415	0.019104	0.008491	0.004776	0.003057	0.002123	0.001559	
250	0.084906	0.021226	0.009434	0.005307	0.003396	0.002358	0.001733	

STRAIN IN BEAM

Note that the entry in cell B9 reads h = 0.1 in. This and the other entries in cells B9 to H9 will be the legends for the seven curves showing the strain that we will superimpose on the X-Y graph. OK, the modification of our previous spreadsheet is complete, and we are ready to begin to prepare the X-Y graph.

Start by marking the data block, namely A9 to H20. Observe that we have included the headings for the seven columns of Y data that are given columns B to H of the data block. Remember that the data for the caption along the X axis is listed in column A. The strain will be plotted automatically by EXCEL along the Y axis. After the data block is marked, click the Chart Wizard button, and reserved an area for the graph in the region below the spreadsheet. The page for Step 1 confirms that we have already marked the correct data. If there is an error you have the opportunity to correct it in the dialog box. The next page (Step 2) offers you the options for the type of graphs. Do not select the line graph, because it often distorts the data along the X axis. Instead you should select the X-Y scatter graph type.

Step 3 provides several options for scaling the axes and the appearance of the plot. We selected option 2, because it provides both lines and data points on the graph. This option does not provide for the major grid lines, but we can add them later as we modify the appearance of the graph. Note also that the 4th and 5th options provide scales for semi-log and log-log X-Y graphs. We will not demonstrate the use of these options, but you should know that they exist.

Step 4 is the important part, where we define the axes and the legends. We have made our task easier by the modifications made to the spreadsheet that properly arranged the data block used in the X-Y graph. We select the column dot, since the numerical data for Y (the strain on the beam) is arranged in columns B to H. We select row 1 in the top dialog box, because the scale on the X axis is listed in column A which is the 1st column in our data block. Finally, we select row 1 (row 9 on the spreadsheet) which is the 1st row in the data block because it contains the legends needed to identify the seven curves that we will plot using EXCEL. In making these selections, we preview the appearance of our X-Y graph. It looks good to me.

The final page (Step5) is to add the title, and to caption the X and Y axes. We added the title STRAIN IN BEAM, and captioned the X and Y axes as WEIGHT W_u (LB) and STRAIN respectively. The draft of the graph is completed when we click the Finish button. We click on the graph to activate it, and then drag the corner and/or side markers of the graph to enlarge it. With the graph still active, we click on the menu item Insert, Gridlines, and add major gridlines for the X and Y axes. Before printing the graph, click on File then Print Preview, and check to see that you have arranged the display so that it conveys the numerical data as well as the X-Y graph.

The combination of the spreadsheet and the graph, which is displayed either below the tabular data or beside it, is an effective technique for presenting the results of theoretical or experimental data. However, in some instances we want to show the graph without including the spreadsheet. We can show the graph alone without the spreadsheet by copying it. Click on the graph to activate it, then click on copy. Next click on the sheet 2 Tab to bring a new spreadsheet onto the screen. Position the cursor on cell A1, and click on the Paste button. The graph from Sheet 1 is copied onto Sheet2. Click on the graph to activate it, then drag its corners or sides to position and size it on the sheet. Examine the appearance of the graph with Print Review prior to printing. The result is shown in Fig. 2.

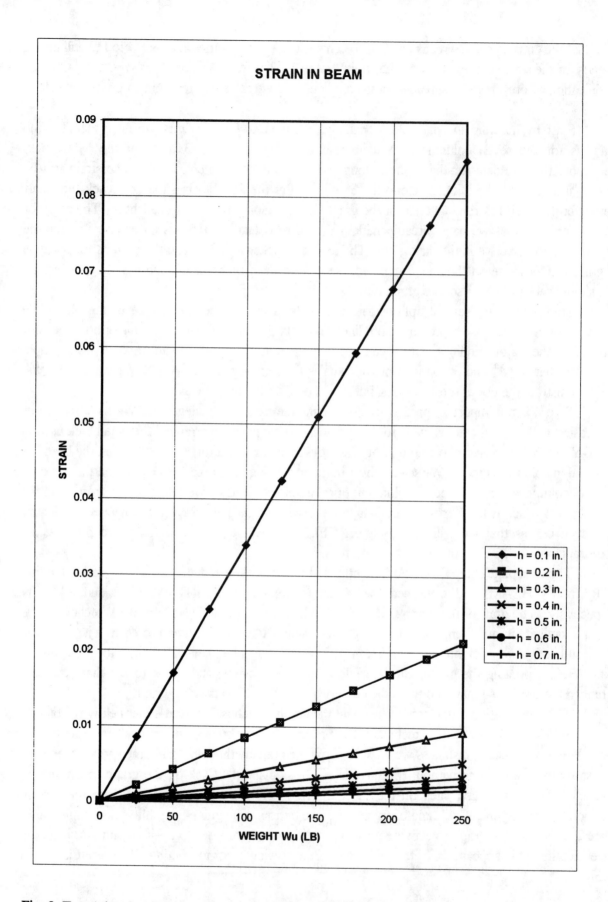

Fig. 2 Example of an X-Y graph printed without showing the spreadsheet.

SUMMARY

We have introduced you to EXCEL one of the popular spreadsheet programs in use in the United States. The introduction was limited because our purpose was to develop only entry level skills. You should know how to interpret the screen, the menu and the toolbars. You should also know how to navigate on the spreadsheet with the keyboard or the mouse, and to mark cells or blocks of cells by dragging with the mouse.

We showed three examples related to the weighing machine being developed in this course. We first illustrated the technique for preparing a parts list. In this instance, the spreadsheet was used to organize a table containing both descriptive and numerical data. The second example showed how to perform calculations using EXCEL to operate on simple formulas. We first illustrated how to perform a single calculation in a cell, then a line of calculations down a column, and finally a sheet full of calculations when we considered two variables. The third example was related to preparing graphs. We demonstrated the techniques involved in preparing a pie chart, a bar or column chart, and a X-Y graph. We made extensive use of the Chart Wizard in EXCEL, that leads you through the technique in a five step process.

EXCEL is a tool, somewhat like a hammer. With time and practice, you learn to use a hammer, you hit the nail more squarely, and with more force. The same is true for any spreadsheet program. You will grow more proficient with time. The results that you can produce with a spreadsheet throughout your tenure in college and in your professional life make the effort well worthwhile.

REFERENCES

1. Gottfried, B. S., Spreadsheet Tools for Engineers, Mc Graw Hill, New York, NY, 1996.
2. O'Leary, T. J. and L. I. O'Leary, Microsoft Excel 7.0a for Windows® 95, Mc Graw Hill, New York, NY, 1996.

EXERCISES

1. Explore the Excel screen and learn the purpose of the icon buttons.
2. Explore the Excel menu and learn the options available to you as you employ this spreadsheet program.
3. Prepare a parts list for the weighing machine that your team is developing.
4. Suppose that you have a rich grandmother, who has provided an inheritance for you. Unfortunately it is in trust, and you cannot get your hands on the loot until you are a mature 50 years of age. If the amount of the inheritance was $ 7,000.00, and the trust yields 7 % per annum, determine the value of the trust each year until your fiftieth birthday. Also prepare a X-Y graph showing the value of the trust with time. Neglect the effect of taxes on your analysis.
5. Duplicate the work involved in producing the spreadsheet shown on page 10 of this chapter.
6. Prepare a bar chart comparing the percentage of men and women in your high school graduating class with the percentage in the freshman engineering class in 1996.
7. Suppose that you decide that the chart on page 18 of this chapter covers too large a range of height of the beam. Prepare a new chart covering the range in height from 0.3 in to 0.7 in.

NOTES

CHAPTER 10

MICROSOFT PowerPoint

INTRODUCTION

During this course your team will be responsible for two design reviews, and you will have to make a presentation before the class and your instructor. In the presentation, you will describe the product specification, the important features that your team is incorporating in the design, and illustrate your concepts with suitable drawings. If you want to make a very good presentation, it is important that the audience follow your descriptions of the activities of the team. To help in this regard, use graphics which enable you to transmit information through two senses, --- audio and visual. This chapter provides instruction for a graphics presentation program, (GPP) that facilitates the preparation of the visuals needed for your presentations.

Microsoft PowerPoint is a graphics presentation program, that markedly reduces the time needed for you to prepare professional quality visual aids. It has a wide range of capabilities from simple overhead transparencies to sophisticated on-screen electronic displays. GPPs assist you in producing either black and white or colored transparencies, 35 mm slides, full screen projected electronic slides, and valuable support materials such as hard copy printouts, notes and outlines.

Before beginning to learn about PowerPoint, you need to understand that preparing a presentation is a five step process.

1. Plan the presentation to provide the information necessary for a design review. In planning, determine the amount of time that you have to speak, the size and layout of the room, and the type of equipment available for visual aids.

2. Formulate the presentation by preparing an outline that identifies all of the topics, that you need to cover.

3. Prepare the slides needed in your presentation, using a GPP such as PowerPoint to guide you. As you compose the slides, edit them in the process, making certain that there are no

misspellings. After you have completed the slides, review them, and make modifications for their enhancement. Can you add graphics or bullets, to better capture the attention of the audience? Is the font size correct, and have used the bold, underline or shadow options to the best advantage? If you are using electronic displays, you have significant latitude for enhancing the visuals, and controlling the pace of the presentation.

4. Finally, you must rehearse the presentation. It is necessary that you know the material on your slides. Nothing is more deadly than a speaker reading his or her material from the slide to the audience. Rehearse until the presentation is smooth and polished, and you feel confident about the message to be conveyed, and your ability to handle the slides and the projector.

THE PowerPoint WINDOW

Let's begin by loading PowerPoint. If the Tip of the Day dialog box shows on the screen, read and then close it, to reveal the PowerPoint Window as shown in Fig. 1. There are four rows of controls located above the body of the window. The top row, called the title bar, displays the file name that you assign to the presentation when you first save your work. At the right side of the title bar, you will find the three buttons to minimize, restore/maximize, and close the file. The second row contains Microsoft's standard menu items. Click on one of the items, and a pull down menu is displayed. Click an explore the menu items until you are comfortable with the menu, and have a good understanding of their purpose.

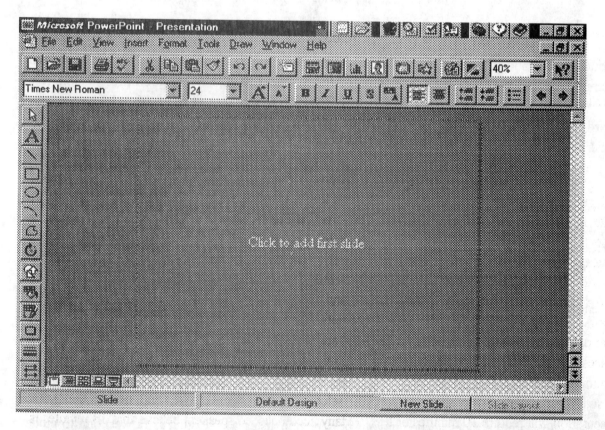

Fig. 1 PowerPoint display showing toolbars and working window.

The third row is the standard toolbar that displays 21 buttons, and a zoom control arrow to change the size of the display. Point to any button, and note that its purpose is soon defined in a small box attached to the pointer. The buttons on the left side of this toolbar are common to all of the Microsoft applications; however, the buttons on the right side of the bar are unique to PowerPoint. We will introduce them later, as we prepare an example presentation for a design review. The fourth row down from the top is the formatting bar. Again, most of the buttons are common with other Microsoft applications.

The center area of the window is the workspace, where you will prepare the slides. On the left side of this workspace, you will find a vertical drawing toolbar. We can prepare drawings in PowerPoint, to enhance the visual impact of the slides. The vertical scroll bar is located on the right side of the window. The two buttons at the bottom of the scroll bar permit us to move from one slide to the next, either forward or backward.

At the bottom of the window, we have two more rows. The top one is for the horizontal scroll controls. To the left of the scroll bar, you will find five buttons that control the presentation views. These buttons are very helpful, as we work to develop a multi-slide presentation, because they:

1. Present the slides individually.
2. Present the outline of all the slides, showing all of the titles and text.
3. Permit you to rearrange the slides with a slide sorter.
4. Permit you to add notes to the speaker's copy with a note page.
5. Present a slide show enabling you to review the complete presentation.

These buttons will be described in more detail later when we edit a presentation prepared as an example of a design review. The bottom row contains the usual the status bar where helpful messages are displayed. To the right of the status bar are shortcut buttons that you use to introduce a new slide into the presentation, and to select a specific slide layout.

PLANNING AND ORGANIZING WITH PowerPoint

The PowerPoint program has built-in features to assist you in planning, organizing and timing your presentation. To show these features, let's load PowerPoint, and click on File then New. A folder is displayed, as shown in Fig. 2, with three tabs named General, Presentation Design, and Presentations. Select presentations, and then note that you have 20 choices for different types of presentations. Select the first one offered, namely the Auto Content Wizard. The Auto Content Wizard page, which appears on your screen, indicates that wizard provides ideas and organization for your presentation. Click on the Next button, and the first of five dialog boxes associated with the wizard is displayed. This first dialog box prompts you for information pertaining to the title slide. After providing this title information, click Next, and the second dialog box provides you with a many choices for the type of presentation that you are creating. Let's hope that you do not need to Communicate Bad News. If so, select Reporting Progress. The next dialog box gives you choices on style and timing. We suggest that you select the professional style, and decide that your presentation will require 30 minutes or less. (Many accomplished speakers believe that 18 minutes is the ideal length for a presentation. This time is long enough to include substance, but short enough to avoid boring the audience).

The fourth dialog box gives you four choices for the type of output, and a choice of preparing materials to hand out to the audience. For this first example, select Black and White Overheads and indicate that you will print handouts. Click on Finish, and PowerPoint provides you with nine slides organized in sequence to provide the necessary information for your design review.

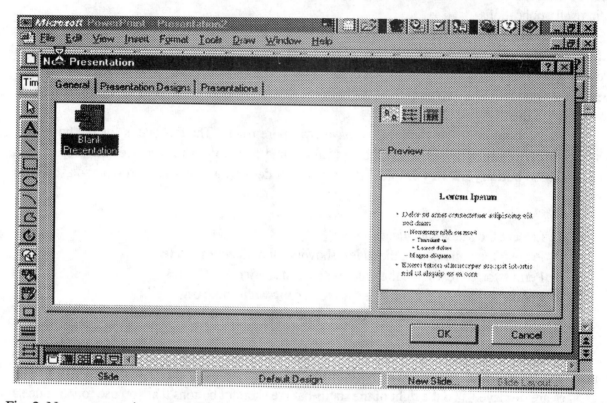

Fig. 2 New presentation page where selection of the Auto Content Wizard is made.

We will use these nine slides and prepare part of a preliminary (first) design review for the weighing machine. Begin with the title slide. Note that you have already provided the title in responding to the questions raised by the Auto Content Wizard. You may want to edit the title, or provide additional information in a subtitle. To edit the title, double click on the upper box to activate it and you are ready to make the changes needed to improve its appearance. Keep the title and the subtitle information very brief. You will have the opportunity to expound later in the presentation. Typing information onto the slides is the same as typing in a word processing program. An example title slide is shown in Fig. 3a.

Click on the lower button on the vertical scroll bar to advance to slide 2. In the organization provided by PowerPoint, the purpose of slide 2 is to define the subject. This slide prompts us to divide the subject into topics which we plan to discuss, and then we list these topics. In other words, we tell the audience in the beginning what we intend to tell them later in the presentation. It is a good idea, because no one likes surprises. We show an example of this slide in Fig. 3b. The information contained in slide 2 was entered on the presentation outline, as shown in Fig. 4. You can convert from the slide format to the outline format, by clicking on the outline button that is located to the left of the horizontal scroll bar. Under slide 2 on the outline, you can respond to the prompts provided by wizard, and add bullet items to define the topics that you intend to discuss. When you have completed the outline for slide 2, click the single slide button on the horizontal scroll bar and review the appearance of the slide. You may want to edit the slide, by changing the font size to better use the space available.

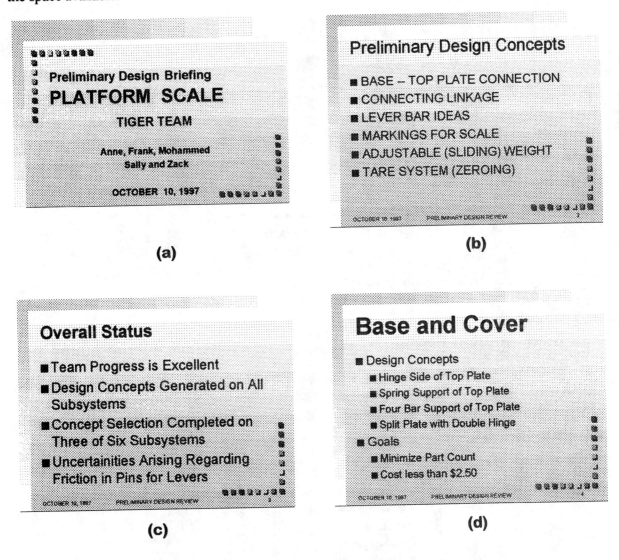

Fig. 3 Example showing content of four slides for a design briefing (before enhancement).

Click on the next slide button, and edit the entries that the wizard provides for slide 3. The title of this slide is overall status, which we believe is appropriate for a design review. You should report on the current status of the team. Describe the tasks that have been completed, report if you are behind schedule on any task, and raise any issues that concern the team. The example slide, presented in Fig. 3c, describes uncertainties regarding friction in the pins used in the lever system for the platform scale.

Wizard has titled slide 4 as Component One: Background. This title lead us to think about reporting on the first subsystem, and so we double click on the title to activate it. We then edit the title, changing it to Base and Cover. In the large area below the title block, we describe our progress in generating design concepts, and show two key criteria for the selection of the concept. You will note that we have made use of bullets and sub-bullets in the descriptions. We offset one bullet relative to the other by clicking the arrow button, that is located on the right side of the formatting toolbar. Try clicking on the arrow buttons, and observe the changes to the position and style of the bullets.

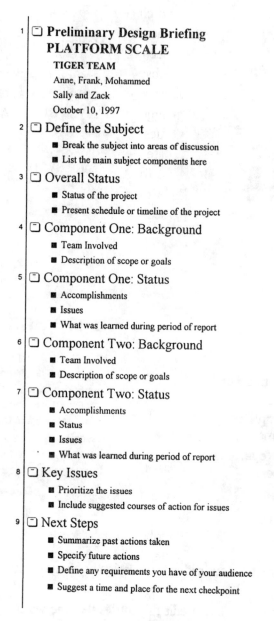

Fig. 4 Example of the outline page which can be used in preparing entries for the slides.

Slide 5, in the sequence suggested by wizard, is to report our status on the design of Component One. We modify the title to indicate that we are reporting on the Base and Cover, and then click on the lower block of the slide. In responding to the prompts provided by wizard, you should provide information describing accomplishments, issues, and lessons learned. These are all excellent topics to include at this point in the design review.

In preparing a complete design review, you will cover each subsystem using two slides similar to Slide 4 and 5. The wizard has only provided for two components (subsystems), so you will have to insert addition slides to cover all of the subsystems in your design. Adding new slides is easy. With slide 5 displayed on the monitor, click on the new slide button on the right side of the status bar. A new slide page is displayed on the screen, that gives you a choice of 12 different formats. A format consistent with that employed in slide 5 is highlighted. Since we want to retain this format, we click on OK, PowerPoint introduces the new slide, and changes the entry on the status bar to read slide 6 of 10. In the top block of your new slide, type the title, and in the lower block type the information describing the design concepts and goals for the second subsystem.

After you have briefly reviewed the subsystems in you development, you will want to conclude the presentation. The last two slides in the wizard sequence provide a template for preparing your concluding remarks. We show suggestions for these two slides in Fig. 5. The next to last slide in your presentation is titled "Key Issues". This is your opportunity to raise any issues (questions), that your team has discovered to date. Do not hide uncertainties; come forward, and seek help. The design review is a formal venue for this purpose. The example shown in Fig. 5a indicates that we are concerned with the effects of friction at the pins connecting the levers, and friction in the linkage connecting the base of the scale to the lever system.

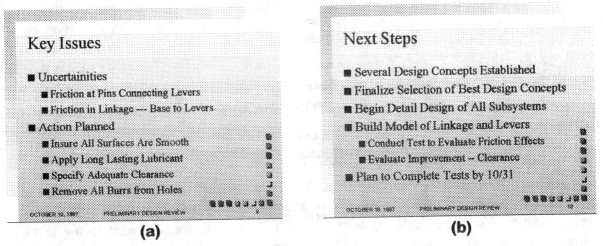

Fig. 5 Example of the two slides used to conclude the design review.

The final slide, illustrated in Fig. 5b, describes your plans for action in the near future. Wizard titles this slide as Next Steps, which is appropriate. Wizard then prompts you to:

- Summarize past actions taken by the team.
- Specify future actions that are planned.
- Define any requirements that you have of the audience.
- Suggest a time and place for the next design review.

These are excellent suggestions for the final slide concluding your design review.

You are now ready to print the output. Click on File, and select Print. The print page that is displayed is adapted for PowerPoint. It contains a selection box labeled Print What. You have several choices including: slides, handouts with 2, 3 or 6 slides per page, note pages, and an outline view. You can prepare the overheads for your presentation, and the handout that you intend for the audience. Print your presentation using these options, and examine the output to decide which style is the most suitable for the handout that you intend for the audience. The black and white paper versions of the slides can be copied to give the transparencies necessary for the presentation. If you have your own laser or ink jet printer, the transparencies can be printed directly. Be sure that you save the file before exiting the program.

ENHANCING THE SLIDES

Suppose you decide to use color slides, instead of black and white. It will cost a lot to use color if you take the job to a commercial copy center, but you may decide that it is worth it. How can you change the black and white slides to color slides? It is easy, and you do not have to retype all of the information into a new version of wizard. To change the color and/or the design of the slides, click on the Apply Design Template button on the right side of the PowerPoint toolbar. The design template page, that is displayed on your screen, has a menu listing of 30 templates, which start with Azure and end with Wet Sand. With the mouse, select one of the templates and note that it is displayed for your review in a small preview box. Try several templates before selecting the one that you like the most.

If you would like to change the color of the background, go to the menu bar and click on Format, and select Custom Background. The Color page that is displayed gives you several options for both the style of the background and the color employed. Click on Options and review the results until you get the style and the colors that you believe will be the most effective in your presentation. Be careful and stay with relatively light colors, because they are more visible to the audience. Dark slides do not project, well unless the room is extremely dark. And dark rooms are an invitation for your audience to sleep.

Suppose that you like the background style and color, but are unhappy with the color of the text and the bullets. Change them by clicking on Format and select Slide Color Scheme. The Color Scheme page has two tabs, namely standard and custom. The standard tab illustrates four choices of color for the titles and bullets. If they are not to your liking, click on the custom tab and prepare to become a painter. You have the opportunity to select the colors, that you want to use in the background, shadows, fill, text and title and bullets. The colors are changed by selecting a feature on your slide, such as the title text. Click the Change Color button, and a multicolored hexagon appears with many choices of different shades of the basic colors. Select the color that you wish to employ, for each of the features on the slide. You can have a single design for all of your slides (recommended), or you can have a different color scheme for each slide. Have fun, because PowerPoint gives you the means to make great slides.

Some folks like to use either a header or footer on their slides. A footer is helpful, if you give many presentations, and want to retain the slides for your files. The information in the footer can include date, occasion for the presentation, location, and etc. For example, you may decide to use a footer stating October 10, 1997 --- Preliminary Design Review and slide number. How do we incorporate this footer on our PowerPoint slides? Click on View on the Menu bar, and select Header and Footer. The Header and Footer page is displayed with two tabs; one for the slide and the other for the notes and handouts. Let's select the slide tab, and add a footer to our slides. We click on the Date and Time square, select Fixed, and type the date of the planned presentation. We mark the slide number, because it will help us to time the delivery of our presentation. We click on the footer, and type in a short message to appear in the center region of the footer. We also click on the box to indicate that we do not want the footer to appear on the title slide. We delete the footer from the title slide, because we want the audience to remain focused on our subject in the very beginning of the presentation. The Header and Footer page has a small preview window, that permits us to review the input prior to printing. We illustrate the footer developed with this procedure in Fig. 6. The slide was in beautiful color on the monitor, but the figure is in black and white. Our apologies, but the use of color in printing in a low-volume, low-cost textbook is prohibitive.

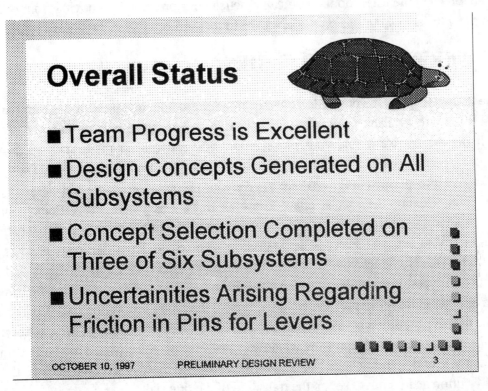

Fig. 6 Example of an enhance slide showing an interesting background, footnote and clip art.

Still not happy with the impact that your slides will make. What about adding some clip art? Select a slide that you want to enhance with the addition of some photo or sketch which is available in the Microsoft Clip Art files. You don't know what's available. Not to worry; click on the Insert Clip Art button that is located on the right side of the toolbar. (The button with the funny face on it.) There is a pause while the clip art files are loaded into the active PowerPoint program. A page entitled Microsoft Clip Art Gallery 2.0 appears on the screen. On the left side of this page, you will find a listing of clip art categories. Select the Animal category and illustrations of the selections of clip art available under this category appear in the large preview window. Click on the turtle, and then click the insert button.

Now look at the slide that you had previously selected for enhancement. We have a large picture of a turtle, smack in the middle of the slide. To move the picture and change its size, we drag it by the one of the little squares either on the corners or the middle of the sides to position the insert appropriately on the slide. We show the turtle in the upper right hand area of the slide in Fig. 6. Note, if you push while dragging the object, it gets smaller, but pulling makes it get larger. Try it, and you will soon manage to both move and size the turtle.

REHEARSING

Rehearsing is an essential part of preparing for a design review. Fortunately PowerPoint has a neat feature that will help you handle the visuals as you rehearse. Go to the first (title) slide, and then click on the Slide Show button to the left of the horizontal scroll bar. The screen changes with all of the toolbars disappearing. The slide occupies the entire screen and the PowerPoint presentation simulates the screen in the class room. You can practice your presentation moving from one slide to the next by using the arrow buttons on the keyboard to forward or reverse the slides.

COMPUTER PROJECTED SLIDES

The ability to project your slides directly from the computer gives you several significant advantages which are available in PowerPoint. First, you can store a large number of slides in memory. If questions arise during your presentation, you can call on these back-up materials to improve your response. Second, you do not need a color copier to make the slides. Third, you can use special effects incorporated in PowerPoint to give your presentations some real class.

Let's first explore the special effect know as transitions where PowerPoint controls the manner in which we switch from one slide to the next. Go to the title slide in your presentation, click on Tools in the menu bar, and select Slide Transition. The Slide Transition page appears with an Effect box that states No Transition. Click on the arrow located beside this message, and a list of about 50 different types of transitions is given. Select one of them, and watch the slide switch from one view to another in the small preview window. Select the type of transition that you like the best. We think the dissolve transition is an excellent choice. Next, select the speed of the transition. Again you can review your choice in the preview window. Finally, select the advance method. Until your a real pro at delivery of design briefings, we suggest you advance the slides on the click of the mouse. Click the OK button and you have set up the transition of the title slide.

Click on the Slide Sorter View button near the horizontal scroll bar, and the first six slides of your presentation are displayed. The first slide has a small slide transition icon visible near its lower left corner. A new toolbar also appears at the top of the window replacing the formatting toolbar. You are now ready to select the type of transition for the remaining slides. Select slide 2, and click on the transition arrow button on the new tool bar. You can select the type of transition from the long list for slide 2. Proceed to mark any slide that you wish to transition in the Slide Sorter View, and to select the type of transition that you want for that slide. After you have made your choice, it is confirmed by the presence of a small transition icon just below the slide when you have the Slide Sorter View active. If you employ the same transition for all of your slides, mark all of them (Edit/Select All), and then select the type of transition of your choice.

PowerPoint also permits you to build your slides progressively for your audience. It is a bit like watching the construction of a building from the foundation to the roof. For slides, we start with the title, and add one line at a time to the text in the block below the title. To modify your slide to incorporate this feature, repeat the slide transition procedure described in the preceding paragraph. The new tool bar that is evident just above the window has a box that is titled No Built Effect. Click on the arrow beside that box, to reveal a wide choice of techniques for building you slides. Pick an option; fly from the left is very common. When you select an option for building the slide, a second small icon shows under the view of the slide in the slide sorter view. You can proceed to build your slides, one by one, or you can use the same style on all of them. Select all of the slides on the slide sorter view, and depress the Control + A keys.

When you have completed the transitions and the building effects, you are ready to review the dynamic nature of your electronic presentation. Real high tech. Go to slide 1, and click on the Slide Show button. As you advance from slide to slide with arrow keys on the keyboard, you will see the transitions and the lines of text for each bullet item being moved onto the screen.

If you have computer facilities available in the classroom, electronic presentations should be mandatory. Creating electronic slides and using them in a professional presentation are skills that are important for you to master. The transitions keep the audience on their toes, and the building effects permits you to pace the flow of information. It takes less than an hour to learn how to prepare a very professional, dynamic presentation using PowerPoint. Take this opportunity to develop your computer graphic presentation skills.

SUMMARY

We have introduced you to PowerPoint, a graphics presentation program that is an extremely useful tool in preparing for design reviews or any other type of presentation. We strongly suggest that you take an hour or two to acquire PowerPoint skills. You will master the skill necessary to produce professional quality overheads at the least, and, if the equipment is available in the class room, the ability to produce computer projected slides.

The screen for PowerPoint has been briefly described. As with all Windows® based programs produced by Microsoft, many of the buttons on the toolbars have common uses. The coverage given here is brief because learning is more rapid if you click, study, and understand. Hopefully, our limited coverage and examples will be helpful.

We suggest that you start to learn how to use PowerPoint by preparing your preliminary design review using the Auto Content Wizard. The wizard organizes your presentation, arranges the sequencing of your slides, and provides excellent suggestions for the content in each slide.

After you have completed the standard presentation with the Auto Content Wizard, you should enhance the slides to improve their appearance. Procedures have been described for changing the design template, the colors of the background and the text, adding a footer, and clip art. Clip art is particularly impressive and easy to implement.

Finally, we introduced techniques for preparing electronic presentations. These presentations depend upon the availability of a computer controlled projector for the classroom. If this equipment is available, the use of a computer in your presentation has very real advantages. Employing transitions, as you move from one slide to the other maintains the audience interest. Utilizing build effects, where the text is presented one line at a time, has the significant advantage of controlling the flow of information to the audience. The procedures for preparing and rehearsing a computer aided presentation are described in this chapter.

EXERCISES

1. Prepare a title slide suitable for a final design briefing using PowerPoint.
2. Prepare a subject slide that provides an overview of a final design briefing using PowerPoint.

3. Prepare a status slide using the slide outline format in PowerPoint. After completing the slide in the outline format, edit the slide, and print a paper copy of it.

4. Prepare a pair of slides describing your progress in the development of one of the subsystems for the weighing machine.

5. Prepare a slide that describes the key issues facing your team as you design the weighing machine.

6. Prepare a slide showing your plans for action in the near future.

7. Using clip art enhance a slide by inserting an animal in the lower right hand corner of the slide.

8. Prepare a transition from the title slide to the second slide. What type of a transition did you select?

9. Prepare a slide using the building effect, where single lines of text are added to the slide one by one as you click the mouse button.

NOTES

PART IV

PRODUCT DEVELOPMENT PROCESSES

CHAPTER 11

DEVELOPMENT TEAMS

INTRODUCTION

The development of quality, leading-edge, high performance products requires the coordinated efforts of many skillful individuals from different disciplines over an extended period of time. This group of individuals is organized within a corporation to act in an integrated manner to successfully complete the product development. The organizational structure that is employed varies from one corporation to another, and it also depends to a large degree on the size of the company that is involved. For relatively new firms in the entrepreneurial stage, team structure is usually not an issue. These firms are usually quite small with only a few products, and only a small group of people are involved. Everyone that is on the payroll is deeply committed to this product, and communication is frequently accomplished around the lunch table. On the other hand, for very large corporations with thousands of employees, the company is organized into divisions or operating groups. These large companies have many products or services that are offered by each division, and several new products are introduced each year. Communication is often very difficult because personnel involved in the development of a given product are sometimes separated by hundreds or even thousands of miles. In a large corporation, organization of the product development team and the physical location of the team members becomes extremely important.

You might be wondering if this introductory material about organization and communication of a development team in a corporation is important in this course. After all you are just beginning your studies of design. The answer is a resounding **YES**! In this class, we attempt to simulate in so far as possible an industrial design experience. Accordingly, we have assigned you to a small group of five or six students. This group is a development team, and the product that you will design will be an analog and a digital weighing machine. Above, we discussed the problems of a large corporation in organizing their development teams and insuring effective communications. You will also encounter problems of organization and problems of communication even though your team is small, and all of you are located on the College Park campus.

In a corporation, cross disciplinary teams are necessary in staffing a product development team. Usually, at least five disciplines are represented on the team including, marketing, finance, design engineering, operations and/or manufacturing, and quality control. When we assigned you to a team, it was not possible to follow the corporate procedures for staffing. You have yet to establish your disciplines; however, you do have interests and talents. In so far as possible, we have used the information that you provided on your application for team membership in the team assignments. We trust that you will find a diversity of interests, talents, and skills in the members assigned to your team. It is suggested that you organize the team, and assign work to individual team members to take advantage of these skills.

TEAM LEADERSHIP

Experience and leadership skills are required of the product manager (team leader) in industry. The product manager is a seasoned staff member well known for his or her expertise, and respected for performing at a level that exceeds expectations. The appointment as a product manager sometimes represents a promotion for the individual involved, and a first experience in managing a technical group. But what about you? You are not "seasoned," and you do not as yet understand our level of expectation for your performance or that of your team. Nevertheless, your team must have a leader. We usually encourage you to meet for a few hours as a team to exchange information about yourselves. From this interchange, some bonding between team members should occur, and the team should eventually develop a consensus regarding the team leader. We will expect that the team will announce its leader before the end of the second week of class. The selection of the team leader is entirely up to the team members. The class instructor will not be a part of this decision. The class instructor will serve as a consultant to the team, but he or she will never attempt to provide leadership.

In industry, the responsibility and authority given the product manager is significant. His or her career depends strongly on the success of the product in the market and on the productivity of the development team. We recognize that you and your team leader do not have the experience needed to insure a successful development. For this reason, we do not expect every team to develop a weighing machine that meets every requirement listed in the product specification. However, we do expect that the team leader will guide the development team by:

1. Providing a development schedule.
2. Assigning tasks to the most qualified team member.
3. Insuring that every team member is working productively.
4. Chairing and organizing team meetings.
5. Maintaining "good" behavior at each team meeting.
6. Insuring that each team member is treated with respect.

We also expect that each team member will work diligently with the team leader, cooperate completely at team meetings, and commit to holding the development schedule.

TEAM MEMBER RESPONSIBILITIES

Members of a product development team in industry have two different sets of responsibilities-- one to his or her disciplinary function and the other to the team. A team member, from a specific engineering department, provides the technical expertise in his or her specialty, and ensures that the product is correctly designed with leading edge state of the art technology. The team member acts to bring all of the important functional issues that affect the product performance to the attention of the team. As a beginning student in engineering, you have not yet established an expertise in some discipline. However, you do have natural talents, skills, and interests. We want you to represent these skills on your development team, and discuss issues in your area of interest at your team meetings.

The most important team activity for any individual member is to share responsibility. A team member shares responsibilities with other team members to increase the team's effectiveness, and to insure its success. Next, the team member must recognize and understand all of the product features, and fully participate in the methods and techniques that are employed to meet the design objectives. Individual team members must assess team progress, and participate in improving team performance. A team member must cooperate in establishing all of the relationships that are required to maintain communication between the team and the instructor. Flexibility and cooperation on the part of the individual team members are required for the team to be effective. Finally, the team members must be able to communicate in a clear concise manner in all three modes: **writing, speaking, and graphics**.

TEAM MEMBER TRAITS

When a group of individuals work together on a team to achieve a common goal, they can be extremely effective. The team interaction promotes productivity for several different reasons. First, meeting together is synergistic. One's ideas, freely expressed, stimulates additional ideas by other team members. The net result is many more original ideas than would have been possible by the same group of individuals working independently. Another advantage is in the breadth of knowledge available in the team. The team is cross disciplinary with different interests, talents and skills. The very wide range of skills necessary for the product development process are usually inherent within the team. Each member of the team is different with some combination of strengths and weaknesses. Acting together, the team can build on the strengths of each individual and eliminate the weaknesses. The grouping of individuals provides social benefits that are important to the individual team members, and to the corporation. There is a bonding that takes place over time, and a support system develops that is appreciated by the team members. Feelings of trust and mutual understanding, so important to everyone involved, develop over the course of the project. The team with its strength in numbers develops a sense of autonomy that is valued by its members. The team develops solutions, implements them, and then monitors the results with little or no direction from management (the instructors). Empowered teams assume ownership of the project, and they totally commit to the product development process. Successful teams consistently perform beyond expectations.

While cross disciplinary teams have been employed in industry for a decade or more, little has been done by our secondary school system to develop team skills. On the contrary, the secondary schools and even some college systems encourage competitive attitudes and independence. Students compete for the few A grades that might be given in the course, and they are discouraged from cooperating on the assignments. The system trains a student to work independently, but team building skills are not addressed. The work place in industry is much different. Team members do compete, but not within the team. They compete with another development team from a rival corporation. Team members bond together, cooperate and consistently help each other. In a team, we cooperate as much as possible and work together by sharing assignments. Competition between its members is discouraged. Recognition and rewards go to the team much more often than to select individuals.

There is a set of characteristics that describe a good team member and another set that depicts an individual that can destroy the efforts of a team. The characteristics of a good team member are:

- Respects other team members without question.
- Listens carefully to the other team members.
- Participates but does not dominate.
- Displays self-confidence but is not dogmatic.
- Is comfortable with his or her disciplinary skill level.
- Communicates effectively by speaking, and writing and with graphics.
- Disagrees, if it is important, but with good reason and in good taste.

You should evaluate your behavior as you work with fellow team members. Judge how well you follow the guidelines for "good" team member behavior. Are you helping or hindering the progress of your team?

The characteristics of a destructive team member are:

- Shows lack of respect for others.
- Tends to intimidate.
- Stimulates confrontation.
- Dominates discussion.
- Talks all the time, but does not listen.
- Criticizes unnecessarily.
- Does not communicate effectively.

As you compare your team traits with those listed above, you will note characteristics from both lists. To be effective an team members, you must work to enhance the favorable traits and to suppress those that are destructive to the efforts of the team.

EFFECTIVE TEAM MEETINGS

Teams are effective because they meet together to focus their wide range of disciplinary and individual skills toward the solution of well formulated problems. The team meeting is the format for the synergistic efforts of the team. Unfortunately, not all teams are effective because some members are disruptive and they destroy the cohesiveness of the team. Bonding of the team is vital if it is to successfully solve the multitude of problems that arise in the product development process.

Teams fail for three main reasons which are:

1. They deviate from the goals and objectives of the development, and cannot meet the milestones on the development schedule.
2. The team members become alienated, and the bonding, trust, and understanding, critical to the success of a team never develops.
3. When things begin to go wrong, adversity almost always occurs, and "finger pointing" starts. Fixing blame replaces creative actions, and the team is destroyed.

Quality leadership ensures that the team remains focused on the overall goals and objectives of the project, and on the agenda of each meeting. Creating the correct team environment or atmosphere is essential so that the team members cooperate, share ideas and support each other to achieve the many solutions. It is important for the team leader to assess the performance of the team as a whole at each meeting. When the team fails to make progress, it is critical for the team leader or possibly another team member to make the changes necessary to enhance the teams effectiveness. There are several measures of team effectiveness that can be observed by everyone on the team during the course of a typical team meeting. Corrective measures to improve team performance are not difficult to implement. We will identify some of the difficulties encountered during team meetings in subsequent paragraphs.

When the goals and objective of the meeting are not clear or they appear to be changing, progress is impeded. While in many cases team meetings are planned well in advance with a well organized agenda, sometimes new topics are introduced, and the meeting drifts from one item to another. This team behavior indicates that either some of the team members do not understand the goals; or worse --- they do not accept them. Instead, they are marching to their own drummer and trying to take the team with them. Unless corrective actions are taken to focus the team on the original goals, this team meeting will fail resulting in lost time and effort. A team that is focused stays on the task. The team members accept the goals, and their discussions deal with all of the issues important in considering a wide set of solutions to the problem being considered. The discussion continues until the team members reach a consensus, and all of the members accept the solution. This type of a meeting is successful because a problem has been solved, and all of the team members concur with the solution. The team can then move forward, and focus its talents and energies to address the next problem.

Team leadership is an essential element for success. The most effective teams are usually democratic with a large amount of shared leadership. While there is a recognized leader with a considerable degree of responsibility and authority, he or she is supported by each member of the team. The team member with the appropriate expertise (relative to the problem being considered) will usually lead the discussion, and effectively act as leader of the team during this period. However, in some instances, the team leader maintains strict control of the meeting and does not share the leadership role with any member of the team. Some team members resent this style of leadership, and they may not participate as fully as possible. The result is that complete utilization of the resources of the team does not occur. The "tightly controlled" team leader will almost always insist on sitting at the head of the table.

Attitude of the team members is an important element in the success of the team. Are the team members committed? Commitment is evidenced by attitude. If team members come to the meeting exhibiting interest and willingness to participate, they are committed and will make a positive contribution. However, if they are bored and indifferent, they will not become involved in a meaningful way. In fact, if they are sufficiently disinterested, side conversations or arguments may develop and actually destroy the atmosphere necessary for an effective meeting. Every team member should evaluate each person on the team, assessing his or her commitment to the goals and objectives of the project and to the meeting's agenda.

As the meeting progresses, it is important to assess the discussions that are occurring. For a meeting to be successful, the interactions should involve everyone on the team. Discussion should focus on the agenda items with few if any deviations to unrelated topics. Team members must listen carefully to one another. All of the ideas presented are given a serious hearing. No one should be intimidated or made to feel foolish for making suggestions that appear to be too radical. An informal and relaxed attitude exhibited by the participants is the key to free flow of ideas. Teams tend to accomplish less when the discussion is dominated by only a few of the participants. Some team members speak well, but they do not bother to listen to others. They may introduce discussion that is off the topic and ignore suggestions for getting back to the agenda. These members tend to intimidate others in their efforts to control and dominate the team. The result is disastrous because they essentially eliminate the contributions of the other team members.

Teams do not operate with total agreement on every issue. There must be accommodation for disagreements and for criticism. All of the ideas or suggestions made by every team member at any meeting will not be outstanding. In fact, some of them may be ridiculous, and criticism of these ideas must occur. Criticism should be frank, but without hostility. Personal attacks must not be a part of the critique of an idea. If the criticism is to improve an idea or to eliminate a false concept, then it is of benefit to the progress of the team. The criticism should be phrased so that the team member advancing the flawed idea is not embarrassed. When disagreements occur, they should not be suppressed. Suppression of disagreements breeds hostility and distrust. It is much better for the team to deal with the disagreements when they arise. When the root cause of the disagreement is ascertained, the team can take the actions necessary to resolve the problem.

On some occasions voting is a mechanism used to resolve open issues. This procedure must be used with care, particularly if the vote indicates that the team is split almost evenly in their opinions. A better practice is to discuss the issues involved, and to try to reach a consensus. More time is required to reach a consensus than is necessary for a simple vote, but the results are worth the effort. Consensus implies that all members of the team are in general agreement and willing to accept the decision of the team. When the team votes, a simple majority is sufficient to resolve an issue. However, the minority members may become resentful if they are always on the short end of the voting. In a very short time, they will not accept the outcome, and they will not commit to the actions necessary to implement the decision.

The team meetings must be open to all of the members. The agenda should be available in advance of the meetings and subject to change during the "new business" agenda item. Team members should believe that they are important, and have the authority to bring new issues before the team. They should feel free to discuss procedures used in the team's operation. Hidden agenda items detract from the harmony and trust developed through open and fair operation of the team. Secret meetings of an inside group should be carefully avoided. When the fact leaks that secret meetings are being held and that issues are prejudged by a select few, the effectiveness of the team is destroyed.

At least once during a meeting, it is useful to evaluate the progress of the team. Are we still following the agenda? Is someone dominating the meeting? Is the discussion to the point and free of hostility? Is everyone properly prepared to address their agenda items? Is the team attentive, or are they bored and indifferent? Is the team leader leading, or is he or she pushing? If a problem in the operation of the team is identified during the pause for self appraisal, it should be resolved immediately through open discussion.

Finally, the team must act on the issues that are resolved and on the problems that are solved. To discuss an issue and to reach a consensus is part of the process, but not the closure. Implementation is the closure, and implementation requires a plan for action. When decisions are made, team members are assigned action items. These action items are tasks to be performed by the responsible individual, and only when the tasks are completed can the problem be considered solved. It is important that the action items be clearly defined, and that the role of each team member in completing their respective tasks be evident. A realistic date should be set for the completion of each action item. The individual responsible for a task should be clearly identified, and he or she must accept the assignment without objection. A checking system must be employed to follow up on each action item to insure timely completion. If there is a delay, the schedule for the entire product development cycle may be at risk. Delays should be dealt with immediately, and the entire team should participate in the development of modified plans to correct the situation.

PREPARING FOR MEETINGS

Effective team meetings do not just happen. To insure success, it is necessary to prepare for the meeting and to execute post meeting activities. The preparation usually involves selecting, arranging and equipping the meeting room, scheduling the meeting in advance so that the necessary personnel are in attendance, and conducting the meeting following some set of rules accepted by the team. Also critically important are actions taken by the team members following the meeting to implement the decisions.

The meeting room and its equipment are essential to the progress made by the team. The room should be sized to accommodate the team and any of the visitors that have been invited. Rooms that are too large permit the members to scatter, and the distance between some members becomes too long for effective and easy communication. The chairs should be comfortable, but not so soft as to promote napping. The seating arrangement should be around a table so that everyone involved can see each others face. It is very difficult to engage anyone in a meaningful conversation if they have their back to you. In industry, water, coffee, tea or soft drinks are often available if the meeting duration exceeds and hour. (Sorry, but we cannot manage this nicely in the typical college classroom). Smoking is strictly prohibited. The equipment that will be needed for presentations must be available, and its operation should be checked prior to the start of the meeting. Flip charts and white boards are useful for recording the key ideas generated during the meeting. Tape, pens, markers, and other supplies necessary for making and mounting charts and lists should be available. A computer with projection capability is of growing importance in a well-equipped meeting room. The availability of a computer during the meeting permits one to draw from a large data base, and to modify any solution in real time. Also, the results can be displayed in a graphical format for the entire team to review. Finally, make sure that the room temperature is comfortable, and that the light intensity can be controlled from low for projecting overheads to high for round table discussions.

Prior to the scheduled meeting, it is important to make careful preparations. Minutes from the previous meeting and the agenda should be distributed with sufficient time for review before the next meeting. The detailed agenda includes the topics and/or problems that will be addressed in the next meeting. Individual team members responsible for specific agenda items are identified. The details of the agenda are clear to all, and the responsibility that is shared by the individual members is defined. In some instances, information will be needed from corporate members or visitors that are not members of the team. These people should be invited to the meeting so that they can provide the necessary expertise, and respond to questions from all of the disciplines represented on the team.

In conducting the meeting, it is important to start on time. It is very annoying to the majority of the members to wait five or ten minutes for a straggler or two. The team leader must make certain that everyone involved understands that 8:00 am means 8:00 am, and not 8:08 or 8:12 am. The objectives of the meeting should be displayed, and the time scheduled for each item should be estimated. A team member should be assigned as the timekeeper, and another should act as a secretary to record notes and to prepare the minutes of the meeting. If team meetings are frequent and are held over a long period of time (several months), the duties of the timekeeper and the secretary should be shared with others on the team. The meeting should follow a set of rules that govern the behavior of the individual members. We have assigned, in Exercise 11, the preparation of a list of rules which should be followed in conducting an effective meeting. As the meeting draws to an end, the team leader should take a few minutes to summarize the outcome of the discussions. This summary is the ideal place to insure that assignments are understood, responsibility accepted, and completion dates are set. The meeting must be completed on time and everyone's schedule should be respected. If the meeting is not periodic (i. e. every Tuesday at 8:00 am), then it is important that the time and place for the next meeting be scheduled before the meeting is adjourned.

After the meeting, the team leader checks to make certain the minutes have been prepared, that they are complete and correct, and that they have been distributed. If any member was not able to attend the meeting, the team leader briefs that person on the important happenings. Finally, the team leader checks on each action item to make certain that progress is being made, and that new or unanticipated problems have not developed. It is clear that effective meetings do not happen by accident. Many members of the team work intelligently before, during and after the meeting to make certain that the goals and objectives are clearly defined, that the issues and problems are thoroughly discussed, and that the solutions developed result in assignments which are executed with dispatch.

POSITIVE AND NEGATIVE TEAM BEHAVIOR

All the members on a team and particularly the team leader must behave in a positive manner for the team to function effectively. Frequent negative behavior is detrimental and inhibits the progress that can be made by the team, and delays the product development cycle.

Positive mannerisms for the team leader, and team members are:

1. Treat every team member with respect, trust their judgment and value their friendship.
2. Maintain an inquiring attitude free of predetermined bias so that team members will all participate and share knowledge and opinions in a free and creative manner.
3. Pose questions to those team members who are hesitant to encourage them to share their knowledge and experience with the team. Share your experiences and opinions in a casual easy going manner. Help other members to relax and enjoy the interactive process involved in team cooperation.
4. Observe the body language of the individuals to determine lack of interest, defensive attitudes, hostility, etc. The team leader should act to defuse hostility and to stimulate interest.
5. Emotional responses will occur when issues are elevated to a personal level. It is important to accept an emotional response even when you are opposed to the viewpoint presented. It is also important to control you emotions, and to always think and speak objectively.
6. Seating arrangements, once established, tend to become permanent. It is productive to break these patterns changing the seating arrangements to mix the team members. The result, with time, is to enhance bonding, and to stimulate new ideas generated by different combinations of members.
7. Allocate time for self-assessment so that the team can determine if it is performing up to expectations. The team leader may make suggestions during these assessment periods to build team skills.

Negative behavior:

1. The team leader must be extremely careful about intervention. If the team is moving and working effectively, the leader should be quiet because intervention in this instance is counter productive. If the team is having difficulty, intervention may be necessary depending on the problem. If an individual becomes hostile, intervention should be quick and the disagreement producing the hostility dealt with immediately. If the team is seeking consensus, the process may require time and the leader should

wait 10 to 15 minutes before intervening. Under no circumstances should the leader intervene if he or she cannot make a positive contribution.

2. The leader should never argue with a team member. The leader can introduce a different opinion, and participate in a discussion with a different viewpoint. However, the discussion should never deteriorate into an argument. It is the leader's responsibility to quickly resolve arguments that arise among the team members.

3. The team leader cannot be opinionated. The leader is not a judge determining the right solution. The leader facilitates, so that the team can reach a consensus on the correct solution. It is only when the team accepts the solution that implementation can begin.

4. No member of the team should ever participate in a conversation that is derogatory about a person on the team or in the corporation. If you can not make complimentary remarks about a person, keep your thoughts to yourself. Remember respect, trust, and friendship are vital elements in a successful product development team.

5. The leader is responsible for attendance and punctuality at the meeting. If a member is absent, the leader should know the reason beforehand, and explain it to the team to avoid resentment among members present. Every effort should be made to prevent erosion of mutual respect among team members.

REFERENCES

1. Smith, P. G., D. G. Reinertsen, <u>Developing Products in Half the Time</u>, Van Nostrand Reinhold, New York, NY, 1991.
2. Barra, R. <u>Putting Quality Circles to Work</u>, Mc Graw Hill, New York, NY, 1983.
3. Barczak, G. and Wilemon, D., "Leadership Differences in New Product Development Teams," Journal of Product Innovation Management, Vol. 6, 1989, pp259-267.
4. Clark, K. and Fujimoto, T., <u>Product Development Performance</u>, Harvard Business School Press, Boston, MA, 1991.
5. Wheelwright, S. C. and Clark, K. B., <u>Revolutionizing Product Development</u>, Free Press, New York, NY, 1992.

EXERCISES

1 Write a paragraph describing why team structure is so important in the product realization process which takes place in a large corporation. Add a second paragraph indicating why it is much less important in a very small company.

2 Write a letter to a prospective employer explaining why you are a good team member.

3 Write a letter, with the best of intentions, to a friend explaining why you think he or she is not an effective team member.

4 You have been a member of a functional team for two months. The team is not doing well, and it is beginning to miss milestones in the schedule of the development plan. You are to meet with your instructor for a formal performance review. Write a script for a two person

play with your instructor. Questions and comments, and your answers and comments covering what you expect to occur during your review should be included in the script.

5   You (a male/female figure) are attending weekly team meetings, and are always seated next to Sally/Bill who is young and very attractive. Sally/Bill is apparently interested in you because she/he frequently involves you in side conversations that are not related to the ongoing team discussions. Write a plan describing all of the actions you will take to handle this situation.

6   Brad, the team leader, is a great guy who believes in leading his team in a very democratic manner. He encourages open discussion of the issue under consideration for a defined time period, usually 5 to 10 minutes. At the end of the time period, he intervenes and calls on the team to vote to resolve the issue. Write a critique of this style of leadership.

7   You, together with Harry and Mary, are senior members on a development team with 14 other members. The three of you are really very knowledgeable with extensive experience. You get along very well, and have developed a habit of gathering together at the local watering hole the evening before the schedule team meeting. During the evening you discuss the issues on the agenda reaching some decisions between the three of you. Write a brief describing the consequences of continuing this behavior.

8   Sue has a wonderful personality, and, as a team leader, is skillful in promoting free and open discussion by all of the team members. After the team reaches a consensus, she calls for volunteers to implement the required action. When two or more people volunteer, she assigns them as a group to handle the action item. When no one volunteers she assigns the least active person on the team to the action items, and moves promptly to the next item on the agenda. Write a critique of this leadership style.

9   Efficient Edith prepares a meeting agenda that includes 16 major items. She schedules each item on the agenda with a 15 minute discussion period, and she leads the discussion on every topic. The meeting is scheduled to begin at 8:00 am and adjourn at 11:59 am. Write a description of what you imagine happened during the meeting. If you were the team leader, how would you handle the 16 agenda topics that had to be resolved in a short time?

10 You are the leader of a team of eight members that is meeting to develop a new hair dryer. During the first meeting the following events occur:

- The temperature the meeting room is 88 °F.
- The room is set up for a seminar speaker.
- The meeting is scheduled to begin at 9:30 am and at 9:35 am only five of the eight members have arrived.
- During the meeting Horrible Harry begins to verbally abuse Shy Sue.
- At 9:50 am Talkative Tom begins to describe his detailed positions on the world situation and is still going at 10:10 am.
- The team breaks for coffee at 10:15 am and has not returned at 10:30 am.
- Sleepy Steve begins to snore.
- Procrastinating Peter refuses to accept an action item after leading the discussion on the issue.

Describe your actions as a team leader in dealing with these events.

11 Prepare a list of ground rules that should be followed by all the members of a team in conducting an effective meeting.

12 You are a member of a team with an ineffective leader. What can you do to improve the productivity of the team? In responding to this question consider the irresponsible actions on the part of the team that are listed in Exercise 10. Remember in your response that you are not the leader and the most that you can do is to share the leadership role from time to time.

CHAPTER 12

A PRODUCT DEVELOPMENT PROCESS

INTRODUCTION

The design and development of new products and services are essential for the welfare of most industries operating in the world. The only industries that can operate today, disregarding the marketplace and continuing to ignore to the consumers' needs, are those protected by government regulations (i. e. the U. S. Post Office). For private enterprise to be successful, their product line must be competitive in every category. The products must be attractive, functional, efficient, durable, reliable, affordable, and most importantly they must satisfy the needs of the customer. How do we systematically design products that will be a success in the marketplace and generate profits for the company? Over the years companies have evolved a product development process, that is intended to provide a steady stream of new successful products. During the past decade, there have been many changes, and new approaches to product development have been advanced by several of the more successful companies. A general framework for a product development process which is followed, with some modifications, by many companies in the U. S. today is described below:

1. Identify the customer needs.
2. Establish the product specification.
3. Define alternative concepts for a design that meets the specification.
4. Select the most suitable concept.
5. Design the subsystems and integrate them.
6. Build, and test a prototype, and then modify it as required.
7. Design and build the tooling for production.
8. Produce and distribute the product.

The listing is shown in sequence, but we emphasize that we must attempt to conduct many of these phases concurrently. The phrase "concurrent engineering" has been defined to indicate that the eight phases, in so far as possible, are conducted in parallel and not in series.

In this one semester course, we do not have sufficient time to simulate all eight phases of the product development process. We essentially give you the product specification, and you begin the product development process with Step 3 and continue through Step 6. The final two phases of the product development process have not been included, because we are not stressing mass production at this early stage of your studies. While the course does not cover the complete development process, it is a good start and a significant challenge early in your program. You will get a broader exposure in other courses scheduled later in the curriculum.

THE CUSTOMER

We develop a product so that it may be sold at a profit. If a product is to be sold, someone must like it well enough to spend his or her money to purchase it. How do we design a product that the customer will like? The answer is to talk to the customer. Yes, design engineers do go into the field and talk with the customers one on one. These talks (interviews) are conducted at the customer's location, and usually last 30 minutes to an hour. Another approach, for learning about the customer's needs, is to quietly watch the customer use the product over an extended period (several hours). By observation, we determine all of the ways the product is used and record the customer's reactions in each application. Individual interviews and observation periods with the customer represent very cost effective methods of establishing what improvements the customer prefers in a new product.

Focus groups similar to those used in the news programs on TV to determine political preferences are also effective in determining the customer's needs. The focus group (eight to ten customers) is assembled in an off-site location to discuss the product. A moderator with group dynamics skills leads the discussion. The focus session, which lasts for about two hours, is video taped. The development team reviews and analyzes the tape. They gain insight regarding the needs of the customer, and the value that customers place on features that may be included in the development of a new product. Focus groups have the advantage that they encourage cross communication among the customers which often reveals information that is not discovered in the individual interviews. However, the disadvantage of this approach is higher costs because of the need to hire a moderator, tape the session, and rent the meeting room in a hotel that is convenient for the members of the focus group.

Who is the customer? Clearly the person buying the product in the marketplace is a customer. But usually there are others involved with the product that we need to consider as customers. To illustrate this point, let's consider Black & Decker (B&D), a company that designs, manufactures, and sells moderately priced power tools. Clearly, Harry the happy home owner is a customer because he goes to the store and buys power tools. B&D will certainly interview Harry to get his input, but they will also interview the store managers at retail outlets like Home Depot, Walmart, Lowes, etc. These store managers are stakeholders and they share in the success of the product. They are concerned with attractive packaging, the size of the package, the lead time for the delivery, the market introduction date, price, etc. The retail store managers needs are at least as important as the needs of the individuals buying the product. They control the entry to the marketplace.

THE PRODUCT SPECIFICATION

Once the customers' and stakeholders' needs are established and prioritized, we must convert the vague statements gleaned from the customer interviews into meaningful engineering terms. If the customer states that the tool vibrates too much, we must conduct experiments and measure the vibration level. Only then can we set a realistic vibration level in the product specification that will satisfy the customer. We need metrics (measurable quantities) describing each product feature that can be employed as targets to include in the product specification. We establish these metrics by testing not only our own products, but those of the competitors. In this extensive testing program (called benchmarking in industry), we establish the market norms for the product. If our new product is to be successful, we must improve on these norms to produce a superior product.

The benchmark testing provides the basis for writing the product specification. In preparing the product specification, we attempt to provide metrics associated with each feature or subsystem in the product. We establish a target for a certain feature in the specification with a lower limit that is marginally acceptable and a higher limit that is essentially the best that we can achieve. Design is a compromise particularly when several metrics are to be satisfied simultaneously. We design to meet all of the subsystem targets, but, on some features, it may not be possible to achieve the highest level of performance while maintaining targets for cost and development time.

There is a systematic procedure for generating a product specification, where the vague statements made by the customer are converted to well defined metrics that are then included as design targets in the product specification. This procedure called the "house of quality" is outside the scope of this course, but the interested reader is referred to the excellent book by Dr. D. Clausing [1] for a complete description.

DEFINING ALTERNATIVE DESIGN CONCEPTS

Let's suppose that we begin an assignment on a development team the day after the product specification has just been completed. Our next task is to develop design concepts. A design concept (an idea) determines the approach taken to develop a specific feature or subsystem in the product. For example, in the development of a power tool for sanding flat wooden surfaces, we must attach the sandpaper to a sanding pad. Do we attach the sandpaper by clamping, gluing, vacuum, snaps, zipper or Velcro? We have presented six different design approaches (concepts) for attaching the sand paper to the sanding pad. Some of these ideas may be well known, and others may be new but wild. Hopefully one of the design concepts will be new and feasible. In the initial stages of generating design alternatives, we consider all ideas acceptable (the good and the bad). We will evaluate all of them, and select the better ideas later in the concept development process.

Let's consider an example pertaining to the design of a weighing machine (scale). We clearly need to support the person on some sort of a platform which transmits the force (weight) to a mechanism used to measure the weight. There are many ways to design this platform. We can use a simple flat plate that bears on supporting bars like the spring type bathroom scale described in Chapter 3. Another idea is to use a plate that is hinged. What about a split plate with two sides, one for each foot? A sealed two sided rubber pad capable of supporting air pressure backed with a solid plate might be a viable option for a hydraulic weighing machine. In this phase of the product development process, it is important to generate as many design ideas as possible to consider in designing each product feature or subsystem. Do not worry about their feasibility in this first effort to generate ideas. We will sort the good from the crazy later in the development process.

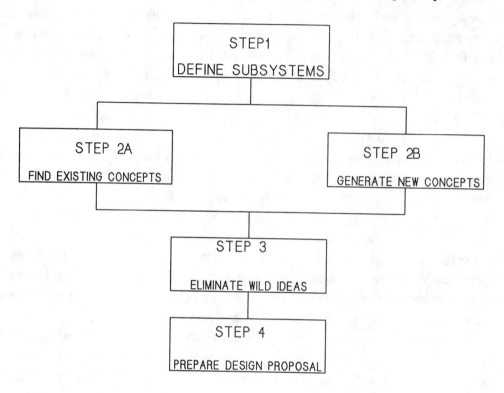

Fig. 1 A systematic approach to the concept generation process.

A systematic approach to the concept generation phase of the development process, is shown in Fig. 1. The first step in this approach is to define the subsystems involved in the operation of the product. In our case, the design of a weighing machine will depend on the overall approach. If we decide to develop a platform scale, we will have several subsystems including:

1. The base supporting the entire scale.
2. The cover plate supporting the item to be weighed.
3. The mechanism for transmitting force from cover plate to linkage.
4. The linkage connecting platform to the lever system.
5. The vertical member supporting the lever system.
6. The lever system.
7. The adjustable weight.
8. The tare (zeroing) system.

We have examined the platform scale, and have subdivided it into eight different subsystems. This is not a unique list. In fact, your team may decide to divide the many functions of the scale differently. The exact list is not important as long as it is complete. We divide the product into its major parts to simplify the design problem. It is easier to think about several smaller design tasks involving a more limited number of concepts than it is to handle a giant task involving a very large number of concepts.

The second step is to generate design concepts for each of the subsystems identified above. We approach the concept generation in two different ways. First, look to others for ideas. You are not the first design team in the world to design a weighing machine or any other product. Someone else was first. If you go to the library and search the literature, you will find scores of papers or books describing many different types of balances, platform scales, spring scales, electronic scales, etc. that have been built over the past century or even earlier in history. A careful review of this literature, and a search of the patents should permit you to develop a listing of design concepts that already exist for each of the eight subsystems listed above. Don't worry about using someone else's idea. We do it all the time. We call it by a special name --- "reverse engineering". We perform a detailed examination of the competitor's product, test it, and then adopt the good features and reengineer the features that we believe to be inferior. We also try to improve on the good features by making them better. As long as we do not violate a patent, there is no requirement that the design be original. However, the new product must be superior to the competing products, feature by feature, when we compare them.

OK, we have a complete listing of the existing design concepts used in previous designs of our weighing machine. Let's next generate some new ideas from within the design team. There are two approaches to follow. First, we suggest that each team member think about the subsystems; understand their purpose, function, and the way that they work. Then after developing a thorough understanding, the individual team members should list as many design ideas as possible for each subsystem. The independent thinking by team members is important early in our attempts to generate alternative design concepts. Write your ideas on index cards, with one idea on each card, and then give the cards to the team leader. Keep a copy of the card for yourself because you will use it in the next part of the new concept generation task.

The second approach in generating alternative design concepts involves a collective effort of the team. One of most widely used methods for generating ideas is brainstorming. The development team, five or six people, is a good number of participants in a brainstorming group. The team leader moderates the group, keeping it on task and avoiding any tendency to begin a round table discussion of any idea. The team leader must formulate the questions posed to the team that initiates its response. A typical question might be; present your ideas for the mechanism connecting the platform cover to the linkage. The questions should be focused on a subsystem, but they should not be so narrow as to limit the response from the team.

The team spends the next few minutes quietly thinking about the question. If ideas occur to you during this period for quiet thought, write them on an index card. You may also use the index cards that you prepared when you considered the design concepts independently. The team leader goes around the table in sequence, and each team member briefly presents one idea to the team. After presenting the idea, he or she hands the index card to the team leader. The brainstorming method is very effective when a few simple rules are followed:

1. No criticism is allowed during the session. The critique comes later.
2. The team responses should be short and to the point. Detail is deferred until later.
3. Wild ideas are OK because our goal is to generate as many ideas as possible.
4. Improvements and combinations of ideas are sought.

Brainstorming should be fun with everyone participating. Begin with prepared ideas from your index cards, and then add spontaneous ideas as they occur to you from the synergism of the group. You should, if possible, try to embellish the ideas of other team members with improvements. Humor helps during the brainstorming process because it breaks down barriers and improves the bonding of the team members. As the session moves from discussion based on the index cards to the generation of spontaneous ideas, the team needs a recorder to write the ideas on the blackboard or a flip chart (a very large pad of paper visible to the entire team). The visual display of the emerging ideas keeps the team on track and facilitates more thoughts on improving the ideas already identified.

A brainstorming session should run for 20 to 30 minutes, or it should be concluded when the team has exhausted all of their ideas. The team leader takes the index cards and recorders notes, and is responsible for sorting and classifying the teams contributions. These team generated ideas are added to the list of concepts that were determined from our literature and patent searches. Together they give us relatively complete lists of design ideas for each subsystem. At this point in the process, the lists, one for each function or subsystem, are composed of many ideas with very brief descriptions.

A follow-up meeting of the team is conducted to consolidate the lists. Wild ideas were encouraged in the brainstorming session to create a fun-like, open, and creative environment. The following session is a return to realism. Its purpose is to remove the crazy ideas from our list. The team reduces the list so that it includes only three or four of the most feasible ideas for each subsystem. Reducing the lists means that someone's idea must be eliminated. It is important that the team reach consensus in rejecting ideas. The procedure to cull ideas should not generate ill feelings. Good reasons for excluding an idea should be given in good humor. Considerable time is needed for thorough discussion. No one should feel belittled or the butt of a joke by the idea elimination process.

When the lists of the design concepts have been reduced so that they include only feasible ideas, we expand on the ideas. The ideas are represented by short written statements on index cards or notes without significant detail. Discussion of the concepts occurred in the meeting to reduce the number of alternatives, but we have not prepared detailed written descriptions of the concepts. To proceed with the evaluation of the remaining concepts, we need brief design proposals which include more detailed information, and drawings or sketches showing the key features of each idea. Individual team members should develop a design proposal for his or her concept. The proposal should describe the idea in a few paragraphs, present a sketch if needed, a list of advantages and disadvantages, and a design analysis if possible. The proposal should contain much of the information necessary for the team to select the superior idea from among the few remaining design concepts.

SELECTING THE SUPERIOR CONCEPT

We have reduced the number of design concepts originally generated by a preliminary meeting that weeded out the obviously weak and deficient design ideas. We have also expanded the ideas in the design proposals, providing the team with drawings, descriptions and perhaps even analyses. But, hopefully, we still have too many good ideas. We must now select the best from the good, and that is not always easy to do. Fortunately we have the technique developed by Stuart Pugh [2] to help us in the selection of the best concept. The Pugh selection technique is a team activity that involves the preparation of an evaluation matrix. Keep calm, it's not a math matrix and it is very easy to evaluate. Let's consider the nine steps in the Pugh selection technique that are:

1. Select the alternative design concepts to be evaluated.
2. Define a criteria upon which the evaluation is based.
3. Form the Pugh matrix.
4. Choose a datum or a reference for concept comparisons.
5. Score the concepts.
6. Evaluate the ratings.
7. Discuss the good and bad elements of each concept.
8. Select a new datum and repeat the Pugh matrix.
9. Revise select concepts as needed and evolve a superior concept.

Step 1 concerns the selection of the design concepts to be evaluated; however, we have already selected the most feasible three or four concepts based on previous team discussion. The essential element in successfully completing this step is that all of the team members understand each concept in considerable detail. If all of the team members are to participate effectively in the Pugh selection process, then every one must understand each concept in great detail. For example, consider that we have generated four concepts for the cover for the base of the platform scale, which include:

A. A solid rectangular plate constrained to small vertical motions.
B. A split rectangular plate with two hinges.
C. A solid rectangular plate with one hinge.
D. A split hexagon plate with a central support point.

Do you understand how each of these four concepts can be implemented in the design? The cover for the base is employed to support the weight of an individual standing on the platform scale while transmitting a force to a mechanism located below the cover. Can the concept be implemented in the limited time allotted for prototype evaluation in this course?

Step 2 defines the criteria that will be used in the selection process. This criteria may change in evaluating concepts for different features or subsystems, but it is fixed for any one feature or subsystem. All of the concepts being considered must be judged against the same scale. What are the criteria that should be used to evaluate the four concepts listed above for supporting the weight of the person standing on the scale? In the instructions provided to you for the design of the weighing machine, we specified the performance expected in terms of the accuracy of the weight measurement, we listed some constraints on the size of the scale, the time available for testing, the need to be able to read the output from a distance, and the cost. Clearly the criteria used to judge the concepts must take into account all of the requirements imposed by the course instructors as well as any additional criteria that your team imposes. Let's list some items that you may wish to include in the criteria:

1. Cost of materials
2. Availability of materials or components
3. Simplicity

4. Ease of manufacturing
5. Ease of assembly
6. Ease of disassembly
7. Impact on the environment
8. Weight
9. Power requirements
10. Friction effects on accuracy

Your team may want to add addition items to include in this list of criteria. For instance, do you believe that ease of maintenance should be included as a criterion?

Step 3 involves the development of the Pugh matrix as shown in Fig. 2. The matrix is a simple table with rows and cells. You could use a spreadsheet program to prepare the matrix or a sheet of engineering paper as illustrated, in Fig. 2. The concepts A, B, C are placed in the column headings, and the criteria 1, 2, 3,....10 are placed in the row headings.

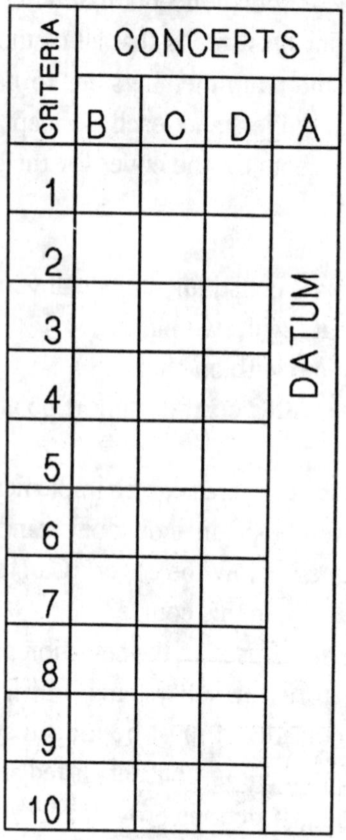

CRITERIA	CONCEPTS			
	B	C	D	A
1				
2				
3				DATUM
4				
5				
6				
7				
8				
9				
10				

Fig. 2 Format for the Pugh chart used for concept selection.

In step 4, you choose the datum or reference against which your comparisons will be made. The datum is one of the concepts that you are evaluating. We initially select what is believed to be one of the better concepts, and this design approach serves as our reference. Suppose that we consider Concept A (the solid rectangular plate with constrained vertical motion) to be the datum. We then compare the remaining three concepts B, C and D to the solid rectangular plate concept.

In Step 5 you score the concepts, and enter the results in the Pugh matrix. The scoring is simple. We consider each criterion separately and compare a selected concept (say B) against the datum concept (say A). The evaluation is a team activity. If the team reaches a consensus and decides that concept B is superior to concept A for criterion 1, then we score the intersecting cell in the Pugh matrix with a plus sign (+). On the other hand the team may decide that concept A is superior to B, and mark the cell with a minus sign (-). If the team decides that there is not a significant advantage of one concept over the other, the cell is marked with the letter S, indicating that the concepts are the same. It is important to recognize that the team discussion leading to the scoring is of great benefit. The team members interchange information in reaching a consensus. This heightens everyone's awareness and understanding of the design issues involved in implementing the concept. The scoring continues with each concept compared to the datum for every criterion. An example of a completed Pugh matrix is shown in Fig. 3.

Let's use the matrix presented in Fig. 3 as an example to illustrate the scoring technique. Note that this scoring was performed by the author without the very significant benefit of a team discussion. In your team's evaluation you may score the items differently, because you are considering different information or you may modify the concept slightly during your discussion. The matrix is not unique, since it is totally dependent on the information available to the team at the time of the evaluation.

CRITERIA	CONCEPTS			
	B	C	D	A
1	S	S	S	
2	S	S	S	
3	-	-	-	DATUM
4	-	-	-	
5	-	-	S	
6	-	-	S	
7	S	S	S	
8	-	-	S	
9	S	S	S	
10	+	+	+	

Fig. 3 Example of a completed Pugh chart where four concepts have been compared.

Let's consider the author's scoring for the 1st criteria in Fig. 3. I was comparing concepts B, C and D against concept A on the basis of cost. First, I estimated that the cost of the datum (the solid rectangular plate). This plate must be about 8 by 12 inch in size to accommodate the 13E shoe size that is required in the specification. This fact lead me to consider a small sheet of ½ inch thick plywood for the manufacture of the cover. The plywood is available at the local lumber supply store for about $2. All of the other concepts B, C, and D are about the same size, and they also could be constructed from the same sheet of plywood. Therefore, there is no difference between the four concepts base on criterion 1 (cost). Accordingly, we place the letter S in cells B1, C1 and D1. DifferenceS do exist between the four design concepts, but criteria 1 is not helpful in drawing a distinction between the different ideas.

We repeat the scoring for each criterion, and mark each cell in the matrix with +, -, or S. Each score represents a consensus, and the discussion leading to that consensus gives the team an opportunity to share information in depth about every concept as it is judged against every criterion. The matrix should not be considered fixed. If a criterion is not effective in the selection process, it can be eliminated from then matrix. Concepts can be modified during the rating process, or elements from one concept can be incorporated in another. The Pugh matrix is not a strict scoring device. Instead, it is a technique that leads to the interchange of information between team members which gives additional opportunities to modify and improve the design concepts.

Step 6 evaluates the ratings, by examining the scores for concepts B, C and D when compared to concept A. In Fig. 3 for concept B, we have one + sign, five negative signs and four Ss. Do not cancel the positive with one of the negatives, because the negatives will be examined later in the Pugh process. For concept C, we have the same score as concept B, and for concept D, we have one positive, two negatives and seven Ss. Concepts B and C seem to be inferior to concept A, but concept D is very nearly the same as concept A. We have evaluated the ratings and summarized the pluses and minuses, but we defer the selection.

Step 7 is inserted in the Pugh selection process to formalize the discussion of the advantages and disadvantages of each concept. Let's examine concept B and focus on the minus sign in cell B3. That minus score indicates that the hinges used to support the edges of the split plate increased the complexity of the design. Can we retain the same concept but reduce the complexity? Think about replacing the hinges with a simple support. The edge of the plate on the bottom surface could be grooved to capture support ridges incorporated in the base. This change in the design would eliminate the two hinges used originally in concept B, and reduce the complexity. The minus sign in the Pugh matrix permits the team to focus its attention on a weakness in a concept, and to improve the design by eliminating the weakness.

Sometimes minus signs lead us to abandon a concept. Let's consider cells B10, C10 and D10, all of which contain a plus sign indicating that we anticipate friction problems if we implement concept A. Friction effects will detract from the accuracy of the scale, and could prevent us from achieving the specified accuracy requirements. Can we do anything to change the concept and eliminate friction effects while retaining the necessary constraints on the motion of the platform? If we cannot determine a technique for eliminating friction effects concept A will not be feasible and will have to be abandoned.

Suppose Shirley Sharp suggests the use of flex plates to constrain the motion of the cover. These flex plates eliminate the friction effects, but increase the complexity of the design and change the scoring on the Pugh chart for some of the other criteria. If we make modifications to improve a concept relative to one criterion, it probably will be necessary to repeat the scoring of the Pugh chart.

Step 8 selects a new datum and repeats the Pugh procedure. We recommend this additional step if there are many concepts that remain after the first evaluation which represent uniquely different approaches. For the example given here, we would not gain much from a reevaluation unless the concepts are modified. Concepts B and C with hinges are very similar and both score poorly. Concepts A and D are nearly equal in scoring. The idea of using flex plates with concept A can be extended to the other concepts replacing either simple supports or hinges. If flex plates are introduced with one or more concepts, it would be advisable to completely repeat the Pugh evaluation process.

Step 9 revises the selected concept to evolve to a superior concept. We have under Step 8 already improved the highest scoring concept (A) by introducing flex plates to constrain the motion of the cover without introducing frictional forces. However, we can consider improvement of the other concepts by working to eliminate the reasons for the negative scores. Keep asking questions about the factors detracting from the merits of an idea. New approaches emerge, negative scores can change to positive scores. Answers to your questions often lead to design modifications that eventually provide the superior concept. When we finally have superior concepts for every feature or subsystem, we can proceed with the detailed design.

DESIGN FOR ????

No, we have not made a typo in the section heading. When we design, we must consider so many different aspects, that we replaced the listing of them with ???? in the section title. Let's begin our discussion of design with the four Fs, namely form, fit, function and finish. The four Fs refer to detail design of piece parts involved a some system. This is the easiest part of the design which is the reason for considering it first. Form refers to shape. Is the part shaped correctly to perform its function? If it is too thin, it may fail by breaking or by deforming excessively. If it is too large or too thick, it will weigh too much and cost too much. If it has a sharp external corner, it represents a safety hazard because sharp corners are cutting edges. If it has a sharp internal corner, high stresses, which may lead to premature failure, will develop.

Fit is involved when we assemble two or more pieces to form a subsystem. Will the pieces fit together without gaps or discontinuities. If holes are drilled in a given component for bolts, are they in alignment so that bolts fit without interference? Fit becomes critical when we deal with rotating shafts that require bearings. Tolerances required for several of the diameters in a housing containing the bearing are very tight and require precise machining of the component parts, if the bearing is to fit properly. If the fit is too tight, the bearing will overheat and seize. If the fit is too loose, vibration will occur which detracts from performance, and shortens the life of the bearing and other components.

Function refers to the ability of the system to perform satisfactorily with its components. If a part fails in service, then the system cannot function. Suppose you design a crankshaft for an auto that resonates at an engine speed of 3800 RPM. Your crankshaft is great. It has the right form, it fits properly and it has the correct surface finishes on the bearing journals. But the crankshaft vibrates at or near highway cruising speed. The design fails because the performance of the auto has been severely compromised by the vibrations induced in the engine by the crankshaft. The crankshaft has not functioned correctly.

Finish refers to the surface finish of the part. Is the surface rough, smooth or polished, or does it matter? Some surfaces are important because of appearance. The sheet metal on our auto is very smooth so that the paint that is applied will produce a high gloss finish. Other surfaces are not important. The aluminum block on an auto engine is die cast, and its outside surface reflects the finish of the die. We rarely look at the engine until it dies, and we accept a relative rough surface finish. Some surfaces are polished or ground to enhance their performance in bearing applications. When we design a part, we must know how it will be used in service and specify the appropriate surface finishes on the detail drawing for that part.

DESIGN FOR MANUFACTURING

In designing a piece part or a subsystem, it is essential that we consider how the parts are to be fabricated. Manufacturing methods are probably as important as function in the design of parts that are to be produced in large quantities. For a component that is mass produced, we must carefully consider manufacturing methods such as die casting, injection molding, forging, casting, drawing, machining, stamping etc. in the detail design of the parts. We have good and bad design rules to follow depending on the production methods to be employed in manufacturing [3].

In the development of the prototype, manufacturing methods remain important, but they are less critical. For the design of the prototype weighing machine that we are developing in this course, you probably will be much more limited in your manufacturing capabilities. Hopefully you will have a model shop with power saws and safety qualified operators to cut wood and plastic according to your drawing. You may also have hand tools such as hammers, saws, chisels, wrenches, etc. available for your use. Small power tools such as the drill, bayonet saw and belt sander, illustrated in Fig. 4, may also be available in the student workshop. Any component that you design must be fabricated with the tools available to you unless the components are procured from a retail outlet in finished form.

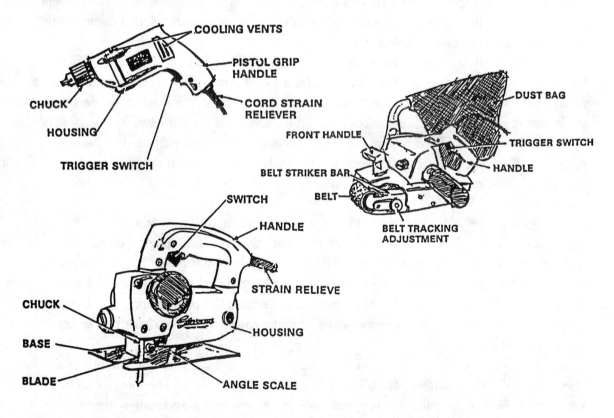

Fig. 4 Power hand tools available in the student workshop.

As an example, let's consider the base of the weighing machine that serves to support the cover. We will assume that your cover is a flat, rectangular plate, 8 by 12 inches in size fabricated from a piece of ½ inch thick plywood. The cover is fabricated by cutting it from a larger sheet of plywood with either a power or a hand saw. The base resembles a shallow box that fits below the cover, as shown in Fig. 5. Each member of the weighing machine is limited to a length of 17 inches, so that they can be packed for storage in the copy paper box. (Remember the constraint that was placed on the design in Chapter 2). We can make the pieces required for the base from strips of lumber (3/4 by 2 inch) cut to the appropriate lengths with a saw. Alternatively, we can make the members from aluminum sheet or angle stock which can also be cut with a hack-saw. However, fastening the aluminum pieces together to form the base will pose many more problems that working with wood which can be nailed, screwed or glued together to form strong joints. It appears that we will not encounter any problems manufacturing the pieces needed for the base. Will we encounter any problems when we try to assemble all of these pieces in short period of time?

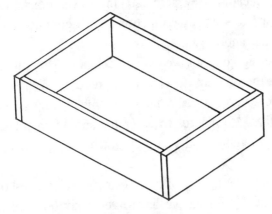

Fig. 5 Illustration of the base used to support the cover of a scale.

DESIGN FOR ASSEMBLY

Products are systems containing several subsystems, and each subsystem contains several components. In the design or selection of each component, we must consider the methods and procedures used to assemble them in the construction of the subsystems. We also must be concerned with the assembly of the various subsystems to form the complete system (product). There are several levels of consideration in design for assembly. The first and most important is whether or not the components can be assembled. Don't laugh. The author has seen designs which were impossible to assemble and had to be scrapped. Second, the assembly operation should be easy and not require special operator skills. Finally, the time required for the assembly should be minimized.

Assembly time is very critical to the success of very complex products that are mass produced. For example, Ford Motor Company leads the big three auto producers in the U. S. in assembly. They can assemble an auto with 24.72 people-hours compared to 26.32 and 27.76 people-hours for Chrysler and General Motors respectively. Nissan required only 17.84 people-hours at its plant in Smyrna, Tennessee. It is estimated that General Motors could save about $3.5 billion each year if it could reduce the time required for assembly to that currently achieved by Nissan. Clearly, savings in billions of dollars can be made by employing design for assembly in the development of mass produced products.

One approach to design for assembly is to minimize the number of parts used in the product. As designers, we tend to use many more parts than are needed based on assembly theory. Boothroyd et al [3] have developed three criteria that are used to justify the use of a part in the design of a product.

1. The part must move relative to other parts in the operation of the product.
2. The part is fabricated from a different material or must for some other reason be separated from adjacent parts.
3. The part must be separate from adjacent parts to permit either assembly or disassembly of the product.

Using these three criteria, we can develop a theoretical part count for any given product. With very rare exceptions, we greatly exceed the theoretical minimum part count when we design. Nevertheless, these rules are valuable, because they stimulate the development team to consider combining parts and to design unique parts to serve dual functions. Part counts are often reduced by 50% or more by applying the design for assembly criteria.

How does this affect you in the design of your weighing machine? After all you are designing a one of a kind prototype to learn about the product development process. Well, you are stuck with an assembly problem, because you have a very limited time to assemble, test and disassemble your weighing machine. If you need more than 45 minutes to assemble and 15 minutes to disassemble, then you will have to use test time for assembly, and you may not have enough time to accurately calibrate and adjust your scale.

There is a lot of work involved in the assembly of the prototype. Can we shorten the time by conducting several tasks at the same time. After all we have five or six team members, and they should all be involved in the assembly operation. The team should divide the assembly operation into several assembly tasks that can be conducted in parallel. For example, the cover and base can be assembled from their individual parts while the lever system (assuming that you are developing a platform scale) is being assembled. The other subsystems can be assembled simultaneously, and the overall time required to complete the assembly can be shortened by working tasks in parallel rather than series. You are not expected to work like a pit crew at the Indianapolis raceway, but every one should stay productively busy.

There are many aspects to design for assembly, and this treatment is only a brief introduction to a very important topic. For a much more complete coverage see the excellent book identified in reference [3].

DESIGN FOR MAINTENANCE

Have you ever changed the oil filter on your auto? On some cars you have to raise the body, crawl under the car, and reach up to unscrew a horizontally oriented filter. When the filter is loosened, it leaks, oil runs down the side of the engine, and onto you. The oil spill is difficult if not impossible to contain. Clearly, the design of this engine, and the placement and orientation of the oil filter did not accommodate the need for periodic and scheduled maintenance. Other cars have the oil filter on the other (front) side of the engine. The filter is replaced from the front without going under the car. The designers have also oriented the filter nearly vertically so that the oil does not spill when the filter is loosened and the seal is broken. This second filter is an example of design for maintenance. The designer recognized the need for periodic maintenance and developed the product to allow easy access, and rapid, inexpensive replacement of the required parts.

In the design of the weighing machine, you should not have to perform any maintenance. However, you should recognize that many commercial products do require some periodic maintenance, and that maintenance is a very real design criterion for many product developments.

PROTOTYPE BUILD AND EVALUATION

There are several important reasons for building a prototype. The first is to determine if all of the component parts fit together. Frequently we find errors, and some of the parts do not fit properly when we attempt to assemble the subsystems. It is much easier to change the drawings and to fix the fit problems when we build the first prototype. Imagine trying to assemble 10,000 units of product using parts that are in error. The prototype, often called the first article build, is vital to insuring parts that fit perfectly. The assembly of the first article also reveals difficulties that may arise at a later date on the production line when the product is mass produced. The idea is to find the problems and fix them immediately while the development is in the prototype stage. Finding problems later is much more expensive. Also problems discovered after the product has been introduced to the market requires a recall and harms the reputation of the company for producing a high quality product.

The prototype also enables engineers to test the product. These tests permit us to measure the performance parameters, and to verify the analysis used in the design phase predicting performance. If any deficiencies in performance are revealed in the test program, design modifications are made to the prototype, and the tests are repeated. The prototype is a vehicle for design change. It is instrumental in enhancing the performance of the product. In some instances, prototype development is nearly a continuous process with one prototype following another. At select times in this process the company will freeze a design, and introduce a new product based on a these prototype experiences.

You are building a prototype of a weighing machine. As you build the first article, you will probably find some mistakes. Do not be depressed. Drawing errors and other analytical mistakes are common. Find the errors and fix them immediately. Make the prototype work as early as possible in the semester. When you fix a problem, find the mistake that was made and learn from it. You should make certain that you revise the drawing to reflect the revisions made in correcting errors. An early test of your prototype, prior to the last week of class is to be encouraged. The workshops and testing facilities should be ready for your early evaluations of the prototype.

TOOLING AND PRODUCTION

In many instances we design and develop products for a very large market with sales of 100,000 units or more each year. Also, the life cycle of the product may range from five to ten years, so that the total production over the product life is huge. The production of large quantities requires extra care in the design phase. It is widely accepted that 70 % of the final product costs are determined by the design [4]. Decisions made in the design affect the cost of producing the product over its entire life. Selecting a fastener that has a premium in cost of only 2 cents adds $20,000.00 to the cost of producing a product if the lifetime volume is one million units. Designing a component that requires an extra machining operation could easily add more than several hundred thousand dollars to the life cycle costs for high volume products. Clearly, when designing components intended for use in high volume products, we need an excellent understanding of costs, manufacturing process design rules, and assembly theory.

We trust that the procurement of the materials and the components for your weighing machine will provide a small lesson in cost control. For lessons in manufacturing processes and assembly theory, we defer to later courses in the curriculum or independent study on your part. The references at the end of the chapter will get you started.

MANAGING THE PROJECT

When a development team works on a project, they are charged with the responsibility for completing the project according to a schedule and a budget. The team leader should manage time and costs. Development costs can be extremely large. Chrysler spent about one billion dollars to develop the Concorde auto. The project took 3.5 years and involved a total of about 2250 people [5]. As you move up in a corporation, your position will likely involve some degree of management. You will need to learn about scheduling and project costs relatively early in your career. Let's start now with scheduling.

You know something about scheduling already, because the University imposes a schedule on you from day one. Classes start on September xx or January xx and end on December yy or May yy. The final exam schedule is fixed before the semester starts. You have a schedule of classes --- M, W and F at 9, etc. The instructor has imposed some deadlines for design reviews that you must meet in this course. You will need to incorporate these deadline dates into the Gantt schedule, that we will develop later in this section.

Scheduling starts with listing the tasks that you need to formulate in order to develop the prototype weighing machine. The list of tasks should be in reasonable detail to enable you to realistically assess the time that you can allocate to each task. Let's begin by setting up this list, presented in Table 1, that shows all of the tasks that must be completed during the development of an electronic scale. Note that the listing is in sequence according to time except for the design reviews.

Your team should review the list and the time estimates, and make the modifications necessary to conform with your design. The time estimates are particularly important. We have scheduled 4 weeks for the team to generate, evaluate and select design concepts. That is a lot of time (five or six people-weeks). Do you need that much time, is it the correct amount of effort, or will it prove to be insufficient? Estimate the time carefully, and then check to see if your estimates are correct. Estimating the time required for team members to complete a task is very difficult. Errors in underestimating time lead to delays in design releases and product introductions. These delays are costly to the corporation, and they are serious detriments to career advancement. The author after many years of experience still underestimates the time needed to properly complete a task. Take this opportunity to start developing your time estimating skills.

TABLE 1
Task List for Developing an Electronic Scale

TASKS	TIME (weeks)
CONCEPT DEVELOPMENT	
Receive and understand specification	1
Concept generation	2
Concept evaluation and selection	2
DETAIL DESIGN	
Spring element	1/3
Strain gage --- bridge circuit	1/3
Cover	1/3
Base	1/3
Amplifier and digital display	1/3
Drawing reviews	1/3
DESIGN REVIEWS	
Preliminary review	1
Final review and design release	1
MANUFACTURING	
Procure materials and components	1
Manufacture piece parts	1
Inspect piece parts and rework as required	½
Prepare assembly kit	½
ASSEMBLE AND TEST	
Pre-assembly	1
Final assembly and test	1
TOTAL	14

After completing the listing of tasks, we construct a Gantt chart as shown in Fig. 6. The Gantt chart shows a listing of the tasks in the rows down the left side. The time line, in this case expressed in terms of calendar months during the fall semester, is displayed along the X axis. We draw bars to represent the time required to complete a specific task. The left edge of the bar corresponds to the time when the task is scheduled to begin. The right edge of the bar indicates the completion date. Of course, the length of the bar shows the number of weeks scheduled to complete the task. When the tasks are completed the bars are darkened. In the example presented in Fig. 6, the bars for the concept development tasks are darkened indicating that these tasks are complete. A vertical line is shown in Fig. 6 to indicate a current date, that is, in this example, sometime in mid October. The vertical line intersects the horizontal darkened bar for concept evaluation and selection. This configuration indicates that the team is ahead of schedule because the bar is darkened for a week ahead of the current date.

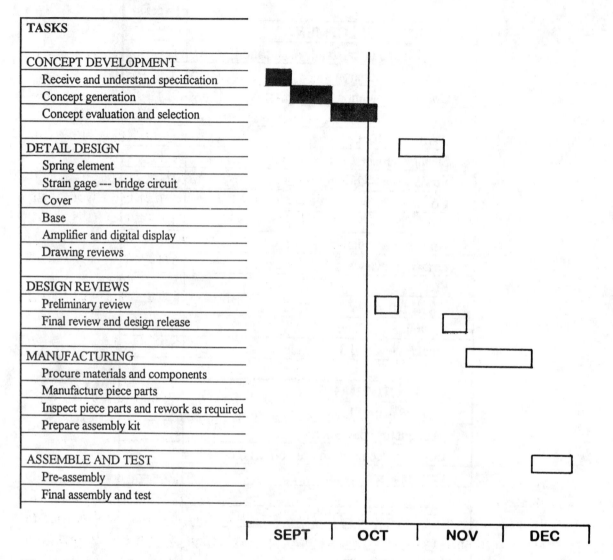

Fig. 6 Example of a Gantt chart for the development of an electronic scale during the fall semester.

The placement of the bars in Fig. 6 is typical of what is called series scheduling. We schedule one task after the completion of the previous task. However, it is possible to overlap the bars along the time scale. Overlapping the bars (parallel scheduling) implies that we will start a new task before the current task is completed. In many instances, we can effectively pursue two tasks simultaneously, and important time savings can be achieved whenever parallel scheduling is possible. In procuring materials, we often encounter delays of weeks or even months in delivery of components necessary to build the prototype. In procuring these long lead time components, parallel scheduling is mandatory because it is necessary to place the purchase order long before the design is complete.

In Fig. 6, we have provided your team with an example Gantt chart for developing an electronic scale over a 14 week semester. You should consider your team's schedule, and change the beginning and completion dates for each task to correspond with your team's plan. Prepare the Gantt chart early in the semester, and revise it periodically to reflect changes to the plan as the design of the scale evolves. Be sure to add the deadline dates imposed by your instructor as milestones on the Gantt chart.

SUMMARY

A complete product development process has been outlined in this chapter. We very briefly described the beginning of the process, where the customers are identified, and interviewed to determine their needs and the amount that they would be willing to pay for new product features. The vague customers' comments are converted into meaningful engineering targets. The targets are incorporated into an engineering specification that defines the goals and objectives which the designers are expected to meet in the development of the product.

Methods for generating alternative design concepts were covered in detail. The most essential element of this process is the synergism of the team members in generating new ideas for design. The approach is to divide the product into subsystems, and then to generate as many design ideas for each subsystem as possible. From these many concepts, we select the best one and then improve it by introducing modifications that eliminate its disadvantages. A selection technique based on the Pugh chart has been described in considerable detail with an appropriate example.

The four Fs of design were introduced to show the importance of form, fit, function and finish. We then briefly discuss several important aspects of design for manufacturing, assembly and maintenance. These are all extremely important subjects and the treatment here is only an introduction. Several references are cited for independent study.

Reasons are given for the importance of building the prototype. The prototype is critical to insuring the success of a development. It verifies that all of the parts fit, that they can be assembled with ease, and that the system performs in accordance with the product specification. The prototype is a vehicle for design change, and it affords a critical opportunity to correct the design errors prior to the introduction of the product to the market.

Finally, we introduce a method for managing the project. The method involves a listing of all of the tasks that must be accomplished to complete the prototype development, and an intelligent estimate of the time necessary to complete each task.. The data from the listing is then used to construct a Gantt chart that enables the team to track the completion of each task relative to a predefine time schedule.

REFERENCES

1. Clausing, D. Total Quality Development: A Step by Step Guide to World Class Concurrent Engineering, ASME Press, New York, NY 1994.
2. Pugh, S., "Concept Selection---A Method that Works," Proceedings of the International Conference on Engineering Design (ICED), Rome, March 1981, pp. 479-506.
3. Boothroyd, G, Dewhurst, P. and W. Knight, Product Design for Manufacture and Assembly, Marcel Dekker, New York, NY 1994.
4. Anon, Company Literature, Munro and Associates, Inc., 911 West Big Beaver Road, Troy, MI 48984.

5. Ulrich, K. T. and S. D. Eppinger, <u>Product Design and Development</u>, McGraw-Hill, New York, NY, 1995.

6. Cross, N., <u>Engineering Design Methods: Strategies for Product Design</u>, 2nd ed., Wiley, New York, NY, 1994.

7. Carter, D. E. and B. S. Baker, <u>Concurrent Engineering: The Product Development Environment for the 1990s</u>, Addison-Wesley, Reading MA, 1992.

EXERCISES

1. Consider a household refrigerator as a product. List the customers that you should consider in any redesign. Also give reasons for including each person or organization as a customer.

2. Interview a family member, Mom, Dad, Sister or Brother about some product that they own and use frequently. Try to list all of the favorable comments as well as their complaints. Are they willing to pay for product improvements? If so, how much?

3. Working together with your fellow team members divide the weighing machine that you are designing into sub systems. Briefly describe each sub system and define all of the interfaces that exist between sub systems.

4. With the team conduct a brainstorming session to generate design concepts for each of the sub systems defined in exercise 3. List all of the concepts generated, the good and the bad, for each sub system.

5. Define the criteria that you plan to use to evaluate one of your subsystems in the Pugh selection process.

6. Construct a Pugh matrix and evaluation several design concepts using the criteria defined in exercise 5.

7. When the Pugh analysis of exercise 6 is complete, discuss the implications of the positive and negative scores that appear in the Pugh matrix.

8. Write an engineering brief describing design for manufacturing.

9. Write an engineering brief describing design for assembly.

10. Write an engineering brief describing design for maintenance.

11. Why do we invest funds and time in building prototypes before we release a product to the market?

12. What was your reaction when your auto or a friend's auto was recalled to replace a design defect?

13. Discuss the implications of mass production on the development process.

14. Prepare a task list for your development of the prototype of a weighing machine.

15. Prepare a Gantt chart for your weighing machine development.

NOTES

PART V

COMMUNICATIONS

CHAPTER 13

TECHNICAL REPORTS

INTRODUCTION

We hope that you like to write, because engineers usually have to prepare several hundred pages of reports, theoretical analyses, memos, technical briefs, and letters during a typical year on the job. We trust that you will learn how to communicate in every way, ---by writing, speaking, listening, and by employing superb graphics. Communication, particularly good writing is extremely important. Advancement in your career will depend on your ability to write well. We recognize that you will be taking several courses in the English Department and other departments in Social Sciences, Arts and the Humanities. These courses will require many writing assignments, and they should help you immensely with the structure of your composition. However, most of the assignments will be essays, term papers, or studies of different works of literature. There are several differences between writing for an engineering company, and writing to satisfy the requirements of courses like English 101 or History 102.

In school you write for a single reader, namely the instructor, and he or she must read the paper to grade it, and to determine how well you are doing in class. In industry your report may be read by many people, within and outside the company, and with different backgrounds and experience. In class the teacher is the expert, and in industry, the writer of the report ---YOU--- are suppose to be the one with the knowledge. Many of these folks in industry don't want to read your report. They are busy, the phone is ringing continuously, meetings are scheduled back to back, and they have lots of other things to accomplish. They read the report only because they must be aware of the information that it contains. They want to know the key issues, why these issues are important, and who is going to take the action necessary to resolve them. In writing reports in industry, we cut to the chase. There is no sense in writing a 200 word essay, when the facts can be given in a 40 to 50 word paragraph that is brief and cogent.

In the following sections, we will cover some of the key elements of technical writing including, an overall approach, report organization, audience awareness, and objective writing techniques. Then we will describe a process for technical writing that includes four phases including, composing, revising, editing and proof reading.

APPROACH AND ORGANIZATION

The first step in writing a technical report is to be humble. Realize that only a very few of the many folks, that may receive a copy of your report, will read it in its entirety. Busy managers, and even your peers read selectively. To adapt to this attitude, organize your report into short, stand-alone sections that attract the selective reader. There are three very important sections, namely the title, summary, and introduction. You locate them at the front of the report, so that they are easy to find. Up front they attract attention, and are more likely to be read. In college you call the page summarizing your essay an abstract, but in industry it is called an executive summary. After all if it's prepared for an executive, perhaps a manager will consider it sufficiently important to take time to read it. Follow the executive summary with the introduction, and then the body of the report. A common outline to follow in organizing your report is:

- Title page
- Executive summary
- Table of contents, list of figures and tables
- Nomenclature only if necessary
- Introduction
- Technical issue sections
- Conclusions
- Appendices

The title page, the executive summary and the introduction are the most widely read parts of your report. Spend a lot of time polishing these three parts, as they provide you with the best opportunity to convey the important results from your investigation..

The title page should give a concise title of the report, usually in less than ten words. Keep the title short, as you have an opportunity to give detail in the body of the report. The authors give their names, affiliations, addresses, and often numbers for the phone, fax and e-mail. The reader, who may be anywhere in the world, should be able to contact the authors to ask questions. For large firms or government agencies, a report number is formally assigned, and it is listed with the date of release on the title page. An example of a title page of a report prepared for a government agency is shown in Fig. 1.

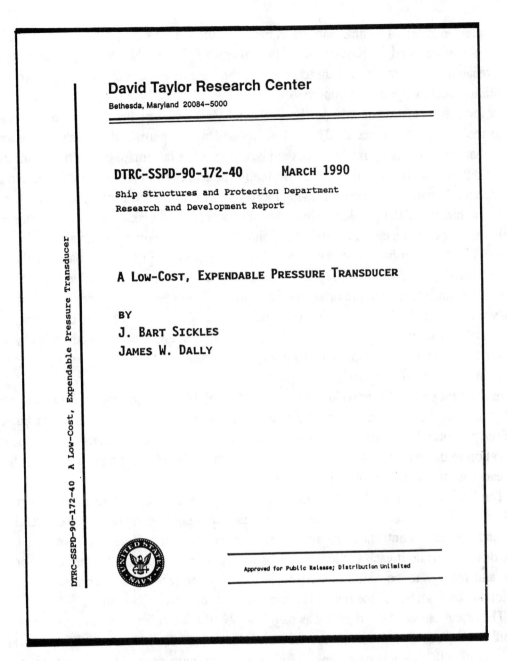

Fig. 1 Illustration of the title page of a professional report.

The executive summary should, with rare exceptions, be less than a page in length (about 200 words). As the name implies, the executive summary gives in a concise and cogent style, all the information that the busy executive needs to know about the content of the report. What does the big boss want to know? Three paragraphs usually satisfy the chief. First briefly describe the objective of the study, and the problems or issues that it addresses. Don't include the history leading to these issues. There is always a history, but the introduction is a much more suitable section to develop history and other background facts related to the issues or problems. In the second paragraph, describe your solution to the problem, and/or the resolution of the issues. Give a very brief statement of the approach that you employed, but reserve the details of the approach for the body of the report. If actions are to be taken by others to resolve the issues, list the actions, those responsible, and the dates of implementation. The final paragraph indicates the importance of the study to the business. Cost savings can be cited, improvement in the quality of the product, gain in the market share, enhanced reliability, and etc. You want to convince the executive that your work was worthwhile to the corporation, and that you did a superb job.

In the introduction you restate the problem. But you say that the problem was already defined in the executive summary. That's OK, redundancy is permitted in technical reports. We recognize that we have many readers, so we place important information like the definition of the problem in different sections of the report. In the introduction, the problem statement can be much more complete. Expand the statement giving more information related to the problem. A paragraph or two on the history of the problem is in order. Folks reading the introduction are more interested than those reading only the executive summary, and they want to know more detail.

Following the problem statement, establish the importance of the problem to the company and to the industry. Again this is a repeat of what was included in the executive summary, but you expand your arguments for the importance of the study. You can cite statistics, briefly describe the cost analyzes that lead to the cost savings claimed, give a sketch of customer comments indicating improved quality, etc. In the executive summary, you simply stated why the study was important. In the introduction, you develop arguments to convince the reader that the investigation was important and worth the time, effort and cost.

In the next part of the introduction, you briefly describe the approach followed in addressing the problem. You began with a literature search, followed by interviews with select customers, a study of the products that failed in service, interviews with manufacturing engineers, and the design of a modification to decrease the failure rate and improve the reliability of the product. In other words, tell the reader what you did to solve the problem.

The final paragraph in the introduction outlines the remaining sections in the report. Yes, we know --- that information is in the table of contents, but again we repeat for the selective reader. Also placing the report contents in the introduction gives us the opportunity to provide a one or two sentence description about each section. Perhaps we can attract some reader interest in one or more sections, and manage to convince a few people to study the report more completely. The added description for each section in the report is not possible in the table of contents.

The organization of the report was suggested by the bullet list shown previously. However, you should recognize that there is an organization for every section, and every paragraph in the report. In writing the opening paragraph of the section, you must convey the reason for including the section in the report, and the importance of the content covered in the section. You certainly have a reason for including the section in the report, or you would not have wasted your time in writing it. Make sure you share this reason with the reader, and convince the reader that the section is important before you go into the detail of the content.

A paragraph is written to convey an idea, thought or a concept. The first sentence in the paragraph must convey that idea, clearly and concisely. The second sentence should describe the significance or importance of the idea. The following sentences in the paragraph support the idea, expand its scope, and give cogent arguments for its importance. Give the idea in the first sentence, and then add more substance in the remaining sentences. A speed reader should be able to read only the first two sentences of every paragraph, and glean 90% of the important information conveyed in the report, if it written properly.

As you add detail to a section or a paragraph develop a pattern of presentation that the reader will find easy to read and understand. First, tell the reader what you will tell him or her, then tell the story, then follow with a summary indicating the action to be taken to implement the solution. Do not try this approach in the theme that you have to write for English class, because the instructor will clean your clock for being redundant. But technical writing is not theme writing. In technical writing, we pound on the reader to get the point across, and to take corrective actions for the benefit of the company and the industry.

We often include sections in engineering reports on analysis. In these sections, we state the problem, and then give the solution. After the problem statement and the solution, we introduce the details. The reader is better prepared to follow the details (the tough part of the analysis) after they know the solution to the problem.

KNOW YOUR READERS AND YOUR OBJECTIVE

Another important aspect to technical writing is to know something about the readers of your report. The language that you use in writing the report depends on the knowledge of the audience. We are writing this book primarily for 1st year engineering students. We know your math and verbal skills from the SAT scores required for admission. We know that you have relatively good language skills, and very good math skills. You are bright and articulate. Most of the chapters in this book are in line with your current abilities. However, there are one or two chapters in the book that may stretch your technical background. These have been added as a challenge for some of the brighter or more interested readers, and as theoretical background for all of the readers.

As you write a report for this class, you know that the instructor is the only reader with which you need to communicate. You also know that he or she is knowledgeable. In that sense the report writing assignments in this class are not realistic unless they are modified. Suppose we asked you to write an assembly routing (a step by step set of instructions for the assembly of a product) for the production of your weighing machine in a typical factory in the U. S. What language would you use as you write this routing. Remember that a significant fraction of the factory workers in the U. S. are functional illiterates, and many other detest reading. You would use few words, and many cartoons or pictures.

Know the audience before you begin to write, and adapt your language to the audience. Consider four different categories of readers:

1. Specialists with language skills comparable to yours.
2. Technical readers with mixed disciplines.
3. Skilled readers, but not technically oriented.
4. Poorly prepared readers, who read with difficulty if at all.

Category 1 is the audience that is the easiest to address in your writings, and the audience in the last category is the most difficult. Indeed with poorly prepared readers, it is probably better to address them with visual presentations conveyed with TV monitors and video tapes.

A final topic, to be considered in planning, is the classification of the technical document that you intend to write. What is the purpose of the task? Several classifications of technical writing are listed below:

- Reports
    - Trip, progress, design, research, status, etc.
- Instructions
    - Assembly manuals, training manuals, safety procedures.
- Proposals
    - New equipment, research funding, money for construction, development funding.
- Documents
    - Engineering specifications, test procedures, laboratory data.

Each classification of writing has a different objective, and requires a different writing style. If you are writing a proposal for development funding, you need persuasive arguments to justify the costs of your proposed program. On the other hand, if you are writing an instruction manual to assemble a product, arguments and reasons for funding, are not an issue. Instead, you would be trying to prepare complete and simple descriptions, that precisely explain how to accomplish a sequence of tasks involved in the assembly.

THE TECHNICAL WRITING PROCESS

Whether you are writing a report as part of your engineering responsibilities in industry, or as a student in ENES 100, you will face a dead line. In college, the deadlines are imposed well in advance and you have a reasonable time to prepare you report. In industry you will have less time, and you will have to get approvals from management before you can release the report. In both situations you have some limited period of time to write the report. The idea is to get started as soon as possible. Waiting to the day before the deadline is a recipe for disaster.

Many professionals do not like to write, and they suffer from writer's block. They sit at the keyboard hoping for some ideas to occur. Clearly, you do not want to join this group, and there is no need for you to do so. There is a technique for you to follow that helps to avoid writer's block.

Start your report by initially following the procedure described in the previous sections. Understand the task at hand, and classify the type of technical writing that you have to produce. Define your audience, so that you establish before hand the sophistication of the detail, and the language that you may employ. Prepare a skeleton outline of the report. Start with the outline presented previously, and add the section headings for the body of the report. This outline is very brief but it provides the organization of the report. That organization is important because it has divided the big task (writing the entire report) into several smaller tasks (writing the report a section at a time).

It is impossible to write a report without information. You must generate the information that goes into the report. You can use a variety of sources, interviews with peers, instructors, or other knowledgeable folks that are willing to help. Go to the library and execute a complete literature search, and then read the most suitable references. Take notes as you read, gleaning the information that you can use in your report. Be careful not to plagiarize. You can use material from published works, but you cannot copy the exact wording. The statements must be in your own words. If the report has an analysis, get to work and prepare a statement of the problem, execute its solution, and make notes about the interpretation of the solution.

When you have collected most of the information that is to be covered in your report, organize your notes into different topics that correspond to the section headings, and incorporate them in your initial outline. This initial outline of the report organization will grow from a fraction of a page to several pages as you incorporate your notes in an organized format.

You are now ready to sit down and write. Writing is a tough task that requires a great deal of discipline and concentration. We suggest that you schedule several blocks of time, and reserve these exclusively for writing. The number of hours that you should schedule will depend on the rate at which you compose, revise, edit and proof read. The author can compose at about a page an hour, but most students need more time.

In scheduling a block of time for composing, you should know your productive interval. Most writers take about a half an hour to get into the swing, and then they compose well for an hour or two, before their attention and/or concentration begins to deteriorate. The quality of the composition begins to suffer at this time, and it is advisable to quit. This fact alone should convince you to start your writing assignments well before the deadline date.

While you are writing avoid distractions. Writing requires deep concentration. You must remember your message, the supporting arguments, the paragraph and sentence structure, grammar, vocabulary, and even spelling. Find a quiet, comfortable place, out of the mainstream and focus your entire concentration on the message, and the manner in which you will present it in your report.

Some sections of a report are easier to write than others. The easiest are the appendices, because they carry factual details that are nearly effortless to report. The interior sections of the body of the report that carry the technical details that are also easy to prepare. The writing describing the detail is less concise and less cogent. Don't get careless on these sections, because they are important, but each sentence does not have carry a knock-out punch.

The most difficult sections are the introduction and the executive summary. It is better to write these two sections, after the remainder of the first draft of the report is completed. We suggest that the introduction be written before the executive summary. The introduction contains much of the same information as the executive summary, although it is more expansive. The introduction can be used later as a guide in preparing the executive summary.

REVISING, EDITING AND PROOFREADING

Writing is a difficult assignment. Don't expect to be perfect in the beginning. Practice will help, but for most of us it takes a very long time to improve, because it is such a complex task. Expect to prepare several drafts of a paper or report, before it is ready to be released. In industry several drafts are essential, since you often will seek peer reviews, and manager reviews are mandatory. In college you have fewer formal requirements for multiple drafts, but they are a good idea if you want to improve the document and your grade.

The first draft is focused on composition, and the second draft is devoted to revising the initial composition. We try to get our ideas down in reasonable form, and in the correct sections of the report. In the second draft, we seek to revise the composition, and do not concern ourselves with editing. Concentrate on the ideas, and their organization. Make sure the message is in the report, and that it is clear to all of the readers. We will polish the message later.

Several hours should elapse between the first and second drafts of a given section of the report. If we read a section over and over again, we soon lose our ability to judge its quality. We need a fresh, rested brain, for a critical review. In preparing this textbook, the author composed on one day, and revised on the next day. Revising is always scheduled for the early morning block of time, when the brain is rested and concentration is at its highest level.

Let's make a clear distinction between composition, revision, editing and proofreading. Composition is the writing the first draft, where we get our ideas down, and we organize the report into sections, subsections and paragraphs. Unless you are a super person, your first draft is far from perfect. The second draft is for revisions, where you focus on improving the composition. The third draft is for editing where we correct errors in grammar, spelling style and usage. The fourth draft is for proofreading where we polish the manuscript.

As you revise your initial composition, you are concerned with the ideas and the organization of the report. Is the report organized so that the reader will quickly ascertain the principal conclusions? Are the section headings descriptive? Sections and their headings are helpful to the reader and the writer. They help the reader organize his or her thoughts. They aid the writer in subdividing the writing task, and keeping the subject of the section in focus. Are the sections the correct length? Sections that are too long tend to be ignored.

Question the premise of every paragraph. Are the key ideas presented together with their importance, before the details are included. Is enough detail given, or have we included too much trivia. Does the paragraph contain a single idea or have we tried to pass two or even three ideas in the same paragraph? It is better to use a paragraph for every idea, even if the paragraphs get short.

Have you added transition sentences or transitional phrases? The transition sentences, usually placed at the end of a paragraph, are designed to lead the reader from one idea to the next. Transitional phrases, embedded in the paragraph, are to help the reader place the supporting facts in proper perspective. You contrast one fact with another, using words like however and although. You indicate addition facts with words such as also and moreover.

When the ideas flow smoothly, and you are convinced that the reader will follow your ideas, and agree with your arguments, you can begin to edit. Run the spell checker, and eliminate the typos and most of the misspelled words. Try to find the remaining misspelled words, because the spell checker does not detect the difference between certain words like grate or great, or say like and lime. If you wanted to say that you like something, and typed the word lime instead, the spell checker doesn't help find your typo.

Look for excessively long sentences. When sentences get to be 30 to 40 words in length they are going to begin taxing the reader. It is better to use shorter sentences, where the subject and the verb are close together. Make sure that the sentences are sentences. Remember, that a sentence needs a subject and a verb, and they should have the same tense. Have you used any comma splices, which is attaching two sentences together with a comma? Examine each sentence, and eliminate unnecessary words or phrases. Find the subject and the verb, and attempt to strengthen them. Look for redundant words in a sentence, and substitute different words with similar a meaning to eliminate redundancy.

The final step is proofreading the paper to eliminate all of the errors. Start by running the spell checker for the final time. Then print out a clean, hard copy to use for proof reading. Check all the numbers and equations in the text, tables and figures for correctness. Then read the text for correctness. Most of us have real trouble reading for correctness, because we read for content. We have been trained since 1st grade to read for content, but we rarely read for correctness.

To proof read, you read each word separately. You are not trying to glean the idea from the sentence, so stop reading the sentence as a whole. Read the words as individual entities. If it is possible, arrange for some help from a friend. One person reads aloud to the other. The listener concentrates on the appearance of each word, and checks the text against the spoken word to verify it correctness.

WORD PROCESSING

The word processor is a great tool to use in preparing you technical documents. The very significant advantage of the word processor is the ability to revise, edit and make the necessary changes without excessive retyping. You can mark text, delete, copy or move it. You can insert words, phrases, sentences and/or paragraphs.

While working with a word processor, it is very difficult to revise, edit, and proof read on the screen. Much better results are obtained if you have hard copy, and can view the entire page. With the entire page, you can see several paragraphs and check that they are in the correct sequence. Use double spacing when printing the first few drafts to give space between the lines for the modifications that you make to the hard copy.

After you have completed the revisions on the hard copy, make the required modifications to the text on the word processor, and save the results on your own floppy disk. We recommend that you keep only the most recent version (draft) of the document. If you save several versions of the document, it is necessary to keep a log book to record the changes to each version. If the writing takes place over a few week period, it is very easy to lose track of what changes you made, and which version of the hard copy goes with which electronic copy. We find it easier to keep one electronic file (on your floppy disk) of the most recent draft with a hard (paper) copy to serve as a back-up.

The word processor has several features that are helpful in editing. The spell check program finds most of our typos and provides suggestions to correct many of the words that we misspell. The search command permits you to systematically examine the entire document, so that you can replace a specific word with a better substitute. The thesaurus is available on command to help you with word selection, but you must be careful when using it. Make sure you understand the meaning of the word that you select, and don't try to impress the reader by using long or unusual words. Short words that are easily understood by the reader are always to be preferred in technical writing.

Formatting the report is another significant advantage of word processing. You can easily produce a document with a very professional appearance. The formatting bar permits you to select the type of font, the point size (the height of the characters), the pitch (the number of characters per inch), and emphasis such as bold, underline or italic.

You can also format the page with four different types of line justification commonly available. We are using the word processor in preparing this textbook, to justify alignment for both the right and left margins. The word processor automatically adds the spaces in a line and aligns both margins. In designing your page use generous margins. One inch margins all around are typical. If the document is to be bound, the left margin is usually increased from 1.00 to 1.25 inch.

Tables and graphs can be inserted in the text. Take advantage of the ability to introduce clip art into the text, or to transfer spread sheets and drawings produced in other software programs into the word processing program. Position the tables, and figures as close as possible to the location in

SUMMARY

Writing is a difficult skill to master, and most engineers experience a number of different problems early in their careers. Unfortunately, the writing experiences in college do not correspond well with the writing requirements in industry. In school we write for a knowledgeable instructor. In industry we write for a wide range of people, with different reading abilities, that vary from assignment to assignment. In both college and industry, we write to meet deadlines imposed by others. While writing is not much fun, at least early in your career, there are many techniques that you can employ to make writing much easier and more enjoyable.

The first technique is to organize the report, and we have suggested an outline for a typical report. Gather information for the report from a wide variety of sources, and generate notes that you can use to refresh your memory as you write. Sort the notes and transpose the information on them to expand the outline for your report.

As you organize the outline, before you begin to write, determine as much as possible about the folks who will be reading your document. They can be as technically knowledgeable as you are, but they can also be functionally illiterate. The language that you use in the report will depend on the readers ability to understand. Also be clear about the objective of your document. Is it a report, an extended memo, a proposal or an instruction manual. Styles differ depending on the objective, and you must be prepared to change accordingly.

There is a process to facilitate the preparation of any document. It begins with starting early, and working systematically to produce a very professional document. Divide the report into sections and write the easy sections (appendices, and the technical detail portions) first. Defer the more difficult sections, such as the executive summary and the introduction, until the other sections have been completed.

The actual writing is divided into four different tasks, namely composing, revising, editing and proofreading. Keep these task separate, and compose before revising, revise before editing, and edit before proofreading. Multiple drafts are necessary with this approach, but the results are worth the effort.

Finally, word processors do not substitute for a clear brain, but they are extremely helpful in preparing professional documents. They save enormous amounts of time in a systematic editing process. They enable a mix of art, graphics, and text neatly integrated into a single document. Word processors have a thesaurus and word search features that are helpful in editing. They essentially turn a computer into a print shop, so that you have a wide latitude in the style and appearance of your professional documents.

REFERENCES

1. Eisenberg, A., Effective Technical Communication, 2nd ed., McGraw Hill, New York, NY, 1992.
2. Goldberg, D. E., Life Skills and Leadership for Engineers, McGraw Hill, New York, NY, 1995.
3. Elbow, P. Writing with Power, Oxford University Press, New York, NY, 1981.

EXERCISES

1. Prepare a brief outline of the organization of the final report for the scale development.
2. Prepare an extended outline of the final report for the scale development.
3. Write a section describing one of the subsystems in the weighing machine.
4. Write a section covering the calibration of the scale during the prototype test period.
5. Write the Introduction for the final report on the scale development.
6. Write the Executive summary for the final report on the scale development.
7. Revise the Introduction that another team member wrote.
8. Edit the Introduction after it has been composed and revised by others.
9. Proof read the Introduction after it has been composed, revised, and edited, by others.

NOTES

CHAPTER 14

DESIGN BRIEFINGS

INTRODUCTION

We communicate by writing, speaking and graphics. All three modes of communication have their place as we try to convey our ideas and concerns to others. All three are important. Let's focus our attention on speaking in this chapter. We use speech almost continuously in our daily life, so why do we need to study about design briefings? There are several reasons. We usually speak with our friends and family in an informal style. We know them and feel comfortable with them. They know us. They are concerned with our well being, genuinely like us, and are usually interested in what we have to say. A professional presentation is different. It is formal event that is usually scheduled well in advance. The audience may include a few friends, but mostly they will be strangers. The time we have to convey our thoughts is very limited. The audience may not be very interested in our message. A person or two in the audience may be managers who control our advancement in the company. There are lots of reasons for tension headaches as one prepares for the professional presentation.

The design briefing is extremely important to both the product development and to your career. Information must be effectively transmitted to your peers, the management, and to any external parties involved in the project. Clear messages, that accurately define the problems that the development team can effectively address, are imperative. On the other hand, ambiguous messages are often misunderstood, and they lead to delays in defining problems and implementing timely solutions. The design briefing is an opportunity for a healthy review. It permits your peers to share your ideas, and it affords management an open forum for assessing your work and progress. Since the professional presentation is critically important, let's learn some great techniques for accurately conveying our messages to a mixed group of strangers and acquaintances.

SPEECHES, PRESENTATIONS AND DISCUSSIONS

To begin, let's distinguish between three types of formal methods of oral communication, namely the speech, the presentation, and group discussion. The speech is the most formal of the three. Last year, 1996, was a good year to refer to in describing speeches, because we were besieged with political addresses that clearly illustrate their key features. Speeches are given to large audiences in huge rooms, stadiums or coliseums. The setting is usually not appropriate for visual aids to be used. The audience rarely has very much in common. They are of widely different ages, with different interests and persuasions. The speech is carefully scripted, and the speaker usually reads or very closely follows the script. Ad-libbing is avoided. Communication is one way --- from the speaker to the audience. Questions are usually not appropriate. Time is strictly controlled. Fortunately engineers are rarely called upon to make speeches, so we will not dwell on this topic.

Professional presentations differ from speeches. Presentations are made to smaller audiences, in smaller rooms with electric power and light controls. We depended on visual aids, demonstrations, simulations and other props to help convey our message. The audience is knowledgeable (about our topic), and usually has many common characteristics. The presentation is carefully prepared. It has order and structure, but it is not scripted. The flow of information is largely from the speaker to the audience, but questions are permitted. The speaker is considered the expert, but discussion of questions gives the audience a chance to share their knowledge of the topic. The time is carefully controlled, and it is usually insufficient from the speakers viewpoint. Professional presentation is an important mode of oral communication that you must master.

Group discussions are also very important to the design engineer. The audience is smaller with much more in common. They may be all be members of a development team. The topic being discussed will be narrowly focused. The speaker serves as a moderator and is an expert on the topic being considered. However, members of the discussion group (audience) may be as knowledgeable as the speaker (moderator). The moderator works in two modes. He or she may act as a presenter, giving brief background information to frame an issue that leads to discussion from any member of the discussion group. The moderator may also direct questions to a member of the group know to be the expert on the issue being addressed. The speaker (moderator) controls the flow of the information, but the flow is clearly two ways --- from speaker to the group and from one group member to another. In group discussions, time is difficult to control and the content and the range of coverage is strongly dependent on the skill of the moderator. Group discussion is very important in industry because the leader of a development team will often use this method of communication to identify problems, and to initiate an effort directed toward their solution. We will not address discussion group methods in this course; however, we recommend that you watch Washington Week in Review on PBS to gain some insights. While the participants are journalists, we can adopt many of their clever techniques in engineering.

A design briefing is a type of professional presentation. It is the method that you use in speaking and the use of visual aids which we will emphasis in this course. Indeed, you probably will be required to make presentations describing your development of the weighing machine on at least two occasions. The first likely will be the preliminary design review, and the second may be the final design review.

PREPARING FOR THE BRIEFING

A design briefing is far too important to take casually. You should prepare very carefully to insure that you will convey the information necessary to accurately report the status of the development, and to identify problems or unresolved issues facing the team. There are three aspects that you should consider in your preparations:

1. Identify the audience.
2. Planning the organization of the presentation and your coverage.
3. Prepare the visual aids.

In this class, it is too easy to identify the audience. You have your classmates, perhaps an undergraduate fellow, and the instructor. In industry the situation will be a lot different as the audience will be much more diverse. The size and diversity of the audience depends on the magnitude of the development project. A small briefing will have 10 to 20 in attendance with most participants from within the company. A large briefing may include 50 to 100 representatives that are both internal and external to the company. The characteristics of the audience is important because you must adapt the content of the presentation to the audience. Classify your audience with regard to their status, interest, and knowledge before you begin to plan the style and content for your presentation.

The status of the audience refers to their position in the various organizations they represent. Are they peers, managers, or executives? If they are a mix, which is likely, who is the most important? If you are preparing a design briefing for high level management, it must be concise, cogent, and void of minor detail. Executives are busy, stressed, and always short on time. High level managers are impatient, and rarely interested in the technical details that engineers love to discuss in their presentations. Recognize these characteristics, and adjust the content and the timing of your presentation accordingly. The executives are interested in costs, schedule, market factors, performance, and any critical issue that will delay the development or increase projected costs of either the development or the product. Usually you will have from 5 to 10 minutes to deliver convey this information at an executive briefing.

First and even second level managers are more human, more interested in you, and in your designs. If you prepare the presentation for this lower level of management, you can plan for more time (10 to 20 minutes). These managers are likely to be engineers, and they will share your knowledge of the subject. In fact, they probably will be more expert or more knowledgeable than you. They will want to know about the schedule and costs, because they share responsibilities with the higher level management for meeting these goals. However, they are also interested in the important details, and you can get into discussion of subsystem performance with them. The first and second level managers usually control the resources for the development team. If you need help, make sure that they get the message in enough detail to provide the assistance required. Do not hide the problems that your team is encountering. Managers do not like to be surprised. If you have a problem, make sure they understand it and are able to participate in its solution. If you hide a problem from management and it causes a delay or an escalation in costs, the manager will take the heat and you will be in one very unpleasant dog house.

Design briefings for peers are less tension producing, and you will have more time (20 to 30 minutes). You will address schedule and cost issues because everyone needs to know if you are meeting the milestones on the Gantt charts. However, the main thrust of your presentation will be on the details. If you are working on one subsystem that interfaces with other subsystems, you will cover your subsystem in sufficient detail for all to understand its geometry, interfaces, performance, etc. It is particularly important that the interface issue be fully addressed in the presentation. If four subsystems are to be integrated in a given product, a lot of details about all four subsystems need to be addressed to ensure that the integration goes smoothly. Suppose for example, that we are developing a power tool with a motor that draws 20 amps, but we employ a 10 amp switch to turn the power on and off. The switch will function satisfactorily in the short term tests with the prototype, but it will malfunction in the field with extended usage sometime after the product has been released to the market. Clearly, the two folks responsible for the interface between the motor and the switch did not communicate. The integration of the two subsystems failed.

The most important purpose of design briefings with peers is to make certain that all of the details have been addressed, and that the subsystems will be integrated without problems occurring when the prototype is assembled and tested.

Peer reviews in the absence of management are very beneficial in the development process. The knowledge of all of the members is about the same, although some team members are much more experienced than others. The topic is usually a very detailed review of one subsystem or another. The audience is fresh, and they can provide a critical assessment of the technology that you employed. They can check the accuracy of you analysis, comment on the choice of materials, and relate their experiences with similar designs on previous products. Peer reviews afford the opportunity for the synergism that make fine products even better. They also provide a forum for passing on important lessons from the more experienced designers to the new designers in a friendly, stress-free environment.

The subject of the presentation is self evident in a design review. We are developing a product, and the content will deal one issue or another regarding the development. There is a choice of specific topics and considerable latitude regarding the content. We have given advice in previous paragraphs about matching the content with the interests of the audience. Executive reviews serve a different purpose than peer reviews, and the content and the time allotted for the presentation is adjusted accordingly.

In all types of reviews there are two absolute rules that you must follow. First, you must know the material cold. It takes an audience about two seconds to understand that you are faking it. When they realize that you are not the expert, your presentation is a failure, and you have lost the opportunity to communicate. Second, be enthusiastic. It is OK to be calm and cool, but don't be dead on arrival. You must command the attention of your audience or they will turn you off. You control the attention of the audience only if you are enthusiastic and knowledgeable in presenting your material (story).

PRESENTATION STRUCTURE

There is a well accepted structure for professional presentations. We discussed this in some detail in a previous chapter dealing with PowerPoint, a graphics presentation program marketed by Microsoft. We show a recommended structure or outline for the presentation in Table 1.

TABLE 1

Recommended Structure for a Professional Presentation

- Title
- Overview
- Status
- Introduction
- Technical Topics
  - First Topic
    - Background
    - Status
  - Second Topic
    - Background
    - Status
  - Final Topic
    - Background
    - Status
- Summary
- Action Items

Let's examine each of these topics individually, and discuss the content that you will include on the visual aids that accompany your presentation. Recognize that the visual aids (slides or overhead transparencies), control the flow and the content of information that you will present. For this reason, we will address the topics listed above in the context of the information to be placed on the slides.

The title slide obviously carries the title of the presentation. But it also gives your affiliation, and the names of the team members that contributed to the work. Remember, you may be reporting on the work of the entire team; it is necessary for you to acknowledge their contributions on the title slide, and in your opening remarks. The title should be brief. Ten words or less is a good rule to follow in drafting the title for your presentation. It is also a good idea to use descriptive words in the title. For example, you could use **Preliminary Design Briefing ----WEIGHING MACHINE**. Six words, and you have told us the topic (a design briefing or review), the type of briefing (preliminary), and the product being developed (the weighing machine). That is a lot of information to convey in six words.

Below the title, the team name is given, and the members involved in the development are listed. It is essential that you share the ownership of the material covered in your presentation. Often the work of several people is covered by a single member of the team. It is not ethical to make the presentation without acknowledging the contributions of others. As students, we typically use first names in identifying team members. However in industry the complete names of the team members are used.

It is also a good idea to date the title slide, and you may wish to cite the occasion for the presentation. This information is useful if you make many presentations, want to refer to your files, and to reuse some of the slides after editing. An example of a title slide prepared with PowerPoint is shown in Fig. 1.

Fig. 1 Illustration of an informative and descriptive title slide.

The second slide is an overview. The overview for a presentation is like a table of contents for a book. You list the main topics that you intend to cover in the presentation. In other words you tell the audience what you plan to tell them in the next 10 to 20 minutes. The number of topics should be limited. In a focused presentation, you can cover about a half of dozen topics. If you press and try to over a lot of topics (10 or more), you will have difficulty retaining the attention of the audience throughout the presentation. Constrain the tendency to tell all and instead focus on the important issues. In a design review, the topics usually are organized to correspond with the subsystems involved in the product under development. We show such an example for a preliminary design review of a weighing machine in Fig. 2.

The third slide is used to present the status of the development. You should report on the progress made by the team to date. It is a good idea to incorporate a Gantt chart on the status slide to show the schedule, and the progress of the team on each task defined on the chart. If there are problems or uncertainties, this is the time to air the difficulties that the team is encountering. An example slide, shown in Fig. 3, illustrates uncertainties that the team has with regard to balancing (zeroing) the output voltage from a Wheatstone bridge circuit.

The fourth slide covers the introductory material such as background, history, or previous issues. It is important that the audience understand the product, the main objectives of the development, the role the team has in that development, and the most pressing of the current issues. This introductory slide permits you to set the stage for the body of the presentation that follows. The audience responds better to your presentation when they know in advance the background and the topics that you will cover. They mentally prepare to receive your message.

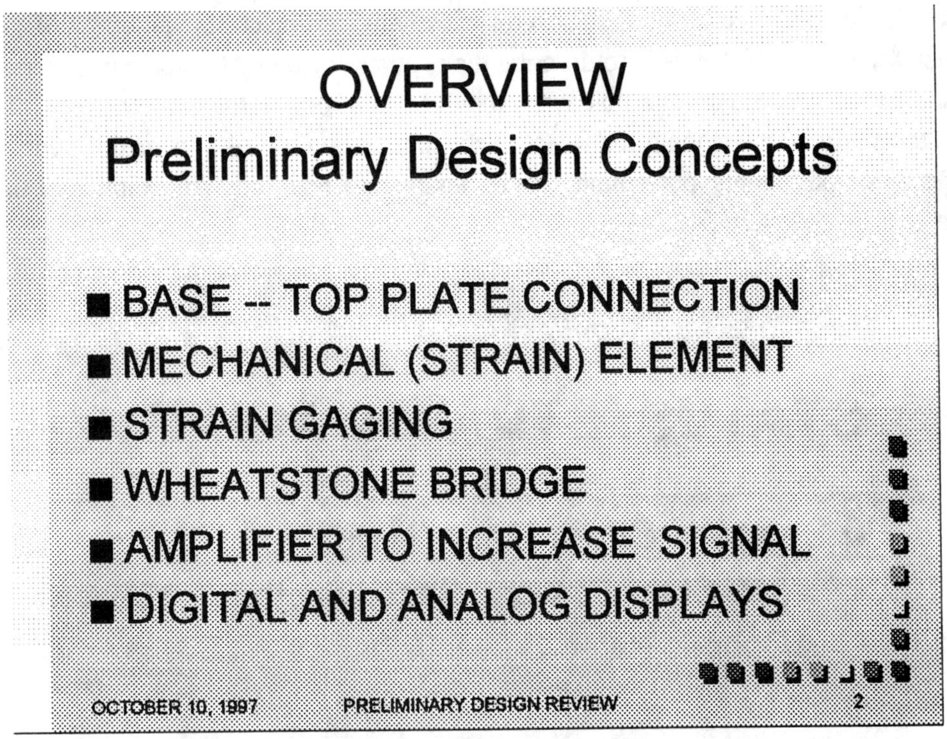

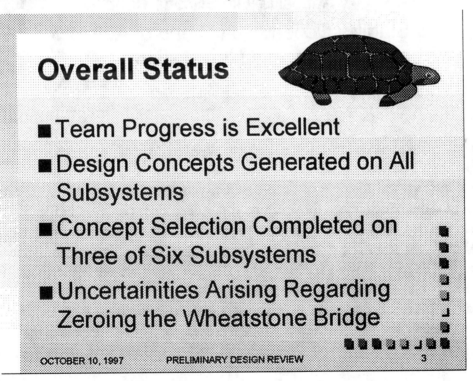

Fig. 2 The overview slide indicates the content in your presentation.

Fig. 3 Status slide indicates overall progress, some accomplishments and signals potential problems.

The body of the presentation usually deals with five or six technical topics. In most instances, the topics selected correspond to several of the more critical subsystems involved in the product. Avoid trying to cover a very large number of topics. It is better to report on a limited number of topics thoroughly than to rush through a dozen topics with incomplete coverage. We suggest two slides for each topic. The first slide is to describe the progress made by the team in generating design concepts, and to indicate the criteria that was employed in selecting the best concept. The second slide covers the status of developing the selected concept. The type of information reported on this slide includes accomplishments, outstanding issues, lessons learned, etc. An example of the status of a subsystem is presented in Fig. 4.

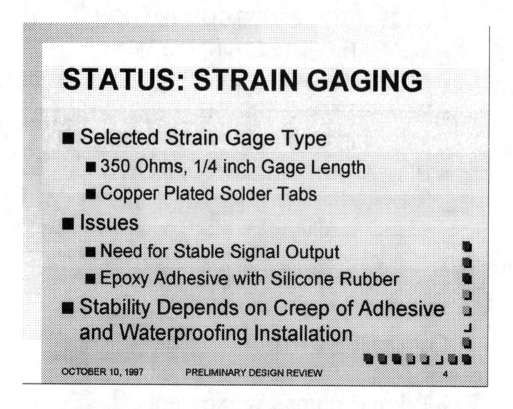

Fig. 4 Example of information presented in describing the development of a subsystem.

After the technical topics have been covered, you must conclude the presentation with a decisive pair of slides. Suggestions for your two concluding slides are given in Fig. 5. The next to the last slide is titled "Key Issues". This is your opportunity to identify very important issues (questions), that your team has discovered. Don't hesitate, as this is the time to introduce the uncertainties, and to come forward and seek help. Design reviews are not competitive. We seek to help, regardless of our status or our role as a member of the audience. Managers will arrange support for your team if it is required. Peers will make suggestions and introduce fresh approaches that the team may find useful. The design review is a formal process for all to participate in the development. Take advantage of the review to reveal uncertainties, and to seek help whenever your team needs assistance.

The final slide is to indicate plans for the future. Management wants to know how you intend to solve some of the issues that you have revealed, and the time required for the solutions. Your peers will be interested in the technical approaches that you intend to follow. You should also define any requirement that you have of anyone in the audience. If you need help from your peers or more resources from the manager, make sure it is clearly indicated on this final slide. Be certain that the individuals with the resources agree to provide the support that is requested. If action items are listed, identify the responsible individual and estimate the time to completion. The planning incorporated in the final slide is very important. It is the prescription for your team's get-well program.

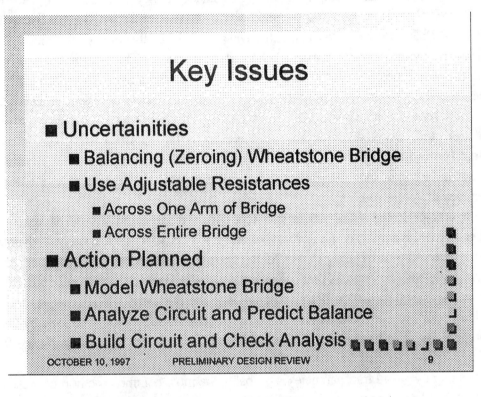

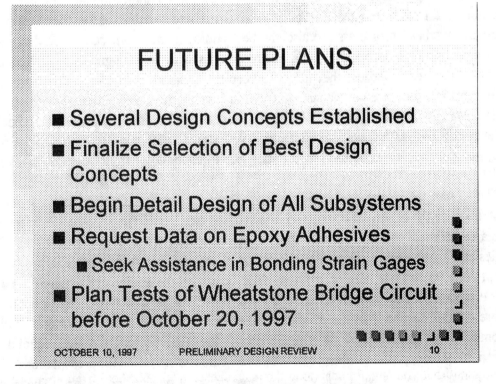

Fig. 5 The final two slides describe the key issues and the future plans.

TYPE OF VISUAL AIDS

Visual aids are essential for a technical presentation, because they control the content and the flow of the information that is conveyed to the audience. They provide visual information that reinforces the verbal information. You communicate using two of the individual's five senses. Take time to select carefully the type of visual aid that will most effectively convey your message.

In selecting visual aids, we are usually limited in two ways. The first limitation is in our ability to produce the visual aids, within the time available and for the amount we are willing to pay for them. The second limitation pertains to the equipment and the room in which the presentation is to be made. Is the room equipped with a 35 mm slide projector, an overhead projector, or a computer controlled projector? Another important consideration is the control of the light intensity in the room. Most rooms have light switch, and you can turn the overhead lights off. However, some rooms do not have blinds, and a bright sunshine will cause real problems in the visibility of 35 mm slides when they are projected.

Computer projected slides are an excellent choice if the room can be darkened, and if the projection equipment is available. (It is very expensive.) This selection has several advantages. The presentation is easy to prepare using PowerPoint. The slides are in color, and you can use both slide transition and slide building features that enhance the visual impact of the presentation. It is also easy for you to copy your presentation onto a floppy disk, use the computer attached to the projection system to load your program, and employ a mouse to click through your slides line by line. Finally, you can prepare low cost hand-outs, with six slides per page, which help the audience follow your presentation.

Projecting overhead transparencies is probably the most common method of presenting visual material. Overhead projectors are relatively inexpensive (a few hundred dollars compared to a few thousand dollars for the computer controlled projectors), and for this reason they are available in most rooms. Overhead projectors are also the best choice if the room cannot be darkened. The projectors are bright enough to give good quality images without extinguishing the lights. Bright rooms have an advantage because they help you keep the audience awake. Overhead transparencies are also easy to make on a copy machine. If you are willing to sacrifice the benefits of color, use inexpensive black and white transparencies. If you want to use color in the presentation because it is more effective, the costs increase significantly if you use a color laser printer or a color copier. However, ink jet printers can be used to produce good quality color transparencies at a reasonable cost.

The 35 mm slide is the third option. It is the preferred option if you have many photographs to project, because the quality of colored 35 mm slides is excellent. The slides are easy to prepare if you have a camera and the time to have the film processed. Their cost is reasonable, if you make many slides and use the entire roll of film (12, 24 or 36 exposures). It is possible to prepare colored slides from your PowerPoint files on a floppy disk; however, the equipment required is not always available. Projectors for 35 mm slides are common, but in an engineering college they usually are much more difficult to find than the overhead projector. The room must be darkened to view the image from a 35 mm projector. Another disadvantage, the relatively long distance required from the projector to the screen, precludes the use of very small conference rooms for presentations.

Sometimes speakers will employ two or three different types of projectors. The switch from one to the other is disruptive. If possible, use only one type of projector. However, if you must use more than one type of projector to properly convey your message, try to minimize the number of times you switch from one to another.

There are several other visual aids that are sometimes used in professional presentations. Video tapes are common, but the size of the TV monitors often make viewing difficult for a large audience. However, if you want to show motion or group dynamics, video clips are clearly the best approach. Video cameras are readily available and, after a bit of practice, we all can become reasonably good video producers.

Motion picture films, particularly of older material with historical interest, is very effective. Finding an 8 mm or 16 mm motion picture projector may be difficult, so allow time in your schedule for making the necessary arrangements.

Hardware or materials to be used in the product is sometimes passed around the audience during a presentation. The hands on opportunity for the audience is a nice touch, but you pay for it with the loss of the attention by some members of the audience during the inspection period. We recommend that you defer passing materials to the audience during the presentation. Instead, invite the audience to inspect your exhibits that are placed on a strategically located table sometime after the presentation. This approach maintains the attention of the audience throughout the presentation, while giving those interested in the hardware time for a much more thorough examination later after the conclusion of your discussion.

DELIVERY OF THE PRESENTATION

Excellent presentations require very well prepared slides, and a very smooth, well paced delivery. There are many aspects to the delivery part of the presentation process including, dress, body language, eye contact, voice control and timing. Let's start with dress. Broadly speaking there are four levels of dress. The highest level, black tie for men and formal gowns for women, fortunately is a rare event and never appropriate for design briefings. The next level, business attire with a suit for men, and a conservative dress for women, is sometimes appropriate for design briefings. A conservative Eastern company may have a dress code requiring business attire, whereas a less formal Western company would encourage more casual attire. Casual attire should not be confused with sloppy attire. Casual attire is neat and tasteful, but without suits, ties and white shirts for men, and suits for women. Sloppy attire like old jeans and a sweatshirt is strictly taboo. We suggest that you select either business or casual attire as your dress for the presentation. We recognize that most students these days prefer the sloppy dress style. Resist the impulse to dress that way. If you look like a bum or a homeless street lady, who will believe your message?

Posture is another important element in the presentation. In the words of former President Reagan, stand tall. Your body language signals your attitude to the audience. You are the presenter, and you are in control. Make sure you are calm, cool and collected. Nervous gestures with your hands, rocking on the balls of your feet, scratching your head, pulling on your ear, etc. should be avoided.

Before you begin your presentation, take control of the audience. One approach is to pause a moment before projecting the title slide, and immediately make eye contact with the entire audience. How do you look at everyone in the house? You scan the audience from left to right and then back, looking slightly over their heads. Occasionally drop your eyes and make eye contact for a second or two with one individual and then another. Pause long enough for the audience to become silent (10 to 20 seconds). If someone is rude and keeps on talking, walk toward them and politely ask for their attention. When you have everyone's attention, project the title slide and begin your delivery.

If you have rehearsed, you will not need notes. The slides carry enough information to trigger your memory. If they don't, you have not rehearsed long enough to remember the issues. Scripting the presentation is not recommended. People who script will eventually start to read their comments, and that is a deadly practice. Rehearse until you are confident. Make notes to help you rehearse, but throw them away before the presentation.

When you begin to speak make sure that everyone can hear you. If you are not sure of how far your voice carries, ask those in the back of the room if they can hear. If you have a microphone with an amplifier and speakers, be careful. The tendency is to speak too loudly. Try to control the loudness of your voice within the first minute of the presentation, as you introduce the topic with the title slide. The audience will read that slide with anticipation, and they will bear with you as you adjust the volume of your voice.

Don't mumble. Speak slowly, clearly and carefully enunciate each word. You don't need to speak as slowly as Vice President Gore, but it is better to go too slowly that too rapidly. If you can improve your enunciation, you will be able to increase the rate of your delivery without loosing the audience.

Don't run out of breath while speaking. Learn to complete a sentence, pause for breath (without a gasp), and then continue with the next sentence. There is nothing wrong about an occasional pause in the presentation provided the pause is short.

If you forget some detail, don't worry. Skip it, and move to the next point that you are trying to make. If the detail is critical, count on one of your team members to raise the issue in the question period. We all forget some of the material that we intended to present, and introduce some additional items on an ad hoc basis. Usually the audience will never recognize our omissions or additions.

Studies have clearly shown that people can listen faster that the presenter can speak. The trouble with following someone who speaks very rapidly is not the rate of speech, but the enunciation. Most folks who speak rapidly tend to slur their words, and we have trouble understanding poorly pronounced words. The trouble in listening to Al Gore is to stay with his message. While your waiting for his next word, your mind goes off on a mental tangent. You anticipate that your mind will return to the speaker in time for the next word, but unfortunately our minds are sometimes tardy and we miss key words or even sentences.

Listening is a skill, and many of us have not developed it properly. As speakers, we have the responsibility to keep the listener, even the poor ones, on the topic. We use several techniques to keep the attention of the audience. Employing the screen is probably the most effective tool for this purpose. As you project the slide, walk to the screen and point to the line on the slide that corresponds to the topic that you are addressing. If a few members in the audience are coming back from their mental tangents, you reset their attention on the current topic.

When you use the image that is projected on the screen, and good speakers use this technique, position yourself correctly. We show the proper position in Fig. 6. Stand to the left side of the screen, being careful not to block anyone's view. Face the audience and maintain eye contact. When you have to look at the slide, turn your head to the side and read over your shoulder. The 90° turn of your head and the glance at the slide should be quick, because you want to maintain eye contact. Under no circumstance, should you stand with your back to the audience and begin reading from the slide as if it was a script.

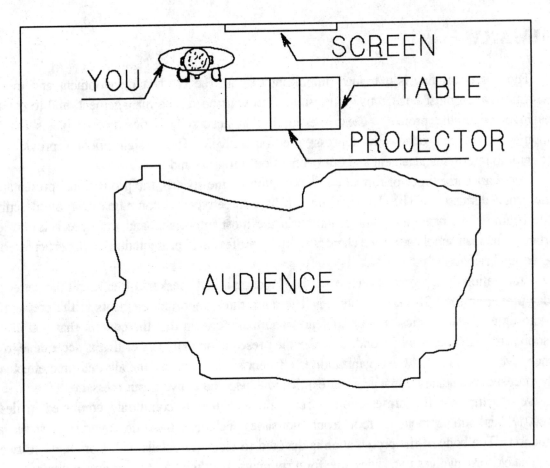

Fig. 6 Typical room arrangement for a presentation. Position yourself to the left of the screen and face the audience.

As you develop the discussion, point to key phrases to keep the audience focused on the topic. Assault both of their senses. Convey your message through their eyes and ears simultaneously. When you point to a line on the slide, point to the left side of the line. People read from left to right, so start them reading from the initial point on the left side.

As you speak modulate your tone and volume. Avoid both a monotone and a sing song delivery. A continuous tone of voice tends to put the audience asleep. A sing song delivery is annoying to many listeners. If you have a very high pitched voice, try to lower the pitch. A high pitch tone is also bothersome to many people and should be avoided if possible.

If a question is asked, answer it promptly, unless you intend to cover material later in the presentation that addresses the question. The answer should be brief, and you should try to avoid an extended dialog with some member of the audience. If a member of the audience is persistent, simply indicate that you will speak with them off-line after the presentation is concluded. The timing of your presentation can be destroyed by too many questions. Some questions are helpful because they permit limited audience participation, but excessive questions cause the speaker to loose control of the topic, the flow and the timing. The format of the communication changes from a presentation to a group discussion. Group discussions have a purpose, but it is different from a design briefing.

SUMMARY

The design briefing is extremely important to both the product development and to your career. Information must be effectively transmitted to your peers, the management, and to external parties involved in the project. Clear messages, that accurately define the problems that the development team can effectively address, are imperative. The design briefing provides the opportunity to inform management and our peers of our progress and problems.

There are three types of formal communication --- the speech, the professional presentation, and the group discussion. The characteristics of these three types of communication are described. For the engineer the professional presentation is the most important, and the speech is the least important. The main emphasis of the chapter is on the professional presentation, with particular focus on the design briefing.

Preparation for a presentation is essential. Accomplished speakers often spend the better part of a day preparing for a 20 minute briefing. There are three important elements in the preparation. First, know the characteristics of your audience, and make certain that the content that you include corresponds to their interests. Second, organize the presentation in a manner that is acceptable to the audience. Make certain that the organization is efficient so that you use the allotted time effectively. Finally, prepare high quality visual aids (slides) that will help you convey your message.

A structure for the presentation is recommended that is commonly employed in design briefings. This outline includes four front-end slides including the title, overview, status and introduction. The body of the presentation is devoted to technical details with from eight to twelve slides. The closure involves two slides, one for a summary and the other for the action items.

The type of visual aid that you select is important because the visual transmission of information will affect the outcome of the presentation. Computer projected slides have some significant advantages, but the availability of the equipment and a suitable (darkened) room often precludes their use. The overhead transparencies are the most common medium. 35 mm slides are the most suitable medium, if you have a presentation that includes many photographs and a large room in which to deliver the presentation.

The delivery is the make or break part of the presentation. We have provided you with two pages of details about do this --- and don't do that; however, the best way to learn to deliver is by practicing your presentation. On your first two or three attempts, have a friend video tape the event. Then review your behavior during your presentation. You will identify many problems of which you were not aware. Another suggestion is to use one of your free electives to take a speech course. This course will not help much in crafting the content to include in a design briefing, but it will help you to develop very important delivery skills.

REFERENCES

1. Eisenberg, A., Effective Technical Communications, 2nd ed., McGraw Hill, New York, NY, 1992.
2. Goldberg, D. E., Life Skills and Leadership for Engineers, McGraw Hill, New York, NY, 1995.
3. Wilder, L. Talk Your Way to Success, Simon and Schuster, New York, NY, 1986.

EXERCISES

1. Write a brief description of the characteristics of your classmates in this course. Focus on the characteristics that will influence the language and content in your design presentations.
2. Prepare an outline for a preliminary design review. Include in the outline the titles of all of the slides that you intend to use.
3. Prepare the title slide for your preliminary design review.
4. Prepare an overview slide for your preliminary design review.
5. Prepare a status slide for your preliminary design review.
6. Prepare the "Key Issues" slide for your preliminary design review.
7. Beg, borrow, but do not steal a video camera. Video tape a practice presentation by a teammate. Critique his or her presentation with good taste.
8. Locate in the College of Engineering all of the projectors that you can borrow for the presentations to be made in your classroom.
9. Practice the delivery of your presentation employing the most suitable projector that you can arrange to borrow.
10. Prepare a group of slides for a design briefing that uses the slide transition and slide building features of PowerPoint.

PART VI

ENGINEERING
AND SOCIETY

CHAPTER 15

ENGINEERING AND SOCIETY

INTRODUCTION

Let's be very clear from the very beginning of this chapter that engineering and society are closely coupled together. Engineering is an integral part of society. We must treat engineering as a social enterprise as we continue in our development of sociotechnical systems.

When we, as engineers, develop a new product or improve an existing system we change society. The magnitude of the change differs from one product to another, and the impact on society may be small or very large, but there is an effect. When a few companies introduced a random orbit hand sanding tool a few years ago, they reduced the time, effort, and cost of producing a flat, very smooth, scratch-free surface. These sanders helped several hundred thousand folks working in the finishing business to produce high quality finishes on furniture and cabinets. The results also benefited many millions of customers who purchase furniture each year. This new product had a modest but beneficial impact on society. It also had a very small negative effect, because some resources were employed in manufacturing the product and small emissions to environment undoubtedly occurred. On balance, society gained by introducing this product.

When transistors were first developed from new ultra pure semiconductors, the impact was revolutionary. While the first transistor was developed in the mid 1950s, the integrated circuit was introduced in 1959, and the microprocessor was released in 1971, the technological revolution in microelectronics is still underway. Micro-electronic based products are markedly affecting the way almost everyone lives in society today. We call it the information age. The new products, which are introduced almost daily, are driven by technology. They will continue to change the way we work and play in a very major way for at least another generation.

We hear a lot about new products and new life styles. Advertising over television, radio and even newspapers keep us well informed of availability and prices of both old and new products.

However, we rarely hear much about the technology involved even when the product is new, novel and revolutionary. The media and the public usually aren't very interested in learning about the technology employed in developing a new product. This is an unfortunate fact because it would be easier to introduce new or modified systems if the public were more technologically literate.

Engineers are in part to blame for this apparent wall between the public and the understanding of technology. We often speak in a language full of technical terms and leave those rare individuals, showing an interest in our work, in a confused state. As engineers, we must recognize our responsibilities to communicate effectively to society. We can begin by learning how to explain why and how our products work in terms easily understood by the general public. We must also be much more enterprising in understanding society and their changing views relative to technology. Finally, we need to think on a more global scale. Engineers work to develop sociotechnical systems which usually are large and complex. These sociotechnical systems rely on hundreds of products and many different types of services to function. Even though, as an engineer, we may be dedicated to developing a single product that is a component of a much larger system, we must understand the entire system. The superior designs are developed by those teams that have a thorough understanding of all aspects of the entire sociotechnical system.

Let's consider an example of a sociotechnical system, namely the commercial air transport system. Do we have technical products involved in this system? YES, literally hundreds of advanced state of the art products. Do we provide technical services for the airlines? YES again. Does an engineer working for General Electric in the development of a new jet engine need to know what Boeing is doing in the design of new aircraft or modifications of existing aircraft? Does that same engineer need to know about the sound levels that the public will accept as the plane lands and takes off? Does the engineer need to understand the causes of each and every airline accident?

From these simple questions it should be evident that our sociotechnical systems are extremely complex. It will require significant effort for you to understand the detail of the interaction of the many products involved in these systems with the public. Nevertheless, the most successful products will be developed by engineers that understand and appreciate the improvements that both the customer and the public seeks.

Finally, as an engineer, you must understand that the systems we develop are adaptive. They are dynamic and change with time in response to several different forces. First, society changes. What society and the customer wanted in the 1950s is a far cry from what either will accept today. The customers, a segment of society (more on this fact later), typically expect that the new model of the product will be much improved over the older model and cost less. Others in society, not necessarily customers, will expect (perhaps demand) that you manufacture the product without your presence being noticeable.

An example is in order to distinguish between the customer and society. Let's consider the automobile as the product. The customer for an automobile demands style, comfort, mobility, convenience, affordability, reliability, performance and value. Society requires (through federal regulation) safety, fuel efficiency and reduced pollution. The fact that the requirements from the customer and society often conflict should be recognized. Conflicting requirements challenge engineers to create improved designs that advance the state of the art in all areas of technology.

The governments (Federal, State, County and City) are also involved in the relationship between engineers and society. The government (really the various agencies) are continually issuing regulations that affect the sociotechnical systems and the product. Consider the commercial air transport system again. Safety is a very serious issue, and the Federal Aviation Agency (FAA) is responsible for issuing regulations to insure safe operations of commercial aircraft fleet. Last year (1996) was not a good year for the FAA or our commercial airlines because several very serious accidents occurred with significant loss of life. As a consequence of the fire in the cargo bay, which caused the crash of a Boeing 737 aircraft (Value Jet) in the Florida Everglades with the loss of all passengers and crew, the FAA is now requiring the airlines to retrofit all of the planes with a fire suppressant system. Regulations from government agencies drive part of the changes to any sociotechnical system.

The understanding of the relationship between engineering and society is relatively new. If we examine the five volume tome, *A History of Technology*, [1] published by Oxford University Press over the period from 1954-58, we find a chronological, deterministic catalog of technological development. The editors almost totally ignore the relation between engineering and society. In his review of this tome, Hughes [2] states that the editors belatedly conclude their five volume treatise with an afterthought essay on technology and the social consequences. Today the relationship is more clearly recognized, although we have not yet managed to include adequate treatment of this very complex relationship in the typical engineering curriculum. We will attempt to show a few examples of the relationships between engineering and society in this brief chapter. For a more complete treatment, we encourage you to read reference [3].

ENGINEERING IN EARLY WESTERN HISTORY

Technology has impacted the way that we live since we began to record history. Long before the first engineering college was founded, we had intelligent, hard-working men[1] working as engineers developing new products and processes. We can look at history and cite many developments that changed society and for the most part benefited society.

Prior to about 6000 BC there was little evidence of technology. Men and women were hunters and gathers. They scrounged for food and lived in caves if they were lucky and shacks or huts if they weren't. The significant development in the period from 6000 to 3000 BC was in agriculture with the domestication of animals and the cultivation of grains. With time, it was possible to insure a relatively stable if limited food supply, and people began to collect together in villages and small cities.

The first technology involved the building of roads, bridges and boats for local transportation. People clustered in villages and small cities needed to transport food, water and other materials over short distances. The geographic regions where technology was evident were extremely local. Early history (Western) involves only Mesopotamia, Egypt, Greece and then Rome. In a few select cities there were aqueducts to transport water and limited sewerage. In most regions, only the ruling class lived well. The great monuments of Egypt and Greece were built with slave labor. In fact slaves (people who were on the loosing side in a war or those born into slavery) provided most of the power required for building, mining and agriculture for many thousand more years. While the rise of religion, after the coming of Christ, is often credited with the reduction in the prevalence of slavery,

---

[1] Historical accounts of engineering achievements indicate that the profession was male dominated. The author is sensitive to the gender issue in engineering education, but it must be recognized that the presence of women in the profession is very recent when considered relative to a historical span of about 8000 years.

the steam engine developed by Thomas Savery, Thomas Newcomen and James Watt in the early 18th century was the beginning of the end of the need for vast amounts of human power.

The Egyptians built many huge monuments and were masters in moving large blocks of stones from distant quarries to the construction site and then to raise the block to its position in the structure. Some estimates indicate that 20,000 to 50,000 men were required drag 20 to 50 ton blocks of stone used in the construction of these monuments. In spite of the majestic grandeur and durability of the monuments, the new innovations introduced by the Egyptian engineers were not outstanding. They constructed temples in a simple fashion using only uprights (columns) and cross pieces (short, deep beams). While the arch was known during this period and used with mud bricks, it was never adapted to stone construction. Classical treatments of the history of technology tells us very little about the impact of these developments on society. However, it is apparent that the construction of a pyramid or a temple that took the efforts of a significant part of the working population for many decades probably did little to improve the welfare of even the free segment of the population. It is also clear that the effort required of the slaves to move the countless large and heavy stones must have been brutal.

The Greeks are better known for their contributions to science, although they had very capable engineers. The contributions of Thales, Euclid and Aristotle to the development of the basics of geometry and physics are well known. The new knowledge in geometry was used almost immediately in architectural engineering. However, the scientific knowledge during this period was so limited that nearly 2500 years elapsed before the body of knowledge was adequate to be useful to the engineering community.

The Greek engineers built cities, roads, water (hydraulic) systems and some machinery. Archimedes and Hero were innovators who used the pulley, screw, lever, and hydraulic pressure to develop cranes, catapults, pumps, and several hydraulic devices. The Hellenistic period brought better buildings, and local roads which improved living conditions for the upper class. However, the construction of the city infrastructure was still performed largely by slaves, because the use of other forms of power was nonexistent. The Hellenistic world was divided into extremely independent city-states. Politics of the day did not foster cooperation between these governmental entities, and as a consequence few long roads were built. Transportation was largely by sea in boats equipped with oars and a galley of slaves. Some of the boats were equipped with a square sail, but these sails were effective only on down-wind tacks.

The Roman engineers followed the Greeks, and while they were less inventive, they were masters of the detail. As the Roman Empire spread from the Middle East to Scotland, the Roman engineers built long roads connecting the distant cities. These roads were so well constructed that they were in service for many centuries. The large cities constructed in the conquered lands had paved streets and heated homes with water supplies and sewer systems. The Roman engineers had little or no theoretical knowledge but they had developed practical methods of construction that served their purposes well. During this period we did not understand statics or mechanics of materials. Beam theory would not be developed for another 1700 years. The Roman engineers based their construction on experience and they used very large safety factors. These empirical methods, while crude by today's standards, sufficed to produce many bridges, roads aqueducts and cities. A great bridge over the Tagus river in Spain supported a road about 200 m long with six spans about 60 m above the river. Considering it was completed at about 100 A. D., it is a remarkable engineering feat. The durability of the Roman construction is well documented by some of their structures which are still in use today some 2000 years after they were placed in service.

The Roman engineers deviated from the Egyptian and the Greeks in their construction by using much smaller building stones and bricks. They were still able to produce massive structures with the smaller blocks by utilizing mortar and cement. Smaller stones, bricks and cement clearly impacted society. Building could be constructed with much less effort. Thousands of slaves were not necessary to move a single block of material. Life for the slaves clearly improved as well as for the upper classes which benefited from the more rapid construction of buildings and houses for their cities.

While the Roman engineers were outstanding in the construction of civil infrastructure, they did very little to advance the use power in driving the few machines in use during that period. Humans were still used in most cases to power pumps, mills and cranes. In rare cases water wheels were used at power mills to grind grain, but these were of the undershoot design and much less effective than the overshoot design that was developed later. While references on the history of technology are silent on the issue, the life of a worker powering a pump or a mill must have been very difficult.

The vast Roman empire crumbled, and we entered a period that the historians classify as medieval civilization from 325 to 1300 AD. The fact that barbarians stripped and destroyed the cities does not imply that progress stopped, but it sure did slow progress to a snails pace[2].

In spite of the slow development of engineering and the lack of the construction of monuments during the middle ages, we did experience an early agricultural revolution. White [4] has describe the changes that occurred as agriculture moved from the dry sandy soil of Italy into the heavy alluvial and wet soil of northern Europe. The plows that were effective in the light sandy soils would not turn the sod in the heavy, moisture laden soil of the north. A new heavy wheel plow was invented that would cut through the grass sod and turn over the soil. However, this plow required eight yoked oxen to provide the large thrust required to pull it through the heavy sodded soil. Very few farmers of the day owned eight oxen. They had to combine their land, share their oxen, and cooperate in plowing and planting. The combination of the farmers and the land holdings led to a manorial society.

It may be difficult to imagine that the invention of a new plow made such a major change on the way people lived because today we are able to produce all of the food that we need in the U. S., and export excesses in significant quantities, with less than 2 % of our population. In the middle ages, 90 % of the population were involved in agriculture, and they had much less to eat.

Oxen were used as a source of power for plowing and hauling on the farms. However, oxen are very slow and consume large quantities of food. Horses were of limited use in this early period because of inadequate harnesses and saddles, and the splitting of their hoofs. (Early harnesses fitted about the horse's neck tended to strangle the animal.) The horse collar which transferred the load to the shoulders of the horse was introduced together with iron horse shoes to solve these problems. Horses gradually replaced the oxen because they were 50 % faster, worked longer and ate less.

---

[2] When Rome collapsed the Byzantine Empire followed and some significant engineering achievements were made in developing dome structures and curved dams. The author has omitted discussion of these developments to keep this historical treatment as brief as possible.

The horse was not very effective in frontal attacks in early warfare because stirrups did not exist. Horse mounted warriors were not effective, because they could not develop adequate thrust on their lances and stay on the back of the horse. With the advent of the stirrup, the horsemen could stand-up, lean forward, and thrust the lance through the defender shields without loosing their seat. The simple addition of stirrups to saddles had a profound influence on society. Feudalism evolved with an aristocracy of warriors conquering land that sustained a new battle strategy based on armored knights on large strong horses [2].

In early history small technological improvements produced major changes in the form of government, the type of society and the way that people of all classes lived. It is remarkable that for about 5000 years humans were the prime source of power for most activity. Agriculture and engineering progress proceeded together to provide more food, better housing, some roads and bridges and eventually horse and oxen powered plows and transportation.

ENGINEERING AND THE INDUSTRIAL REVOLUTION

Civilization has always been power limited, and even today our space travel is severely constrained by the inadequacies of rocket power. Until the last three centuries animals, windmills and waterwheels provided the only power available for agriculture, construction, milling, mining, and transportation. Progress was often slow because power was not available when and where it was needed. Life, in the power starved 17th century, was rather bleak for the general population. They worked from dawn to dark as farmers or craftsmen. Only those endowed with landed estates and/or money lived what we would consider a comfortable life.

The first of several breakthroughs in developing new sources of power occurred in the 18th century. In Great Britain, Savery, Newcomen and Watt developed stationary steam engines. While these early steam engines were woefully inefficient (only 0.5 %), they were produced in large numbers (500 existed in 1800) and were employed to pump water, and power mills (textile, rolling and flour).

Since the steam engines were replacing horses as power sources, James Watt conducted experiments to measure the power that a "brewery horse" could provide. His measurements indicated this particular horse generated power of 32,400 (foot-pounds/minute). The horse's output was rounded to 33,000 (foot-pounds/minute), which is still in use today as the conversion factor for one horsepower (HP).

Most of the 18th century was devoted to many innovations to improve several different models of steam engines. Applications were largely limited to stationary power sources because the engines were large, heavy and inefficient. However, with improved metallurgy and machining methods, it was possible in the 19th century to reduce the size and the weight of the engines permitting them to effectively power river boats, ocean crossing ships, rail locomotives and steam powered cars.

The development of the steam engine and its adaptation to several modes of transportation formed the foundations of the industrial revolution. The life style of almost everyone was markedly changed with the emergence of steam powered factories and a much more effective transportation system. Fewer people were employed in agriculture, but many more were working under appalling conditions in factories and mines. Twelve hour days, seven days a week, with only Christmas as a holiday was the norm. Manufactured products, such as clothing and house wares, were available to a larger share of the population at reduced prices, but the margin between earnings and the money necessary for a factory worker, miner, or farmer to stay alive remained very narrow. Steam power had relieved many thousand of men and horses from brutally hard labor, but they were released from one dull job to another. On a more positive note, the progress in agriculture and engineering enabled the population to about triple during this period (1660 to 1820).

ENGINEERING IN THE 19TH AND 20TH CENTURIES

For the eight thousand years of recorded history prior to about 1800, the advance of technology was extremely slow. We lived in an agricultural society with most of the population involved in growing the food that was consumed locally. Factories were being developed and simple products needed on the farm, in the home, or by the military were produced. Transportation of goods was usually limited to sailing ships on the seas or horse drawn barges on rivers or canals.

During the 19th and 20th centuries technology literally exploded. The many pieces of science developed over the preceding millennium expanded knowledge sufficiently to allow countless engineering applications. To illustrate this explosion, let's define modern technology to consist of the following four elements:[3]

1. Abundant power available where you use it.
2. Transportation including air, land and water craft with associated infrastructure to provide convenient and inexpensive access.
3. Communication between everyone, anywhere, at any time, at reasonable cost.
4. Information (knowledge) storage, retrieval, and processing in seconds at any location, at any time at reasonable cost.

While the advances in technology have been astonishing, particularly in the last 50 years, there are still significant deficiencies. Rocket power for launching space vehicles currently severely constrains space travel. Traffic jams in large cities extend the working day for many millions of people every work day. We still suffer nearly 45,000 deaths by accident on the highways every year. Communication in many parts of the world is still not available to the public. (Two years ago the author vacationed near a small fishing village in Mexico, and learned that the radio phone at the remote hotel was the only phone in the area.) We are just beginning to explore all of the different ways that we can use the information that has suddenly become available to the many people fortunate enough to own a personal computer and a modem.

Take some time during the summer or during a break and read about the technological developments of the last two centuries. Several key developments with approximate dates are listed below. Think about how these inventions have affected your life.

---

[3] We assume in this listing that housing, food, and house wares are available to all from the technological based industrial complex already existing in developed countries. A significant different listing would be necessary for third world countries.

- Batteries 1800.
- Steam locomotive 1825.
- Electric generator 1831.
- Steamship 1835.
- Electric motor 1835.
- Telegraph 1837
- Baltimore to Wheeling railway 1850.
- Telephone 1876.
- Internal combustion engine 1876.
- Incandescent lamps 1880.
- Central electric power generation 1882.
- Steam turbines 1889.
- Alternators (ac generators) 1893.
- Radio 1896.
- Automobile 1900.
- Airplane 1903.
- Vacuum tubes 1906.
- Television 1935.
- Jet engines 1940.
- Atomic bomb 1945.
- Digital computer 1950.
- Transistor 1953.
- Integrated circuit (IC) 1959.
- Atomic power 1960.
- Communication satellite 1970.
- Microprocessor 1971.
- Internet 1990.

The list above is not complete because many more innovations are important and could have been included. The idea is to show you that the period from 1800 to 2000 was packed full of new technology. This technology markedly impacted the way that we live, work and play.

We can also debate the exact year of the invention. The exact year is not very important, since an invention usually evolves into a commercially successful product after several iterations with improvements or modifications by several designers. You should not worry about the exact dates. Instead ,you should recognize the continuous progress and observe the shift in emphasis from one period to another. In power, we moved from animals, to steam, to internal combustion (gasoline), to nuclear energy. Power in the form of gasoline fueled internal combustion engines became mobile and electricity was widely distributed. In transportation, we improved from the horse and buggy and sailing ships to automobiles, airliners, and nuclear powered submarines. In communications, we progressed from the pony express to cellular telephones and the Internet. In information, we are evolving from a hard copy library system to a digital multi-media information bank available at any time, to everyone, at nearly zero cost.

Have these advances in technology changed society? I believe that they have had a profound beneficial effect. We produce an over abundance of food with less than 2 % of our population working in agriculture. We manufacture most of our consumable products with about 20 % of the work force. Our work week is shorter and we have many more affordable products. Most families own at least two cars, and home ownership is at an all time high. We do hear concerns regarding the improvements in our standard of living over the past 20 years, particularly for those folks at the lower end of the income spectrum. Technology has probably not helped this group as much as others. Global trade has permitted many of the semi-skilled tasks to be moved to third world countries where the cost of labor is about 10 to 15 % as much as in the developed countries. Many of the previously well paid factory jobs in manufacturing in the U. S. have been moved off shore where labor costs are much lower.

NEW UNDERSTANDINGS

We have been, in the discussion of the historical development of technology, optimistic about the very positive effects of technology on society. In the past, the public accepted, without serious questions, all forms of technology and the risks that were involved. However, the situation has changed to a remarkable extent in the past 40 years. We (engineers included) have lost the public trust. An excellent example of this loss of trust is in the generation of electricity using nuclear power. Since the Three Mile Island incident in 1979, the general public wants no part of nuclear power. There is a valid public perception that the technological risks associated with nuclear power were significantly understated. John O'Leary, a deputy director for licensing at the Atomic Energy Commission had written prior to the accident at Three Mile Island that "the frequency of serious and potentially catastrophic nuclear incidents supports the conclusion that sooner or later a major disaster will occur at a major generating facility" [6]. In spite of this warning of the very high probability of a serious accident federal, state or local governments have not planned for orderly emergency evacuation of nearby residents or for containment measures to limit the contamination at most of our reactors.

The Chernobyl accident in 1986 proved that O'Leary was correct. The No. 4 reactor at the Chernobyl facility exploded sending a radioactive plume in the air so high that it was detected by an alert nuclear plant operators in Sweden. Many died in this radioactive explosion and many more will die in the future from radiation poisoning. Countries as far away as Italy were affected by high radiation levels.

The nuclear industry and the government agencies regulating it, world wide, have not been candid with the public. Nor has the public been very interested in intelligent debate, resorting instead to large, ugly demonstrations to express their displeasure. We need an accurate and honest assessment of the risks of any nuclear power project by the industry, the regulators, and experts representing the public. The risks must be weighed against the benefits to society. Some time in the future fossil fuels will be come unavailable, scarce or costly. When the fuel crises occurs, we will find that nuclear power is the only viable alternative for stationary power generation.

We are not here to argue the case for or against nuclear power. We use it as an example of the fact that understating or poorly assessing risk brings a severe penalty: Loss of public trust is followed by a ban on technical development in the subject area.

As engineers, we must broaden significantly our perception regarding technological risk. Today many engineers dismiss failure due to human, operator or pilot error because they are not machine errors. This attitude is wrong. The human operator and the machine constitute a system, and affect public safety accordingly. The machine and its operator must be included in system design since they constitute an interacting set of weaknesses and capabilities [6].

Another new understanding pertains to geography. Prior to the second world war (WWII) international trade was very limited. We produced what we consumed, mined our minerals, and pumped our own oil. Geography was very important because it constrained trade by law and by distance. After the war, laws were changed and trade agreements (GATT, NAFTA, and etc.) were arranged. International trade was encouraged so much so that the U. S. now consistently imports more goods and materials than it exports. Our current deficit in the trade balance, in excess of one hundred billion dollars, is the equivalent of 2,500,000 high income jobs at $40,000 a year.

Previously the cost of shipping had limited trade between countries separated by large distances. However, following WWII technological improvements were forthcoming in the shipping business. We developed the super tanker greatly improving the efficiency of shipping huge quantities of oil. We also developed very large containerized ocean vessels reducing the cost and time of loading and unloading cargo.

With the time and costs of shipping greatly reduced and the legal barriers to trade removed, we evolved into a global market place. Engineers compete to develop world class products or services. Products designed in the U. S. may be produced anywhere in the world. Or products designed in Europe or Asia may be produced in the U. S. Design and production are both free to be located to minimize costs and to maximize benefits to the customers and/or to the corporate entities.

The third new understanding is in the role of the governments (federal, state, county and city). Prior to WWII governments were small and taxes were relatively low. The primary role of the federal government was to insure national security. The state governments worried mostly about transportation systems, and the local governments concerned themselves with education. Today the situation is markedly different. The governmental agencies in response to their interpretation of the law (either old or new) issue regulations. The regulations pertain to issues of concern to society which include protection of the environment, safety, health and energy conservation.

We can debate the wisdom of many of the regulations, but that is not the point. The regulations currently exist and new ones will continue to be issued with high frequency. As an engineer, you must be prepared to serve society-at-large as reflected by these governmental regulations. If you question the wisdom of some of the regulations, the best approach is to become involved in the political process so that you may have the opportunity to influence them. Engineers tend to resist constraints because they limit the freedom of their designs and add to the cost of the product. We think of regulations as government imposed constraint, but they are really constraints imposed by society-at-large. Society controls the regulatory process because it has the power (by voting) to change the politicians drafting the laws and the bureaucrats formulating the regulations.

BUSINESS, CONSUMERS AND SOCIETY

Engineers serve three constituencies, namely business, the customer and society-at-large. The business leaders (management) look to engineers to develop products, services, and processes that meet global competitive challenges. Consumers seek more convenient, reliable, enhanced, value-laden products at reduced prices. Society-at-large, through elected politicians and public interest groups, demand action leading to solutions to problems involving safety, health, energy conservation and preserving or improving the environment. It is an overlapping set of constituencies, as shown in Fig. 1, because society encompasses business, the governments, and consumers.

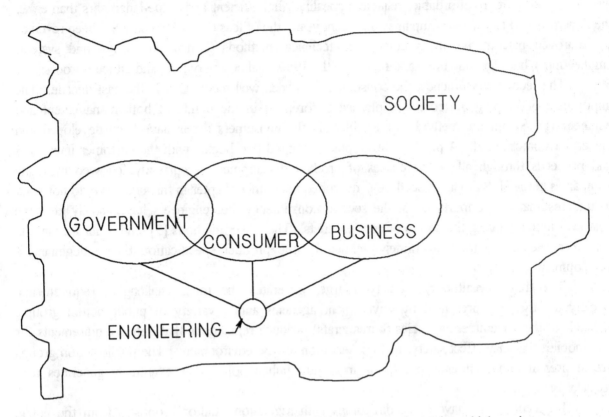

Fig. 1 Schematic illustration showing the overlapping set of constituencies which engineers serve.

To discuss the issues raised by serving three constituencies, we will use the U. S. automotive industry as an example. Marina Whitman has described many aspects regarding the three-way demands, and since we will draw from her excellent treatment, you may want to study her paper [8] in depth.

Business demands are simple although they are often difficult to meet. Any business must be profitable to stay alive. Losses can occur over the short term, but if the red ink is prolonged the company is forced by its creditors into bankruptcy proceedings. Making a well defined profit every quarter is the usual goal of management. For many years following WWII, it was relatively easy for U. S. businesses to make money. Bombing and shelling during the war had destroyed essentially all the manufacturing capability in every country except the U. S. It was easy to be a winner when the competition was without factories or infrastructure. For the automobile industry, the golden years began to erode in the late 1960s and survival became a life or death battle in the 1980s. Their problems came from global competition (mostly Japan). The Japanese firms provided much more reliable automobiles at lower costs than those produced in the U. S.

To show the extent of the market penetration by the imported automobiles, recall that the imports accounted for only 4 % of the market in 1962, but they controlled 30 % of the market by 1990. This devastating loss of market share was not limited to the auto industry. Several other industries including optics, consumer electronics, ship building, steel, etc. suffered even more serious losses.

Engineers and management in the U. S. firms had lost the competitive edge after decades of modest competition from the Europeans. The Japanese automobile companies (Toyota, Honda, Nissan) provided more affordable, reliable, appealing, fuel efficient and value-laden cars than either the American and European companies. In recent years, the of loss of market share has been reversed by improved participative management and technical methods (concurrent design and systems engineering[4]), but regaining the market share of the 1960s will be a very long and difficult process.

The second constituency, the consumers, has been well recognized in the past decade. The importance of keeping the customer pleased is foremost in the minds of both management and engineering. Systematic methods for establishing the consumers needs have been developed and placed into practice [9]. A product development method that begins with the customer interviews and proceeds through all of the phases of product development is currently followed by many companies in the U. S.. This procedure is described in another chapter in this text. We do not want to underestimate the importance of the second constituency, but engineers have usually kept the customer in mind during the design of new products. The newer methods [9] tend to insure that the customers needs are more accurately defined by the product specification before beginning a development.

The third constituency, society-at-large, generates the most challenging requirements. Society at large is represented by government agencies and a variety of public action groups. Officials of the government agencies (bureaucrats) produce regulations that define requirements for automobiles that may affect safety, fuel conservation, or the environment. The public action groups create pressure and influence politicians in setting public policy that eventually generates new requirements.

Engineers often have some difficulties with regulations and/or proposals from the public action groups. The laws are written by lawyers and the regulations are written by public servants. Neither lawyers or public servants are usually technically literate. Sometimes they recognize this fact and seek technical support prior to drafting binding regulations. At other times the regulations are placed into effect prior adequate study by knowledgeable representatives of the public and engineering communities. The difficulty is that the demands of the customer, including style, comfort, affordability, reliability, performance, convenience, etc., may be in conflict with societal imposed regulations on safety, fuel efficiency and emissions.

A very recent example illustrates what many believe to be an ill advised regulation [10]. The National Highway Transportation Safety Administration (NHTSA) mandated requirement for dual airbags in every vehicle licensed in the U. S. Most people consider air bags as a valuable safety feature, and many lives (1600 is the current estimate) have already been saved in serious frontal accidents. However, during the summer of 1996, 10 children were killed by deploying airbags, bringing the total to 32 children killed by the safety feature since their usage was mandated on the passenger side. More recent estimates [11] which include small women and children indicate that 54 deaths are attributable to air bags.

The killing of women and children by a safety device is a very serious **OOPS**. Moreover, the regulation prohibits anyone other than the car's owner to disable even the passenger side air bag. If

---

[4] Systems engineering means many different things to different groups. We define it as a cost and time efficient method to specify, design, develop, and integrate all of the individual subsystems so that the total system for the product meets the requirements of customers, business and society-at-large.

you are a concerned parent you can do it yourself, but few owners have the required skills. Also if you ruin the mechanism and/or the bag, the cost of replacement is about $1000. Because of their high costs, air bags have replaced stereo systems as the theft of choice if your car is robbed.

The issue is not whether we should have air bags or not. They have proven to be a very effective safety device for most adults either driving or riding as passengers. The issue is with the rigidity of the federal mandates and the failure of the regulators to:

1. Update the regulations to accommodate different behavior of drivers.
2. Provide more design flexibility in the requirements.
3. Provide a means permitting the driver to activate or deactivate the passenger side bag.

The air bag regulation was written in 1977 when only 14 % of the population even bothered to use their seat belts. By 1994, when the mandate was fully effective, 49 of 50 states were enforcing seat belt laws, and seat belt usage had risen to 68 %. While seat belts do not eliminate the need for air bags, they do increase the time allowed for the deployment of the air bag. The NHSTA did not change the regulation to accommodate the increased usage of seat belts. The air bags continued to be deployed at a speed of 200 MPH to protect an unbelted adult male. This is great for the big guys, but for kids or even small women a 200 MPH kick in the face is extremely dangerous. When seat belts are used, there is no need for the 200 MPH deployment of the bag because the belt constrains the passenger providing more time for deployment of the bag.

Industry representatives recognized these dangers and warned the NHTSA of the dangers to children. Rather than change the regulation to provide more flexibility in the design of the air bag systems, NHTSA drafted a warning label, and industry added instructions in the owner's manuals explicitly cautioning against placing a rear facing child safety seat in the car's front seat. The manuals indicate that it is safe to place a child in the passenger seat if the child is buckled-in with a seat belt, **BUT** the seat must be in the full rear position. The manual also states that children should never lean over with their faces near the air bag cover while the car is in motion. The warning labels and instructions in the owners manual are a result of poor design that was forced by an inflexible regulation. We now have in place a safety device (air bag) that is dangerous to small children. Warning labels and instructions in manuals reduce the consequences of litigation (there is another law indicating that the user must be warned of possible dangers), but they are totally inadequate to protect children, to young too read, riding in the passenger seat.

It would be relatively easy to modify the design of the control system to include an on/off switch with a warning light on the control panel of the automobile. With this feature the driver could deactivate the passenger side air bag whenever he or she wished to do so. Another design approach entails a modification of the pneumatic system to reduce the explosive force with which air bag deploys. This modification would probably have taken place several years ago when seat belt usage finally became widespread. However, the federal regulation did not allow permit modification then, and it does not permit modification even today after the death of 54 helpless women and children.

Today (January 1997), the public is concerned if not angry. The NHTSA has received about 1000 letters from owners requesting the waivers required to have train mechanics at the dealers to disconnect the dangerous air bag. Most mechanics refuse to disconnect the airbag for fear of future litigation. There appears to confusion about the remedy [11]. The agency contends that customers should decide for themselves and not be required to request permission. The car owners would have to read government literature, sign a waiver, and post warning labels in the car.

The big three automakers do not have a ready solution. GM proposes cut-off switches in existing vehicles. Ford is in favor of permitting only certain categories of people to disconnect the system. Chrysler is reconsidering its opposition to cut-off switches. The dealers and independent mechanics do not want to touch the problem because they fear they might get sued sometime in the future.

The current disarray is alarming. The interaction of the government agencies representing the public-at-large, business, and the consumer is woefully inadequate. We need a solution and we need it tomorrow. Yet the regulatory system is too awkward to provide rapid solutions. The awkwardness comes from inflexibility on the part of the government and inadequate cooperation between the automakers. We can understand competition on the part of business to increase market share, but safety features should be excluded from the normal competitive secrecy. The government agencies must become much more flexible, and recognize that they cannot decree away all danger.

The difficulties between the government and business is not new. An interesting summary of the regulatory relationship between government and business was recently published by Jasanoff [12]. We quote her directly:

> Studies of public health, safety and environmental regulation published in the 1980s reveal striking differences between American and European practices for managing technological risks. These studies show that U. S. regulators on the whole were quicker to respond to new risks, more aggressive in pursuing old ones, and more concerned with producing technical justifications for their actions than their European counterparts. Regulatory styles, too, diverged sharply somewhere over the Atlantic Ocean. The U. S. process for making risk decisions impressed all observers as costly, confrontational, litigious, formal, and unusually open to participation. European decision making, despite important differences within and among countries, seemed by comparison almost uniformly cooperative and consensual; informal, cost conscious, and for the most part closed to the public.

The assessment by Sheila Jasanoff gives us clear direction. We must reduce the adversarial relationship between business and the government. As engineers working for both the government and business, we should be an important element in the future to produce a regulatory system with more flexibility, more cooperation, more cost conscious, and more consensus.

CONCLUSIONS

The standard of living in the U. S. will be determined by interplay of three powerful influences:

- New and rapid technological advances.
- Business (management) response to global consumer demands.
- Social demands as evidenced by legislative and regulatory requirements.

Engineers have always provided the leadership in technological advances, and we will continue to provide business and the public at large with cutting edge, world class technology.

Business requires capital, technology and good management to remain competitive. In addition to providing the technological base for a company, engineers frequently serve in management. If this career path appeals to you, plan on extending you education in a business school. The combination of a Bachelor of Science degree in engineering and a Master of Science in Business provides a very solid foundation for a career path in a technically oriented business.

In the past engineers on the whole have not been actively engaged in societal issues. We lack patience in dealing with the public, and become irritated at their lack of technical literacy. We fume at the inefficiencies of government agencies, and their lack of flexibility. We become outraged at the arrogance of bureaucrats, and their lack of concern for time and costs. We retreat into our caves and work on new developments and new products. In the meantime, the public-at-large grows wary of many important sociotechnical systems, and the government resorts to litigious process which impede real progress.

It is unfortunate that engineers have not been a significant force in dealing with societal issues. As society, business and the customer create conflicting demands, engineers are essential in the crafting of well balanced solutions. The conflicting demands provide new opportunities and complex challenges for engineers. To take advantage of them the engineers of tomorrow will require enhanced communication skill, greater disciplinary flexibility, a better understanding of the mechanisms of regulatory agencies, and a much wider prospective relative to societal demands.

REFERENCES

1   Singer, C. E. J. Holmyard, A. Hall, and T. Williams, eds. A History of Technology, Oxford University Press, New York, five volumes, 1954-1958.

2   Hughes, T. P., "From Deterministic Dynamos to Seamless-Web Systems," Engineering as a Social Enterprise, ed. Sladovich, H. E., National Academy Press, Washington, D. C. 1991, pp. 7-25.

3   Sladovich, H. E., ed. Engineering as a Social Enterprise, National Academy Press, Washington, D. C. 1991, pp. 7-25.

4   White, L., Jr., Medieval Technology and Social Change, Oxford: Clarendon Press, 1962

5   Kirby, R. S., S. Withington, A. B. Darling, and F. G. Kilgour, Engineering in History, Dover Publications, New York, 1990.

6   Ford, D. F., Three Mile Island: Thirty Minutes to Meltdown, Viking, New York 1982

7   Adams, R. McC. "Cultural and Sociotechnical Values," Engineering as a Social Enterprise, ed. Sladovich, H. E., National Academy Press, Washington, D. C. 1991, pp. 26-38.

8   Whitman, M. v. N., "Business, Consumers, and Society-at-Large: New Demands and Expectations," ed. Sladovich, H. E., National Academy Press, Washington, D. C. 1991, pp. 41-57.

9   Clausing, D. Total Quality Development: World-Class Concurrent Engineering, ASME Press, New York, 1994.

10 Payne, H. "Misguided Mandate," Scripps Howard New Service, Knoxville News-Sentinel, January 12, 1997, p. F-1.

11 Associated Press, "Government Wants to Let Owners Cut Off Air Bags; Big 3 Fret," Knoxville News-Sentinel, January 16, 1997, p. A-11.

12 Jasanoff, S. "American Exceptionalism and the Political Acknowledgment of Risk,' Daedalus, Vol. 119, No. 4, 1990, pp. 61-81.

EXERCISES

1. Recall a product that you or your family has purchased in the past month or so. Write a brief paper describing both the positive and negative impacts of that product on society.

2. Recently the Federal Aviation Agency enacted a regulation affect the mailing of packages weighing more than 16 ounces. Write a paper covering the following issues:
   - Describe the regulation in more detail.
   - Why is the FAA writing regulations affecting the U. S. Post Office?
   - What effect (cost) does this regulation have on the individual and on small business.
   - Do you believe that the regulation will be effective for it's intended purpose? Please give arguments supporting your viewpoint.
   - What actions will the post offices (40,000 of them) have to take to make the regulation effective?
   - What actions do the post offices actually take with regard to the regulation?
   - Will these actions be costly? Estimate the costs to both the public and the individual. Assume that a postal clerk is paid $10.00/hour and that an individual considers his or her free time worth $10.00/hour.
   - Give your assessment of the cost to benefit ratio for this regulation?

3. Suppose that you were a slave in the time of Rameses II and were one of many slaves assigned to the task of constructing his statue. For those not up to date on Egyptian statues, this one weighs a 1000 tons and was 56 feet tall. Write a paper describing your daily tasks.

4. Horses were not used extensively to relieve man from brutal work or to provide a significant advantage in military efforts until nearly 1000 AD. Write a brief paper describing both the social and technical reasons for the very long time needed to effectively employ the horse in either military or commercial enterprise.

5. Suppose that you were living on a manor in England in about 1300 AD. Describe your life style and indicate how technology affects your work and your play. Select in which of the two classes of society that you existed.

6. Describe where power is available in the U. S. today? Define the type of power in your response. Also include an example where you personally suffered because of a lack of power.

7. Write a paper comparing the lifestyles, as you imagine them, for a man living in 1997 and 1297. Did technology make a difference in the quality of life.

8. Repeat exercise 7 replacing the man with a woman.

9. Why is the public-at-large turned off on power generation with nuclear energy? Is the public-at-large correct in their collective assessment?

10. Why do we have a global marketplace today? Does global trade improve our standard of living? What does our current trade deficit have to do with wages for factory workers? Does technology help or hinder our balance of trade deficit?

11. If you were the director of the National Highway Transportation Safety Administration (NHTSA), what action would you take regarding the public concern about air bags.
   - Explain the reasons for your actions.
   - Explain how you would get the big three automakers to agree with you.
   - Describe the outline that you would follow in the press conference announcing this action?
   - Describe how you would handle the public interest groups that disagree with your ideas.
   - Describe how you would handle the public interest groups that agree with you.

12. What non technical courses can you take during your undergraduate program that will broaden your prospective and aid you in dealing with sociotechnical issues?

NOTES

CHAPTER 16

SAFETY, RISK AND PERFORMANCE

INTRODUCTION

When engineers design a new product or modify an existing product, we often introduce an element of risk in society-at-large. Sometimes the risk is minimal, and the resulting damage from a malfunction or failure is small. But in some instances, the risk may be large, and the damage may be catastrophic.

Some simple examples are in order to illustrate the concept of risk with attendant benefits to society. Suppose we design a new model of an electric pistol grip drill that is powered from the standard 120 volt 60 cycle single phase power supply from the local utility company. The operator is always at risk for an electric shock. It is our responsibility as engineers to minimize this risk while maintaining the advantage in performance of an electric powered drill.

How do we protect the operator from electric shock? The operator is holding an electric motor with 120 volts across the armature coils and the field coils adjacent to his or her hands. The answer in this case is to build the body of the drill from a tough, durable plastic that is structurally strong, but also an excellent electrical insulator. The case keeps the operator from touching any part of the electrical circuit powering the motor. In fact Black & Decker, a major manufacturer of small power tools uses double insulation to provide two independent insulation barriers to keep the operator away from any part of the electrical circuit. Operators will probably still manage to get shocked from time to time, but it will not be easy. Accidental contact of the operator's hands with a live electrical circuit will be a rare event. In fact, the risk of electrical shock is so small that we routinely pick up a power tool and drill a hole without even a passing thought about the possibility of getting shocked. Operating a power tool involves risk, but it is an acceptable risk because the operator is not concerned about his or her safety.

Let's move up the risk ladder and consider flying in a commercial airliner from say point A to point B some 1000 miles distant. Is it safe for us to make this trip? Most of us appreciate that it is reasonably safe to fly commercial airliners, but that there is a slight probability of a crash[1]. The statistics show the probability of a fatality in an airline accident is about 1 in a billion passenger miles. So if you make the round trip from point A to B your probability of getting killed during your trip due to an airline accident is:

$$P_k = (2)(1000)/10^9 = 2 \times 10^{-6} = 0.0002 \%$$

This is a very low probability for a fatal accident, and most people do not worry much about the possibility of dying in a crash when they board an airliner. But what if you are a sales engineer, and you fly 100,000 miles a year, every year for 20 years. Your total mileage accumulates to 2×10^6 miles, and your probability of getting killed in a crash increases to:

$$p_k = 2 \times 10^6/10^9 = 2 \times 10^{-3} = 0.2 \%$$

The probability has increased significantly if you accumulate the miles flown by a traveling professional over a 20 year period. There is one chance in 500 that you will die in an airline crash. Do you still want to be a sales engineer flying weekly to meet with the customer? Can you tolerate this level of risk? Would you be apprehensive?

Everyone has some tolerance for risk. We weigh the speed, convenience, cost and risk of flying against that of traveling by train or by car. In almost all instances, folks select the plane for long distances, and the car for short trips. We select the train only in those rare instances when it provides relatively good service (high speed with frequent trains) to downtown locations. Most of us do not think about the risks involved in travel because fatal accidents are rare events, except for driving a car.

Fatal accidents in automobiles are not rare events. Each year about 45,000 fatalities occur on U. S. highways, and hundreds of thousands of serious injuries occur. It is clearly more dangerous to drive than to fly, when you compare on a per mile basis. How do we handle the higher risk associated with driving? Rationalization --- I am an excellent driver, and it won't happen to me. Not necessarily a true statement, but the rationalization of ones superior driving skills alleviates the worry and concern about a fatality or injury producing accident. The real risk remains.

When we drive, we are an active factor in determining the risk. Our driving habits (high speed, reckless steering, tailgating, etc.) and driving skills affect to a large degree the level of risk involved. When we fly, we are a participant. We have voluntarily decided to fly, and we even have a choice of airlines. We can pick the airline with the best safety record. (Southwest Airlines is the best with no accidents in their entire 25 year history). However, sometimes we are placed at risk, and we have little or no choice in the matter.

Suppose that you live 40 miles Northeast of a nuclear power plant. Is 40 miles far enough away to avoid significant fall out in the event of a serious explosion involving the nuclear reactor at

---

[1] The author is probably more sensitive to the risk of commercial air travel than most people. I knew, rather well, two colleagues who were killed in separate crashes – one on the East coast and the other on the West coast.

the power plant? Remember that the winds are usually from the Southwest, so the fall out plume will be pointed in your direction in the event of an accidental release of radioactive gasses.

The risks associated with radioactive fallout, air pollution, toxic chemicals, polluted ground water, etc., are a serious concern to almost everyone. Folks do not like to be exposed involuntarily. They expect governmental regulators to control the environment, and to reduce the level of the risks of accidental exposure. Unfortunately we do not have a very good record in protecting the population from these exposures. For example, accidents have occurred in the control of reactors. Citations for safety violations by inspectors of the Nuclear Regulatory Commission are commonplace. From a public perception, the risk of a serious malfunction of a nuclear reactor became unacceptably high in the U. S. and many other countries. After the accident at the Three Mile Island facility in 1979, the commercial nuclear industry died in this country. Some plants under construction were completed and others were converted to fossil fuels, but no new nuclear plants have been started since then. With the more recent and very serious accident at Chernobyl in 1986, new plans for power generation with nuclear energy have been placed on hold for the foreseeable future.

There is a very large difference between the acceptable level of risk depending on whether the risk producing activity is voluntary or involuntary. The degree to which we personally control the risk is also extremely important. Activities like skiing, scuba diving, horseback riding, hang gliding, mountain climbing, dirt biking, etc., carry extremely high levels of risk. Yet folks swarm to the ski resorts and pay big bucks to break their bones. Why? They have voluntarily decided that the thrill, or other pleasures, derived from the activity is worth the risk. Also they control the level of risk. They can pick the bunny slope and minimize the risk or the black diamond slope to maximize the thrill.

As engineers we must recognize that some level of risk is involved in almost all the products we produce. It is imperative that we minimize this risk while maintaining an acceptable level of performance. It is imperative that this risk be acceptable to our customers, and to society-at-large. An appropriate balance between risk and performance must be achieved. It is also essential that we cooperate with business and the governmental regulatory agencies, to provide a realistic assessment of the risk level. Finally, the public should be aware of the risks involved, and they should not be surprised by the news releases describing the gory details of the victims of accidents.

MINIMIZING THE RISK

An engineer minimizes risk by preventing failure of each and every component in the system. This is not an easy task because there are several different ways that components fail. Parts fail by breaking, by wearing out, by corrosion, by burning out, by aging, by fatigue, etc. In some instances, we can anticipate these failures and replace the parts before they malfunction for one reason or another. We call this controlled replacement of finite life parts scheduled maintenance. In most cases, we try to design each component so that failure will not occur during the anticipated life of the product.

A complete treatment of methods to avoid failure would require more than a single textbook. We will briefly introduce an analytical procedure for designing tension members to provide you with an illustration of conservative design to insure safety. These tension members will be sized (sufficiently large) so that they will exhibit a safety factor. In other words, they will accommodate a load higher than anticipated in service over their entire life cycle.

Let's begin by introducing a tension member as shown in Fig. 1.

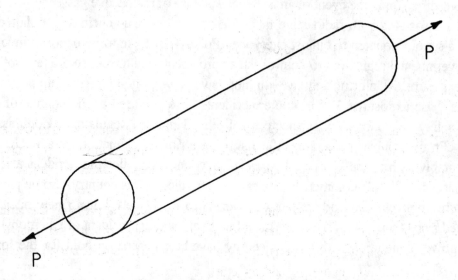

Fig. 1 A tension member is a long thin rod subjected to axial load P.

When the rod is subjected to an axial load P, an axial stress σ develops which is uniformly distributed over the cross section of the rod. We show this stress distributed over a section of the rod in Fig. 2.

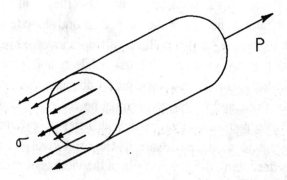

Fig. 2 The stress σ is distributed uniformly over the cross-sectional area of the tension rod.

It is possible to determine the magnitude of this stress from the formula:

$$\sigma = P/A \qquad\qquad (1)$$

where   P is the load in lb. or Newton.
A is the cross sectional area of the rod in in^2 or m^2.

Two simple examples for computing stresses are given below:
If P = 10,000 lb., and the rod has a circular cross section with a radius of 0.25 in., find the axial σ. Note the cross sectional area is given by:

$$A = \pi r^2 = \pi(.25)^2 = 0.1963 \text{ in}^2$$

And from Eq. (1) the axial stress σ is:

$$\sigma = P/A = 10{,}000/0.1963 = 50{,}930 \text{ psi}$$

where the abbreviation psi stands for the units lb/in^2.

At this stage of the analysis, we will not attempt to interpret the significance of our answer of 50,930 psi. We need more information, and we need to make a comparison of the magnitude of the stress with the strength of the material from which the tension member is fabricated, before remarking on it's significance.

In this first example, we have used the U. S. Customary Units expressing the axial load in pounds, the cross sectional area in square inches, and the stress in psi or pounds per square inch. In the next example, we will use the International System of Units (SI) where the load is expressed in Newtons, the cross sectional area in meters, and the stress in Pascal.

If the axial load P is 50,000 Newton (N), and the radius of the circular rod is 5 mm, find the axial stress σ. First find the cross sectional area A as:

$$A = \pi r^2 = \pi (5 \times 10^{-3})^2 = 78.54 \times 10^{-6} \text{ m}^2$$

Then from Eq. (1), we can determine the axial stress as:

$$\sigma = P/A = 50 \times 10^3 / 78.54 \times 10^{-6} = 636.6 \times 10^6 \text{ Pa} = 636.6 \text{ MPa}$$

where the abbreviation Pa stands for Pascal the unit in which the stress σ is expressed. A Pascal is equal to a Newton per square meter (N/m^2). When calculating stresses in the SI system, we usually get very large numbers for the stress when it is expressed in Pascal's. For this reason, we usually use MPa (mega Pascal) where one $MPa = 10^6 \text{ Pa} = N/(mm^2)$.

OK. We can compute the stress on a tension member if somehow we can ascertain the load that it must resist, and if we know its size. The stress that we determine is a number with associated units either psi or MPa. As a number it is not much use to us until we compare it to a strength of the material from which the tension member is fabricated.

Let's suppose that we machine the tension rod used in Example 1 from a very strong alloy steel with a yield strength of 100,000 psi. Do you think that the tension rod is safe? Will it retain its shape under load? Will it fail by yielding (stretching under load and not recovering completely when the load is removed)? We answer these questions by a simple comparison of the applied stress σ, and the strength of the material S. In our first example, the axial tensile stress $\sigma = 50{,}930$ psi, yet the strength of the material from which the rod was machined is S = 100,000 psi. The comparison shows that the yield strength of the rod is greater than the applied stress. The rod will not fail by yielding. But how safe is the rod? We know it will not fail, but we need to establish some measure of the safety for the rod.

To respond to the safety issue for the rod, let's make a ratio of the strength divided by the stress, and define it as the safety factor (SF).

$$SF = S/\sigma \qquad\qquad (2)$$

In our example the safety factor is given by Eq. (2) as:

$$SF = 100,000/50,930 = 1.96$$

The safety factor SF = 1.96 indicates that the tension rod is almost twice as strong as it needs to be to resist yielding under the applied load. We have a comfortable safety factor, and can be confident that the rod will behave safely in service. Our only concern is the accuracy with which the load has been predicted, and the quality control for the material used in the manufacture of the tension rods. If we:

- Have lots of experience with the application;
- Are certain of the magnitude of the load;
- Are confident that our manufacturing division keeps track of their materials;
- And checks on the strength of the materials received from suppliers.

Then we can be satisfied that a safety factor of about two is adequate. However, if we are not so certain about the load and/or the materials, then a safety factor of two is not sufficient. We increase the safety factor to accommodate our ignorance of the applied load, or our lack of control over the material employed. Safety factors ranging from two to four are commonly employed in design. Safety factors of less than two require considerable care and expense. The use of relatively low safety factors is justified only for very high performance applications. In these special situations, the engineering analysis, prototype testing, quality control inspections, maintenance inspections, and documentation need to be extensive.

The margin of safety MS is sometimes used to describe the degree of safety incorporated into the design of a component. The margin of safety should not be confused with the safety factor – they are different quantities. The margin of safety is defined as:

$$MS = SF - 1 \qquad\qquad (3)$$

In our example the margin of safety MS = 1.96 - 1 = 0.96 or 96 %. This computation indicates that our tension rod has a margin of safety of 96 %. That is the strength of the material exceeds the applied tensile stress by 96%.

If you can determine the load on a tension member, it is easy to size (adjust the cross sectional area) the rod to provide any margin of safety that you deem necessary to satisfy management, and to meet your obligations to society-at-large.

FAILURE RATE

Sometimes components fail by burning out, aging or wearing out, and not by breaking. In this type of a failure analysis, the safety factor is not meaningful. We must cope with wear, aging, or burn out by using some other methods. To begin let's recognize that all components do not wear out at the same time. Some people get 50,000 miles on their car before the brakes wear out, and others may get 60,000 miles or even more. Some people can use their personal computers for several years before a component fails. Other folks have problems after only a few months of service. To handle these differences in service before failure, we introduce what is known as a mortality curve as illustrated in Fig. 3.

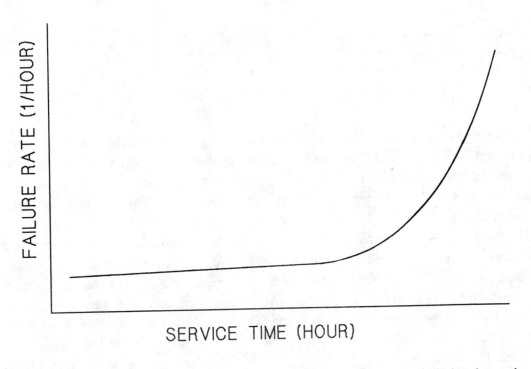

Fig. 3a Mortality curve for mechanical components showing failure rates with time in service.

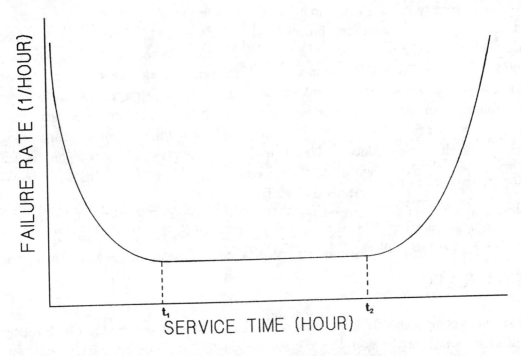

Fig. 3b Mortality curve showing the failure rate for electronic components with time in service.

To introduce the concept of failure rate FR, let's keep records on the performance of a large number of components that are placed in service. We will begin our record keeping with N_0 components that are placed in service at time $t = 0$. After some arbitrary time t, some number N_f will have failed, and the remainder N_s will have survived. Clearly at any arbitrary time:

$$N_f + N_s = N_0 \qquad (4)$$

With time, N_f increases and N_s decreases, but the sum remains constant and equal to N_0. The failure rate FR is determined from:

$$FR = N_f / (N_0 \, t) \qquad \text{when } t > 0 \qquad (5)$$

When we plot the results of Eq. (5) with respect to time, we obtain the component mortality curves presented in Fig. 3. For mechanical components (see Fig. 3a), where initial testing is performed on the assembly line, the failure rate is small when the product is new. However, the failure rate increases non-linearly with time in service because parts begin to wear, or begin to corrode, or are abused.

The failure rate for electronic components can be characterized by the well known bathtub mortality curve presented in Fig. 3b. This curve exhibits three different regions or interest each due to a different cause. The high failure rates in the first region $t < t_1$, are due to components manufactured with minute flaws that were not discovered during inspection and testing at the factory. In the center region of the curve, for $t_1 < t < t_2$, the failure rate FR_0 is very low and nearly constant with time. Clearly this is the best region to operate, if we are to maximize the reliability of the product. Near the end of the service life when $t > t_2$, the failure rate increases sharply with time as the components begin to fail due to the effects of aging.

For high reliability systems it is important to eliminate the inherently defective components associated with early times of operation. For mechanical systems this is accomplished by inspection and testing the performance of the component to eliminate the flawed components produced in manufacturing. Life is not so easy for the electronic component manufacturers, because the feature sizes on the components are so small that inspection requires very high magnification. Even under high magnification, all of the flaws cannot be detected. A better procedure to eliminate the few inherently flawed components is to conduct what is know as burn-in testing prior to incorporating the components into an electronic product. In burn-in testing, we operate all of the components that are produced for a time t_1, and eliminate the flawed components which burn out. The remaining components will then function with the much lower failure rate FR_0.

Clearly, we want to design components with low failure rates. If the failure rates are low, the time between failures will be high. In fact manufactures sometimes cite what is know as the mean time between failure (MTBF) in their product specifications. The MTBF and the failure rate are related by:

$$MTBF = 1/FR \qquad (6)$$

OK. We understand about failure rates of components, and how the failure rates may change over the life of a component. But how do we use this information about the failure rate to establish the reliability of our components to perform safely over the anticipated life of our component or product?

COMPONENT RELIABILITY

To answer the question posed above, we must introduce the concept of probability or chance of occurrence. In considering probability, let's go back to the test that we conducted to determine the failure rate, and make use of the data collected for N_f, N_s, and N_0. We can then determine the probability of survival P_s as:

$$P_s(t) = N_s(t)/N_0 \qquad (7)$$

And the probability of failure P_f is:

$$P_f(t) = N_f(t)/N_0 \qquad (8)$$

Both the probability of survival and the probability of failure are functions of time. With increased usage the probability of survival decreases, and the probability of failure increases.

From Eqs. (7) and (8), it is evident that:

$$P_s(t) + P_f(t) = 1 \qquad (9)$$

And that:

$$P_s(t) = 1 - [N_f(t)/N_0] \qquad (10)$$

If we differentiate Eq. (10) with respect to time t and rearrange the results, we obtain:

$$[N_0/N_s(t)][dP_s(t)/dt] = [-1/N_s(t)][dN_f(t)/dt] = -FR(t)dt \qquad (11)$$

where we consider the term $[1/N_s(t)][dN_f(t)/dt]$ to be the instantaneous failure rate $FR(t)$ associated with a sample size N_s at that instant.

When we integrate Eq. (11) and let $FR(t) = FR_0$, a constant, we obtain a relation for the probability of survival as a function of the failure rate.

$$P_s(t) = e^{-(FR_0 t)} \qquad (12)$$

where e = 2.71828, the exponential number.

Let's consider an example where we can employ Eq. (12) to determine the reliability of a mechanical component. Suppose we have determined from our records of the number of failures of a given component over a long period of time that our failure rate is essentially constant with 1 failure per 10,000 hours of service. Does that sound good to you? Will the reliability be adequate? Let's determine the probability of survival (the reliability) of our mechanical component with time. Substituting into Eq. (12) gives the results for the reliability shown in Table 1.

Table 1
Reliability of a Mechanical Component
for Increasing Service Life with $FR_0 =1/10,000$

Time (1000 h)	Time* (years)	$FR_0(t)$ (unitless)	Reliability $P_s(t)$
1	0.5	0.1	0.905
2	1	0.2	0.819
5	2.5	0.5	0.606
10	5	1.0	0.367
20	10	2.0	0.135
50	25	5.0	6.738×10^{-3}
100	50	10.0	4.540×10^{-5}
200	100	20.0	2.061×10^{-9}

*The conversion from hours to years of service life is based on 2000 hours/year.

The reliability varies significantly depending on the service life. For a short life, say a year, the reliability is reasonable with a 81.9 % chance of surviving. However, for long life, say 10 years, the reliability drops to only 13.5 %, and for very long life 50 to 100 years, failure is almost certain.

While the failure rate of 1 in 10,000 hours looks good in an initial assessment, the reliability which results from this failure rate is disappointing. You cannot be certain that you will have 10,000 hours of service prior to a failure. In fact you have only a probability of 36.7 % to survive for 10,000 hours. If you want a 90 % reliability for a service life of 10,000 hours, the failure rate FR_0 must decrease to about 1 failure in 100,000 hours.

SYSTEM RELIABILITY

The failure of a component may or may not cause the failure of the system. A system may be comprised of several components, and the system reliability will depend on the reliability of the individual components and their arrangement. There are two possible arrangements --- series or parallel. Consider your automobile to illustrate both the series and parallel arrangement of components. In lighting the highway during the night, your auto is equipped with two headlights. This is a parallel arrangement because we have two identical components (the headlights) to perform the same function. If one headlight burns-out, you can still drive. Your visibility is impaired to some degree, but you can still drive. The system has not failed.

On the other hand suppose you want to start your engine. Consider the components that are involved.

- Ignition switch.
- Battery
- Solenoid relay.
- Starter motor.
- Starter motor clutch and gear.
- Engine ignition components (spark plugs and points).

If any of these components should fail, the engine will not start when you turn the key. You need a complete series of successful components for the system to function. Every component must function, and a failure of a single component results in a failure of the entire system.

RELIABILITY OF SERIES CONNECTED SYSTEMS

Let's consider the reliability of a system involving three component which are in a series arrangement as shown in Fig. 4.

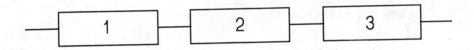

Fig. 4 A system comprised of a series arrangement of three components.

Since all three of the components must operate successfully for the system to perform correctly, the probability of survival of the system (P_s^s) is a product function given by:

$$P_s^s = P_{s1} P_{s2} P_{s3} \qquad (13)$$

where the superscript s refers to the entire system of components.
If we combine Eq. (12) and (13), it is possible to express the system reliability in terms of the failure rate of the individual components as:

$$P_s^s(t) = e^{-(FR_1 + FR_2 + FR_3)t} \qquad (14)$$

As we increase the number of components in a system with a series arrangement, we markedly decrease the system reliability. Note also that the probability of survival continues to decrease with time in service.

Consider the case where we have n components in a series arrangement where n increases from 1 to 1000. Let's also consider that the reliability of each of the n components is the same ($P_{s1} = P_{s2} = \ldots\ldots = P_{sn}$). The results obtained from Eq. (13) are shown in Table 2.

Table 2
Reliability of a System with n Series Arranged Components as a Function of Component Reliability

Number of Components, n	$P_s = 0.999$	$P_s = 0.990$	$P_s = 0.900$
1	0.999	0.990	0.900
2	.0998	0.980	0.810
5	0.995	0.950	0.590
10	0.990	0.904	0.349
20	0.980	0.818	0.122
50	0.950	0.605	*
100	0.905	0.366	*
200	0.819	0.134	*
500	0.606	*	*
1000	0.368	*	*

* System reliability of less than 1%.

Examination of the results presented in Table 2 clearly shows the detrimental effect of placing many components in series. For components with a high reliability (0.999 %), the system reliability drops to about 90 % with 100 series arranged components. However, with components with a lower reliability (90 %), the system degrades to a reliability of less than 1 % when the number of components approaches 50. The lesson here is clear. If you connect a large number of components together in a series arrangement to develop a system, then the individual component reliability must remain extremely high over the entire service life of the system.

RELIABILITY OF SYSTEMS WITH PARALLEL CONNECTED COMPONENTS

Recall our example of a highway lighting system on an automobile that included a pair of headlights. This is a parallel arrangement because both headlights must fail before the system fails. We can drive with one light although the State Troopers might warn us of the dangers to do so.

We represent a system with a parallel arrangement of three components as shown in Fig. 5.

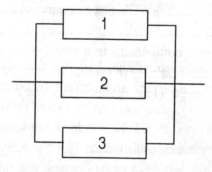

Fig. 5 A system with a parallel arrangement of three components.

These parallel systems are redundant, because all of the components must fail before the system fails. Accordingly, we can write the relation for the probability of system failure as:

$$P_f^s = P_{f1} \, P_{f2} \, P_{f3} \qquad (14)$$

Substituting Eq. (9) into Eq. (14) gives:

$$1 - P_s^s = (1 - P_{s1})(1 - P_{s2})(1 - P_{s3}) \qquad (15)$$

Expanding Eq. (15), and reducing the resulting expression gives:

$$P_s^s = P_{s1} + P_{s2} + P_{s3} - P_{s1} \, P_{s2} - P_{s2} \, P_{s3} - P_{s1} \, P_{s3} + P_{s1} \, P_{s2} \, P_{s3} \qquad (16)$$

Let's examine Eq. (16) to determine how redundancy (the number of parallel components) improves reliability. Suppose we consider P_s of all the components to be the same, but we treat P_s as a variable increasing from 0.1 to 1.0. To show the effect of redundancy we will consider a single component, two components and three components in parallel. The results obtained from Eq. (16) are shown in Table 3.

Table 3
Effect of Degree of Redundancy on Reliability
$$P_{s1} = P_{s2} = P_{s3}$$

Component Reliability	Single Component	Two Components	Three Components
0.1	0.1	0.19	0.271
0.2	0.2	0.36	0.488
0.3	0.3	0.51	0.657
0.4	0.4	0.64	0.784
0.5	0.5	0.75	0.875
0.6	0.6	0.84	0.936
0.7	0.7	0.90	0.973
0.8	0.8	0.96	0.992
0.9	0.9	0.99	0.999
1.0	1.0	1.00	1.000

Examination of the results presented in Table 3 shows that the degree of redundancy improves the reliability of a system with a parallel arrangement of components. If you have a relatively poor component reliability, say 40 %, and you employ two of these components in a parallel arrangement the system reliability improves to 64 %. Add a third component to the parallel connected arrangement, and the reliability increases further to 78.4 %. Keep adding components and you continue to improve the reliability, although it is a case of diminishing returns.

The improved reliability from employing parallel (redundant) components in building a system is a distinct advantage. But we rarely, if ever, get a free lunch. The corresponding disadvantages are the increased costs, and the added power, weight and size of the system. These are significant disadvantages, and redundant design is used only when component reliability is too low for satisfactory system performance, or when a failure produces very serious consequences.

For a more complete discussion of probability of survival and system reliability see reference [1].

EVALUATING THE RISK

When the space shuttle Challenger exploded during launch on January 28, 1986, a Presidential Commission was established to [2]:

1. Review the circumstances surrounding the accident to determine the probable cause or causes of the accident.
2. Develop recommendations for corrective or other actions, based on the Commission's findings and determinations.

Richard Feynman, a Nobel Prize winning physicist from the California Institute of Technology, was appointed to the Commission. Dr. Feynman, near the end of his career and his life, took the appointment very seriously and devoted his entire time and energies to the investigation. One of his many contributions was to determine the probability of failure of the space shuttle system.

During the course of this investigation, Dr. Feynman asked three engineers and their manager to assess the probability of failure P_f for a mission due to the failure of one component --- the shuttle's main rocket engine. Secret ballots by the engineers indicated two estimates of $P_f = 1/200$ and one of $P_f = 1/300$. The manager under pressure finally estimated the risk at $P_f = 1/100,000$. There was such a large difference between management and the engineers at NASA that Dr. Feynman concluded that "NASA exaggerates the reliability of its products to the point of fantasy."

The safety officer for the firing range at Kennedy Space Center, who had been under considerable pressure to remove the distruct charges from the shuttle, did not believe the reliability figures cited by NASA. He had collected data for all of the 2900 previous launches using solid rocket boosters. Of this total 121 had failed. This data provide a very crude estimate of $P_f = (121)/(2900) = 0.042$ or about one chance for failure in every 24 launches. The safety officer considered this high risk of failure to be an upper bound because improvements had been made since the early launches that improved the reliability of the solid booster motors. Also the pre-launch inspections on the shuttle were very thorough. His estimate of risk accounting for taking these improvements was $P_f = 1/100$.

Dr. Feynman [3], after extensive interviews with engineers, reliability experts, managers from NASA, and many subcontractors, concluded that the shuttle "flies in a relatively unsafe condition with a chance of failure of the order of 1 %".

Risk assessment is a difficult task. Each component in a complete system must be evaluated to ascertain its failure rate. Then the components are placed in either a series or parallel arrangement to determine the system reliability which may be lower or higher than the component reliability.

Testing to determine component reliability is possible in some instances where the components are relatively inexpensive. However, establishing reliability estimates by testing requires a very large number of tests which usually destroys the component. If you want to show a probability of failure less than $P_f = 1/1000$, you must test considerably more than 1000 components to failure. Obviously, we could not test more than 1000 booster rockets, where the cost of a single test would be several millions of dollars.

Often the probability of failure of a component must be estimated based on previous experience with similar applications. In some instances, we can compute probability of failure, but these calculations require considerable knowledge of the spectrum of loading and the ability of the material to resist fracture.

When analytical methods are inadequate and engineering judgment is required to assess the probability of failure, the estimate should be made by a senior engineer with considerable experience and expertise. Even then the estimate should be pessimistic rather than optimistic. A frank and honest estimate of P_f, based on all of the data and knowledge available, is much better than unrealistic appraisals that give an unwarranted feeling of safety.

HAZARDS

When we design a product, there usually are risks involved in either the production of the product or its use in the marketplace. The public-at-large is exposed. What can you do to minimize the risk? We have discussed briefly the concept of safety factor, component and system reliability, and evaluating the risk. There is one more area for you to cover --- that is to recognize the hazards involved. You must clearly recognize the hazards before taking the necessary precautions in your designs to minimize the risks associated with them.

Mowrer [4] has a complete list of hazards and an extended discussion of each. You are encouraged to read Mowrer's chapter in reference [4], and to use the extensive check lists incorporated in his coverage. We will provide a very brief excerpt from this material to assist you in recognizing the many hazards that you should consider when you design a product.

The list of hazards includes:

1. Dangerous chemicals and chemical reactions.
2. Exposure to electrical circuits.
3. Exposure to high forces or accelerations.
4. Explosives and explosive mixtures.
5. Fires and excessive temperature.
6. Pressure.
7. Mechanical hazards.
8. Radiation.
9. Noise.

Chemicals can be nasty and you must appreciate the extreme dangers of exposures to certain toxic substances. Do you remember the leak that developed in a storage tank in Bhopal, India in 1984? The storage tank contained the chemical methyl isocyanate, a toxic ingredient, used in the manufacture of pesticides. It was a big leak, and about 80,000 pounds of the chemical was released to the environment. 3,000 nearby inhabitants were killed, 10,000 permanently disabled and another 100,000 injured. The problem was not in the design of the tank, but in the training of the individuals responsible for the plant maintenance and operation. Nevertheless, a catastrophic accident occurred, because several workers and managers, in positions of responsibility, did not understand the dangers of this very toxic chemical.

What about exposure to electrical circuits? Most everyone has been shocked by the standard 120 volt, 60 cycle supplied by the local utility company. Why worry? You should worry because electrical shocks are dangerous. Yes, even the 120 volt supply can cause big problems. Your body acts like a resistor and limits the current flowing from the voltage source through your hands, arms, legs, etc. The problem is the resistance of your body is a variable. It depends on the moisture on your hands or on the type of soles on your shoes, or the moisture on the ground. If your hands are dry, your shoes have rubbers soles, and you are standing on a dry floor when you touch one wire of the circuit with only one hand, then you probably not feel much because you have arranged a very high resistance path to ground. The current flow through your body will be very small. However, if your hands are wet and you touch both the wires from the supply, one with the left hand and the other with the right hand, you have placed 120 volts across your heart. You can be electrocuted quite well with 120 volts under these conditions, where the moisture on your hands greatly reduces the effective resistance of the body.

Do not take chances with electricity. Insulate the operator from the circuits, preferably with two independent layers of insulation. High voltages are even more serious than low voltages because the currents flowing through one's body increase dramatically. When the current flow through the human body increases to about 10-100 milli-amperes, there are very serious consequences to the respiratory muscles. Higher currents of 75-300 milli-amperes produce problems in the operation of your heart. We can increase the electrical current (I) flowing through the body in two ways: First, by increasing the applied voltage (V), and second, by decreasing the resistance (R) offered by the body (I = V/R   Ohm's Law). Using electrical insulation in the design of products with electrical power increases the resistance R, and decreases the current I.

Some folks think that low voltages (5 or 10 volts) are safe, because you barely get a tingle when you touch a low voltage circuit. However, some low voltage circuits particularly on high performance computers carry substantial (100+ amps) currents. If you short a circuit with high current flows, you will strike an arc and generate lots of heat. The flash of the arc can damage your eyes, and the heat can produce serious burns. Insulate and shield even low voltage circuits in your designs.

High forces and high accelerations (or decelerations) go hand in hand. Newton's second law requires the connection (F = ma). If we suddenly apply the brakes on an automobile, the car decelerates and our passenger (without seat belt or air bags) is thrown into the windshield. You must always be concerned with acceleration or deceleration, because of their effects on the human body. Military pilots are trained to withstand high accelerations (high Gs). A good pilot can pull 7 or maybe 8 Gs before they loss consciousness. But a civilian will become irritated at less than 2 Gs. If you want to get an acceleration thrill, go to an amusement park and ride the roller coaster. There is no need to incorporate high accelerations into the design of most new products. A reasonably hot sports car, that accelerates to 60 MPH in six seconds, requires an acceleration of only 0.46 Gs. Be careful and keep both acceleration and decelerations low when you design products that move and change velocity.

With the bombing of the Federal Building in Oklahoma City, we have all become aware of the disastrous effects of a large explosion. The gas pressures that are generated destroy very substantial buildings, break glass in a very large region and kill or injure lots of people. We also are aware of the common explosives like dynamite and ANFO (ammonia nitrate and fuel oil). In most designs we do not encounter a need to accommodate these explosives. What we must be aware of are other less apparent agents that act like explosives under special circumstances. Fuels like natural gas propane, butane, etc. can leak and combine with air to produce an explosion when ignited. Boating accidents are common when gasoline leaks in an engine compartment. The resulting mixture of gasoline fumes and air explode when subjected to a small spark destroying the boat and killing or injuring the passengers. Still another unusual source of fuel for an explosion is dust. When handling large quantities of a combustible solid, dust (fine particles) is generated. If these particles are suspended in air the resulting mixture will explode when ignited. A grain elevator exploded in Westwego. Louisiana in 1977 killing 35 people when an explosive mixture of combustible particles (dust) from the grain and air was ignited.

Over a million fires occur in the U. S. every year. Some are vehicle fires (400,000) and other are structures (650,000) which begin to burn. Some folks die in the fire (4700), and a lot more are injured (28,700). Clearly fires are a serious problem. People get killed, injured or traumatized, and property is lost (eight or nine billion dollars worth per year). Some of these fires are simply stupid. He or she who smokes in bed will some night fall asleep and set the house on fire. A surprisingly large number of fires (about 100,000 per year) are deliberately set, apparently to collect the fire insurance, to take revenge, or as a sick kind of diversion.

Engineers have a responsibility to decrease the number of fires which result from the products which we design. Did these fires start with an appliance or a motor that overheated for some reason or other? Determine the reason for the overheating, and redesign the product so that overheating will not occur. Why did the automobile burst into flames? Did a fitting on the gasoline lines leak? Redesign to eliminate the fittings, or specify a fitting that will not fail under the prevailing conditions. There are many solutions to the problem of fires in the U. S. We, as a nation, are far too casual about fires, and we tolerate carelessness in personal practices, poor design in products intended for the home, and fraud (arson) to collect fire insurance reimbursement for lost property.

We use pressurized fluids for many good reasons, and in most cases our pressure vessels (the containers that hold the fluids) perform very well. It was not always the case. Back in the winter of 1919, in Boston, a huge tank about 90 feet in diameter and 50 feet high fractured. It contained two million gallons of molasses which flooded the local area. Twelve people died and another 40 were injured in the accident.

It is relatively easy to design a pressure vessel today. In fact, the American Society for Mechanical Engineers (ASME) has developed a code that engineers follow in their design to produce pressure vessels that can be certified as safe for service. But today, we still occasionally encounter tanks that fail. The difficulty is usually in the steel plates that are welded together to produce the tank. Both the plates and the welds have to be tough at low temperatures, and the welds have to be free of large flaws. If you have the responsibility of designing a pressure vessel, follow the ASME code, make certain that the welding procedures followed produce crack free welds, and employ a steel which is sufficient strong and tough. If you intend to become a mechanical engineer, you will have the opportunity to learn how to design pressure vessels, select suitable materials for their construction, and specify welding procedures in courses presented later in the curriculum.

Mechanical hazards are features which exist on products which can cause injury to someone nearby. An example might be a fan used to cool the room. Is the fan blade adequately guarded, or can you stick your finger into the rotating blades? The cabinet that you have designed to hold a special tool has a sharp corner at hip level. Can someone walk into the cabinet and break skin on that corner? You have designed a center post crane to lift material, and move it over a 25 foot diameter area. Will the center post be stable under all conditions of loading or will it collapse? You have designed a new pizza machine that rolls the dough into sheets exactly 0.2 inches thick. Have you provided protection for the entrance to the rolls that prevent the pizza person from inserting his or her fingers in the machine? You have designed a wonderful guard that prevents a person from exposing their hands and arms in operating a punch press. However, the guard is attached to the machine with two screws. Have you used locknuts or safety wires to insure that the screws will not loosen during the operation of the press?

There are many mechanical hazards that we encounter in designing equipment and products. Always examine each component and look for sharp points, cutting edges, pinch points, rotating parts, etc. Do not count on peoples good judgment. If they can stick a finger into the machinery, even if it is foolish to do so, you can be certain that sooner or later someone will.

Radiation hazards are due to electromagnetic waves to which we are exposed. The damage produced depends on a lot of different factors such as:

- The strength of the source.
- The degree to which the emitting radiation is focused.
- The distance from the source.
- The shielding between the source and the object being radiated.
- The time of exposure to the radiation.

We have divided the radiation spectrum into several different regions --- very short wave length, visible light, infrared, and microwave radiation. The short wave length radiation is the most dangerous (x-rays, gamma rays, neutrons and etc.) with serious health risks for overexposures. The hazards due to UV, visible light and IR is usually due to excessive exposure where serious burns to the skin can occur. For very intense radiation, even short exposures are detrimental to the eyes. Microwave radiation is absorbed into the body, and can result in the heating of one's internal organs. We are not certain how the organs can dissipate this heat, and the effect of the localized increases in temperature that result because of this heating. It is prudent to avoid exposure to microwaves. In the U. S., OSHA (Offices of Safety and Health Administration) has decreed a power density limit of 10 mW/cm^2 for a period of more than 6 minutes when dealing with microwave energy.

SUMMARY

When we introduce a product in the marketplace, there is usually some risk involved to the workers making the product, our customers, or society-at-large. Hopefully the risks will be small and acceptable to all concerned, because of the significant benefits produced by using the product and/or the sociotechnical system. In determining if the risk is acceptable, the public needs an accurate and honest assessment of the consequences of a failure, and the probability of the occurrence of an accident. It is the responsibility of the engineering profession to assist business and governmental agencies in these assessments.

While we must accept risk, we must also make every attempt to minimize it within reasonable constraints on cost and performance. There are many good, solid engineering methods for ensuring safe design. We have briefly introduced the concepts of stress, strength, safety factor and margin of safety to illustrate one approach to minimizing risk. We have also recommended a range of commonly accepted safety factors, and have provided the issues which should be considered in selecting the safety factor to employ when sizing components.

Sometimes we know that failures will occur, and we need to determine the probability of the failure event. To illustrate methods for determining probability of failure, we first introduced the concept of failure rate and the mean time between failures. We indicated that the failure rate could be determined simply by keeping records of the failures of components as a function of time after they were placed in service. These concepts are important, but they should not be confused with the probability of failure or survival.

We next introduced component reliability, and showed how it could be computed from the failure rate. In developing this relation [Eq. (12)], we assumed the failure rate was a constant over the service life. It is important to observe that the probability of survival decreases as the service life is increased and the probability of failure increases with time in service. For reliable service for a long period of time, the failure rate must be very low.

System reliability is different than component reliability. A system is usually composed of many components. If the components are arranged in series, and they must all function for the system to operate correctly, the reliability of the system is given by Eq. 13. It is evident from the data presented in Table 2 that the number of components arranged in a series markedly lowers the reliability of a system.

Sometimes a system must function all of the time. We simply cannot consider a system failure because of very negative consequences. In these instances, we design with redundant components arranged in a parallel system. Redundancy improves system reliability, as shown in Table 3, but at increased cost, and the need for more power, weight and size.

We have presented the example of the explosion of the space shuttle Challenger to illustrate the importance of evaluating the risk. It is often a very difficult problem and an analytical solution for the risk is frequently not possible. Nevertheless, it is a professional responsibility to prepare intelligent, accurate, honest and frank estimates of the risk, and to make certain that all of the principals involved are aware of the consequences of a failure.

Finally, we have introduced a very brief listing of hazards which cause injury or death. Unfortunately the list is relatively long. It is important that you recognize the hazards, and be vigilant in your designs to avoid them. Avoidance often does not require much in the way of calculation from elaborate formulae, but rather a close inspection of the product and a lot of common sense.

REFERENCES

1. Dally, J. W., Packaging of Electronic Systems: A Mechanical Engineering Approach, McGraw Hill, New York, 1990, pp. 288-296.
2. Lewis, R. S. Challenger: The Final Voyage, Columbia University Press, New York, 1988.

3. Feynman, R. "Personal Observations on the Reliability of the Shuttle," Appendix to the Presidential Commission's Report, Ayer Co., Salem, 1986.
4. Mowrer, F. W., Introduction to Engineering Design: ENES 100, McGraw Hill, New York, 1996, pp. 149-172.

EXERCISES

1. Have you been involved in one or more automobile accidents since you began to drive? Were you or anyone else injured or worse yet killed? Estimate the total mileage you have driven over this period and calculate your accident rate. What were the reasons for the accident or accidents? Comment on your driving behavior and its influence on the accident rate.

2. Are the benefits of driving worth the risks of injury or death? Determine your the probability of being killed in a fatal accident while driving this year. Hint: Statistics on fatal accidents are always listed in the World Almanac. If you do not have a recent edition, use an older edition for your data. Make any assumptions necessary for your analysis, but justify each of them.

3. Your coworker designs a tie rod (a tension member) from a ¾ inch steel bar that has a strength of 60,000 psi. If the team leader has indicated that she wants to maintain a safety factor of 2.8, determine the maximum load that can be applied to the tie bar.

4. If a component is designed with a safety factor of 2.6, what is the margin of safety?

5. If we are so smart and we have all of these gee whiz computer programs to determine stresses, why do we need to specify a margin of safety or a safety factor when we design some structural member?

6. Jane is an engineer in the transmission department of the Fink Motor Corp. Her job is to record data on the mileage prior to failure of the new light weight transmission that has been placed in 50,000 of the new 1998 model of the Clunker. She records the following data:

Mileage (1000 miles)	No. Failures N_f
0 - 5	68
5 - 10	37
10 - 20	20
20 - 30	22
30 - 40	21
40 - 50	35
50 - 60	97
60 - 70	564
70 - 80	1485
80 - 90	5677
90 - 100	13,592

Determine the failure rate as a function of service life expressed in terms of mileage, and prepare a graph showing your results.

7. Using the data from the table in Exercise 6, determine the probability of survival and the probability of failure of the transmissions as a function of service life measured in terms of mileage. Prepare a graph of your results.

8. If the transmission in the Clunker was under warranty for 100,000 miles, what would be the consequences for the Fink Motor Corporation?

9. NASA's space shuttle has two solid propellant rocket boosters to provide the thrust required for launch. Is this a redundant system? State your reason for this conclusion.

10. If the probability of failure of a solid rocket propellant booster on the space shuttle is 1/500, determine the reliability of the solid booster system.

11. You are designing a very large computer system for containing the data base for a very large reservation system. It is estimated that at any instant 600 operators will be accessing the data base, and another 3400 operators will soon be ready with their requests for computer availability. The main frames that you plan to employ have a MTBF of 6,000 hours. Present a design of the computer system that will insure that the 600 operators have a 99.9 % probability of being served. In your design show all the calculations, and carefully list your assumptions. Discuss also the costs involved if each mainframe computer is valued at $400,000.

12. Prepare a paper discussing your reaction to the risk assessment of the space shuttle system by NASA. Give arguments for and against the NASA position regarding their position both before and after the accident.

13. Describe an incident where you received an electrical shock. Something obviously went wrong to cause this incident. Please indicate the problem. Was there a design deficiency? What could be done to prevent the incident from reoccurring?

14. You are designing a cabinet-mounted, self-cleaning oven that requires very high power levels to heat its interior surfaces until they are free of all the splattered grease and grime. What precautions can you take as the lead engineer on this design team to insure that the oven will not be the source of a fire during its anticipated 20 year life?

NOTES

CHAPTER 17

ETHICS, CHARACTER AND ENGINEERING

INTRODUCTION

I am writing this chapter on ethics in February 1997, shortly after the Presidential election of 1996. The newspapers and TV news and talk shows are covering many ethical issues currently emanating from Congress and the White House. Ethical lapses occur on nearly a weekly basis keeping the news services busy reporting on questionable behavior by our elected officials. These officials are the folks that we have empowered to write and approve the laws governing the country. The news media has pointed out numerous instances where our leaders have skirted the truth if not the law. Fund raising by both parties lead to questionable practices that were so extensive that a Governmental Affairs Committee chaired by Senator Fred Thompson has requested $6.5 million to cover the cost of the investigations. (The Committee was finally authorized to spend $5 million on the investigation).

How do you feel when you learn that $50,000 or $100,00 appears to be the price for an invitation to the White House and an over night stay in the Lincoln Bedroom?

The lax ethical practices by many of our elected officials has led to a very serious skepticism about the merits of many of the government institutions and agencies. The fact that fewer than 50 % of the eligible voters took time to cast their ballots in the 1996 elections is testimony to the deep skepticism toward our government and our elected officials.

As engineers, we want to preserve, if not enhance, the confidence of the public. To gain public confidence, it is essential that we all behave in an ethical manner every day of every year. Whether we act as individuals or as professionals, consistent ethical behavior is mandatory. Our primary employer, business, must also act in an ethical manner. Since engineering developments are sponsored, financed and controlled by businesses, corporate and engineering behavior are inseparable. Our actions and words always must be above reproach.

CONFUSED ABOUT ETHICAL BEHAVIOR

Lots of folks are confused about ethical behavior today. My reaction to the results of several polls indicate that Americans are less ethical today than in previous periods. Most folks (64 % of a sample of 5000 individuals) admit to lying if it does not cause real damage. An even large percentage of those sampled (74 %) will steal providing the person or business that is being ripped off does not miss it [1]. Most students (75 % high school and 50 % college) admit to cheating on important exams [2]. Most students have not developed a moral code to use as a guide for their behavior. Poor behavior, when I was in high school (1944-47), included making too much noise, gum chewing in class, talking in class, littering, running in the halls, etc. Today these relatively minor transgressions have been replaced with more serious problems involving drug abuse, alcohol abuse, pregnancy, suicide, rape, robbery and assault. Big negative changes. Why?

There are probably many reasons for the degradation of ethical behavior over the past 50 years. Let's discuss two key reasons. First, institutions (family, church and schools) that once taught ethics are much weaker today than 50 years ago. Many families have been torn apart by a very high divorce rate, and single parent homes are common. The idea of love, honor and obey, in sickness and in health, is often ignored.

The influence of our churches have deteriorated with serious declines in membership. Church finances often limit their activities to current membership. The social problems are often with folks who are not known to the church. The public has also become much more secular with no church affiliation.

Our public school administrators are so concerned with possible litigation that they have largely removed the teaching of values from the curriculum. The consequence is that students graduating from high school have not been given an adequate opportunity to develop a personal code of ethics.

The result of the failure of our institutions to teach ethics is that many of our young people are not aware of what is right and what is wrong. All actions are not gray. Many are plainly right, and many are obviously wrong.

The second reason for the deterioration in ethics on the part of our younger population is the current state of ethical standards in society. We are exposed to cynical behavior and selfish attitudes on TV, magazines, and newspapers every day. The bad guy or gal often gets away with severe crimes with minor penalties. Minor infractions are frequently ignored. Killers and child sex offenders are paroled after remarkably short periods of incarceration. Popular TV programs often show examples of lax ethics and self indulgence to generate a popular but sick kind of humor. The result is to create confusion in our youth as they try to develop some system for judging right from wrong. Our youth mirrors the behavior of society leading to cynicism, selfish attitudes, and dishonesty and irresponsible conduct.

WHITE, BLACK OR GRAY

Christina Sommers [2] develops an interesting viewpoint regarding what is considered right, wrong and controversial by society-at-large. There are some ethical issues that are not clearly defined. Society has not reached a consensus regarding the correctness of these issues. We can develop cogent arguments for and against a specified viewpoint for each of these issues. Examples of these controversial issues include abortion, affirmative action, capital punishment, etc. Sommers refers to these controversial issues as "dilemma" ethics, which should be distinguished from "basic" ethics. With "dilemma" ethics, society-at-large is not certain about what is right, wrong or gray. There are several attitudes prevalent in society, and sometimes the issue is so important to different segments of our people that demonstrations are organized to elevate one view or another.

We will make no attempt in this textbook to resolve any "dilemma" issues. These issues have been unresolved for decades, and they will require much more time and understanding before our society is willing to reach a consensus necessary for resolution. It is much more productive to treat "basic" ethics where right and wrong or white and black are much more clearly recognized by the majority of society.

It is very easy to define right and wrong. In fact the understanding of right and wrong dates back to Aristotle [3], and fundamental doctrine has not changed much since then. Let's start with the four classic virtues [3]:

- Prudence
- Justice
- Fortitude
- Temperance

Prudence refers to careful forethought, good judgment and discretion. In other words, think before you act, exercise care and wisdom. If you act, will your action cause problems, now or later, for you or anyone else? Are you careful about your remarks about others? One of my rules is not to speak ill of others even when conversing with a close friend. Is the action that you are about to undertake in your best interest? Will that action be appreciated by your family, your friends, your manager, and your peers?

Justice involves fairness, honor, keeping your word, honesty, truthfulness, etc. It is clearly wrong to lie, cheat, steal and to break promises. It is also wrong to tolerate those about you who do so. My behavior with regard to justice is governed by the golden rule. Do unto others as you would have them do unto you. It is a simple rule, but it works.

Fortitude is about courage and persistence. We all know or imagine fortitude in regard to warfare and/or battle. Here warriors stand their ground and exhibit a very special brand of courage. But there are other brands of courage which must be exhibited in more typical circumstances. Will you stay with an idea, even if it is not popular, if you know that it is the "right" approach? Showing determination and resolution is sometimes difficult when the risks of failure are high, or when peers encourage you to abandon your idea.

Temperance of course refers to moderation in drinking alcoholic beverages. But temperance has much wider implications with regard to ethics. Temperance means control of human passions such as anger, lust, hostility, exasperation, and lechery. With regard to food and beverages it implies self control and moderation in consumption. Clearly much is to be gained by avoiding overindulgence in food, drink and sex, and in restraining your emotional extremes.

There are other virtues such as loyalty and obedience that are less commonly discussed these days. With the restructuring and massive downsizing that has taken place in the business world in the past two decades, the concept of company loyalty to the employee and vice versa has been destroyed. Hopefully loyalty to the family remains, at least in those situations where a family remains. The concept of obedience was seriously damaged if not destroyed in this country by the Vietnam conflict. The government ordering young people to fight in a war of questionable value was too much for many to endure. Many young men refused to obey the orders to report for duty and fled the country to avoid persecution.

Because of these events (others can also be added), the virtues loyalty and obedience have shifted from the "basic" to the "dilemma" category. Arguments can be advanced on both sides of the loyalty and the obedience questions. What do you think of being loyal to an employer who will eliminate your position during a period of increasing profits? How do you feel about completing your tax return for the Internal Revenue Service? Talk about straining one's patience!

In addition to the four basic virtues, prudence, justice, fortitude and temperance, we have three religious (theological) virtues. These are:

- Faith
- Hope
- Charity

The theological virtues are well known and we will not elaborate on their importance. We will again emphasize that their merits are definite and irrefutable. It is good to be faithful, hopeful and charitable. If you want to be even more complete you can add other ethical constants such as humility and respect to the list.

While the emphasis in most college courses in ethics deals with social dilemmas where arguments and counter arguments are to be developed, almost all of our individual behavior can be judged by very clear and well understood virtues that are part of "basic" ethics. There is no need to be confused, nor is there any reason to impair your "basic" code of ethics with issues raised in the study of moral relativism.

IT IS THE LAW

The governments (federal, state and local) play a role in ethical behavior, because they generate laws and set public policy. Laws (mostly at the State level) are written to aid people in their pursuit of a morally correct life. The purpose of these laws is to prohibit a well defined set of vices. For example, it is illegal to sell drugs, rob banks, commit burglary, kill or abuse another human, engage in prostitution, or pursue deviant sexual practices (rape).

These laws, and many more like them, serve to create a moral ecology in which we exist [4]. The laws require visible or outward conformity to a publicly accepted moral codes of behavior. They do nothing to inhibit unethical behavior if it is not observed and prosecuted. An unscrupulous person can sell drugs if he or she is not observed in the act. Even if they are observed, they can avoid penalty if they are not prosecuted for one reason or another. Even if they are observed, detained, prosecuted, and convicted they may receive a suspended sentence. The law does not insure that infractions will not occur. The law simply indicates that certain behavior may not be tolerated.

In some instances laws are not a suitable approach for creating a moral ecology. An example of an unsuitable law in this country was prohibition. When it became evident that that law was causing more problems than it solved, it was repealed. The government wanted to make it illegal to consume alcoholic beverages, but society-at-large wanted to drink. Bootlegging, racketeers, and speakeasies followed and it soon became apparent that the law was not a suitable approach to solve a temperance problem. In cases like this, the government uses public policy to discourage people from pursuing a vice. Policies are adapted that limit (but do not preclude) the consumption of alcoholic beverages. Hours of sale are restricted, licenses are required, taxes are imposed, the number of outlets are constrained, etc.

Public policies are conveyed by means of rules and regulations, not laws. The policies should be crafted to strengthen families, communities and churches by discouraging, but not prohibiting, irresponsible conduct.

ETHICAL BUSINESS PRACTICES

Corporations are entities that under law are treated as persons. As a person, a corporation has the right to conduct business and the obligation to pay taxes. They are essentially corporate persons, but not natural persons. Just as individuals are judged by their ethical standards, corporations are judged by their ethical conduct [5]. The chief executive officer (CEO) for a corporation sets the moral and ethical tone for the organization. Any large business is organized with several layers of management. If the CEO operates the business with the highest ethical standards, the middle and lower level managers will follow the example set at the highest level in the corporation. However, if lax ethical standards are permitted, the pattern is set and sloppy ethics will become the corporate style and standard.

Baker in an excellent paper [5], has developed a list of issues that occur daily in a typical corporation. The manner in which a corporation deals with these issues establishes its character. We include Baker's list below:

1. How we treat people, employees, applicants, customers, shareholders, and the folks who live near our facilities.
2. Are we free of systemic, or individual, practices of discrimination?
3. How do we spend our shareholders' money on our expense accounts as an institution and as individuals?
4. What is the level of quality that goes into our products? Do we meet our customers' expectations for quality? Do we meet our own standards for quality?
5. What is our concern for safety, not only for our employees, but also for our customers and our neighbors in communities in which we operate?
6. How ethically do we compete?
7. How well do we adhere to the laws of the locales, regions, and nations in which we do business?
8. What is acceptable gift-giving and gift-taking?
9. How honest are our communications to our employees and our public advertising?
10. What are our corporate and personal positions on public policy issues and how do we promote those positions?

This is a long list and I am certain that we could add more items to Baker's catalog of concerns. But adding more items is not as important as recognizing that corporate decisions are necessary on a daily basis by the managers and employees. These day to day decisions determine the corporate character. The ethical judgments that serve as the basis for most of the decisions will depend on the individuals' personal values (virtues) and experience. The quality and consistency of the ethics applied in these decisions will markedly affect the success of the corporation, its management, and its employees.

HONOR CODES

We will not lie, steal or cheat, nor tolerate among us anyone who does [6]. This is a 14 word honor code that cadets learn almost as soon as they step foot onto a military academy or college campus that has strong ties to the military (Virginia Military Institute and the Citadel). Our military academies (Navy, Army and Air Force) are educational institutions that serve to train career officers for the services. Honor and integrity is a very large and an important element in their educational program.

Honor codes are much more widespread than the military institutions. There is a growing trend to introduce honor codes on campuses where they are absent, and to strengthen them on campuses where they already exist [7]. An honor code typically forbids lying, cheating and stealing. When an infraction of the code is noted by a peer, the code requires that student to bring the case forward. The case is heard by a student committee. Faculty members may testify if called upon to do so, but they have no control over the proceedings. The penalties imposed on those found guilty of breaking the honor code are determined by the student committees. University administrators implement the decision of the student discipline committees.

The honor codes are very well respected by the student body, and appreciated by the faculty. Students generally support the honor codes because they are concerned over cheating by their peers. They very much want to play the game fairly and on a level playing field. When the honor system is established, the students handle the responsibility for academic integrity with careful consideration. Punishments are handed out only after complete investigations involving representation of both sides of the case. In some instances lawyers are retained by the individual being judged by the student committee.

The faculty are also very much in favor of the honor system. They never have appreciated proctoring exams. With the increase in cheating over the past 25 years, the proctoring task has evolved into a policing task which is very distasteful. The advantage of the honor code to the faculty is the unproctored examination where the responsibility for policing has been transferred to the students. The students often are permitted to take the subject exam at a time and place of their choosing.

Of course honor codes are not perfect. Some students cheat in spite of the code. The military academies, where the code is the most strict and the most rigorously enforced, have suffered the most notorious lapses in standards. Scandals at the U. S. Naval Academy have been the most prevalent and widespread in recent years with as many as 133 midshipmen involved in cheating on an electrical engineering exam. To make matters worse, the administrators at the academy appear to have messed with the system, and in doing so the administrators lost the trust of the midshipmen.

The lesson from the academies with regard to the effectiveness of the honor codes is that rigid codes and rules reduce cheating. But if the students see a way of beating this very rigid system, they will occasionally take advantage of the opportunity. Honor codes are most effective when they produce an ambience of trust between the students, faculty, and the administration. Cheating, lying, and stealing will generally be reduced. For honest students the scoring on exams will be fairer. For faculty the distasteful task of policing is eliminated. The feeling of trust between the faculty and the students is enabling. With time and prolonged success, the honor code and the trust that it engenders becomes an essential part of the educational process.

CHARACTER

I had the privilege of teaching for a year (1995-96) at the U. S. Air Force Academy. It was a wonderful experience for me for many reasons both personal and professional. One of the most important reason was the opportunity I had in participating in a very well planned approach to incorporate character building into the educational process. Character was a key educational outcome in every class that we taught at the Academy.

It is not possible to transplant character into an individual, and it is not possible to accomplish much in efforts to explicitly "teach" character. Character comes implicitly as we as individuals develop a personal set of ethical responsibilities. In a book like this one, we can list the attributes that form the foundation for a persons character. Let's consider some of these attributes:

- Be committed to excellence. This attribute simply means do the best that you can do in both your personal and professional endeavors. To make this commitment you have to assess your capabilities. How good are your skills? How determined are you in achieving your goals? How much time can you commit without endangering your health? If you know yourself, you can measure your achievements on a realistic scale. Try to use the full measure of that scale.

- Respect the dignity of everyone. Respect is the foundation for all achievements. We work, live and play in a diverse world. We must appreciate everyone that we encounter in life. Race, gender, ethnicity and religion are not criteria for judgment. Recognize the potential of the folks with which you work and study. Support and encourage them and carefully avoid demeaning criticism. Teamwork, so essential in today's workplace, requires that you accept the differences inevitable in our diverse population, and that you fully embrace fellow team members.

- Integrity. A single word describes a vitally important attribute of character. Integrity means that you decide to do the right thing solely because it is the right thing to do. You often face decisions that test your integrity. Do you instinctively make the right (honorable) decision. For example, as you parked last night the bumper of your car scraped the fender of the adjacent car. No one observed your accident. Do you leave a note with your name, address, and phone number? Or do you split and drive to another nearby parking lot? How many parking lot scrapes can you find on your car? Did you find a note from the party inflicting the damage claiming responsibility? A person with integrity will consistently make the correct decision, not because it is right for their personal benefit, but because it is clearly the "right" course of action. They "walk the talk".

- Be decisive. When you encounter a problem in engineering or your personal life, there is an information gathering period followed by a decision. Some folks do not want to make that decision. They prolong the information gathering period; they procrastinate;

they hum and haw; they pass the buck. When you establish the facts, evaluate them and make a timely decision. It will be your decision if you made it in isolation. If you made it within the framework of a team, it should represent a consensus of the members of team.

- Take full responsibility. Decisions are made and actions follow. Often you have a winner, but sometimes it is a loser. If it was a loser and you participated in the decision process, it is your responsibility. Step up and accept that responsibility, and lead the effort to fix the problem. Never under any circumstances begin to participate in a finger pointing exercise. Finger pointing does not resolve problems; it exacerbates them.
- Be temperate. Self-discipline is an essential part of character. Eat, drink and be merry in moderation. Over indulgence is disgusting. No one appreciates a drunk or a glutton. Self discipline insures control of the human passions, sensual pleasures, anger, rage and frustration. With self control it is possible to attain consistently high levels of achievement.
- Exhibit fortitude. On many occasions we encounter significant difficulties in completing the task at hand. It is easy to give up and divert our attention to more pleasurable undertakings. We can quickly forget the missing assignment, the incomplete solution, the late review, or the unfinished drawing. Who is to know? Stamina, mental toughness and discipline are attributes that keep us on task. Stay with the job, and not only complete the work, but do it very well.
- Understand the significance of spiritual values. Most, but not all, of us have a faith. We endorse a set of theological beliefs. Your beliefs and your church may be different than mine. Nevertheless, it is essential that I respect your convictions, and you should respect mine. We all need to be sensitive to the important role that religion occupies in the mental comfort of most of our people. We must accommodate a diversity of beliefs by supporting the right of an individual to choose his or her faith, and to pursue the ceremonies offered by the church representing this faith.

ETHICS OF ENGINEERS

We have devoted most of this chapter to a discuss of issues pertaining to individual behavior, ethics, virtue, morals, and character. A personal code of ethics is the cornerstone. Professional ethics are also important, but they must begin only after a person has developed a well understood personal code of ethics. Hopefully, the preceding sections of this chapter will be helpful in any attempt that you make to establish a sense of right and wrong.

There are several codes of ethics for engineers. Most of the founding societies of engineering have developed and distribute codes, guidelines for professional conduct, and faith statements. The codes of ethics for the different engineering societies are all very similar. Each society has their own code, more to insure a complete distribution of the code than to pursue unique ethical issues.

Let's consider the Code of Ethics of Engineers, presented in Fig. 1, that is sponsored by the Accreditation Board for Engineering and Technology (ABET). The code is divided into two parts. The first part deals with four fundamental principles, and the second part contains seven fundamental canons. This code was first advanced by the Engineers' Council for Professional Development in

1977. The fact that the code is 20 years old indicates that professional ethical values are as constant as personal ethical values.

*Accreditation Board for Engineering and Technology**

CODE OF ETHICS OF ENGINEERS

THE FUNDAMENTAL PRINCIPLES

Engineers uphold and advance the integrity, honor and dignity of the engineering profession by:

I. using their knowledge and skill for the enhancement of human welfare;

II. being honest and impartial, and serving with fidelity the public, their employers and clients;

III. striving to increase the competence and prestige of the engineering profession; and

IV. supporting the professional and technical societies of their disciplines.

THE FUNDAMENTAL CANONS

1. Engineers shall hold paramount the safety, health and welfare of the public in the performance of their professional duties.

2. Engineers shall perform services only in the areas of their competence.

3. Engineers shall issue public statements only in an objective and truthful manner.

4. Engineers shall act in professional matters for each employer or client as faithful agents or trustees, and shall avoid conflicts of interest.

5. Engineers shall build their professional reputation on the merit of their services and shall not compete unfairly with others.

6. Engineers shall act in such a manner as to uphold and enhance the honor, integrity and dignity of the profession.

7. Engineers shall continue their professional development throughout their careers and shall provide opportunities for the professional development of those engineers under their supervision.

345 East 47th Street New York, NY 10017

*Formerly Engineers' Council for Professional Development. (Approved by the ECPD Board of Directors, October 5, 1977)

Fig. 1 Code of ethics recommended by ABET.

The fundamental principles seek to insure that the engineer will uphold and advance the integrity, honor and dignity of the profession. These goals are common to the personal goals covered in the previous sections; however, the approach to achieve the goals is different.

1. We are selective in the use of our knowledge and skill so as to ensure that our work is of benefit to society.
2. We are honest, impartial and serve our constituents with fidelity.
3. We work hard to improve the profession.
4. We support the professional organizations in our engineering discipline.

The seven fundamental canons in the ABET Code of Ethics of Engineers, presented in Fig. 1, are explained in considerable detail in guidelines that are used to clarify the relatively brief statements that represent the "regulations" that shape our professional behavior. With the permission of ABET, we have reproduced the guidelines below:

ACCEDITATION BOARD FOR ENGINEERING AND TECHNOLOGY SUGGESTED GUIDELINES FOR USE WITH THE FUNDAMENTAL CANONS OF ETHICS

1. Engineers shall hold paramount the safety, health and welfare of the public in the performance of their duties.
    a. Engineers shall recognize that the lives, safety, health and welfare of the general public are dependent upon engineering judgments, decisions and practices incorporated in structures, machines, products, processes and devices.
    b. Engineers shall not approve nor seal plans and/or specifications that are not of a design safe to the public health and welfare and in conformity with accepted engineering standards.
    c. Should the Engineers' professional judgment be overruled under circumstances where the safety, health, and welfare of the public are endangered, the Engineers shall inform their clients or employers of the possible consequences and notify other proper authority of the situation, as may be appropriate.
        (1) Engineers shall do whatever possible to provide published standards, test codes, quality control procedures that will enable the public to understand the degree of safety or life expectancy associated with the use of the design, products, and systems for which they are responsible.
        (2) Engineers will conduct reviews of safety and reliability of the design, products or systems for which they are responsible before giving their approval to the plans for the design.
        (3) Should Engineers observe conditions which they believe will endanger public safety or health, they shall inform the proper authority of the situation.
    d. Should Engineers have knowledge or reason to believe that another person or firm may be in violation of any of the provision of these Guidelines, they shall present such information to the proper authority in writing and they shall cooperate with the proper authority in furnishing such further information as may be required.

       (1) They shall advise proper authority if an adequate review of the safety and reliability of the products or systems has not been made or when the design imposes hazards to the public through it use.

       (2) They shall withhold approval of products or systems when changes or modifications are made which would affect adversely its performance insofar as safety and reliability are concerned.

   e. Engineers should seek opportunities to be of constructive service in civic affairs and work for the advancement of the safety, health and well-being of their communities.

   f. Engineers should be committed to improving the environment to enhance the quality of life.

2. Engineers shall perform services only in areas of their competence.

   a. Engineers shall undertake to perform engineering assignments only when qualified by education or experience in the specific technical field of engineering involved.

   b. Engineers may accept an assignment requiring education or experience outside their own fields of competence, but only to the extent that their services are restricted to those phases of the project in which they are qualified. All other phases of such a project shall be performed by qualified associates, consultants, or employees.

   c. Engineers shall not affix their signatures and/or seals to any engineering plan or document dealing with subject matter in which they lack competence by virtue of education or experience, nor to any plan or document not prepared under their direct supervisory control.

3. Engineers shall issue public statements only in an objective and truthful manner.

   a. Engineers shall endeavor to extend public knowledge, and to prevent misunderstandings of the achievements of engineering.

   b. Engineers shall be completely objective and truthful in all professional reports, statements or testimony. They shall include all relevant and pertinent information in such reports, statements or testimony.

   c. Engineers when serving as expert or technical witnesses before any court, commission, or other tribunal, shall express an engineering opinion only when it is founded upon adequate knowledge of the facts in issue, upon a background of technical competence in the subject matter, and upon honest conviction of the accuracy and propriety of the testimony.

   d. Engineers shall issue no statements, criticism, nor arguments on engineering matters which are inspired or paid for by an interested party, or parties, unless they have prefaced their comments by explicitly identifying themselves, by disclosing the identities of the party or parties on whose behalf they are speaking, and by revealing the existence of any pecuniary interest they may have in the matters.

   e. Engineers shall be dignified and modest in explaining their work and merit, and will avoid any act tending to promote their own interests at the expense of integrity, honor and dignity of the profession.

4. Engineers shall act in professional matters for each employer or client as faithful agents or trustees, and shall avoid conflicts of interests.

a. Engineers shall avoid all known conflicts of interest with their employers or clients and shall promptly inform their employers or clients of any business association, interest, or circumstances which could influence their judgment or the quality of their services.

b. Engineers shall not knowingly undertake any assignments which would knowingly create a potential conflict of interest between themselves and their clients or their employers.

c. Engineers shall not accept compensation, financial or otherwise, from more than one party for services on the same project, nor for services pertaining to the same project, unless the circumstances are fully disclosed to, and agreed to, by all interested parties.

d. Engineers shall not solicit nor accept financial or other valuable considerations, including free engineering designs, from material or equipment suppliers for specifying their products.

e. Engineers shall not solicit or accept gratuities, directly or indirectly, from contractors, their agents, or other parties dealing with their clients or employers in connection with the work for which they are responsible.

f. When in public service as members, advisers, or employees of a governmental body or department, Engineers shall not participate in considerations or actions with respect to services provided by them or in their organization in private or product engineering practice.

g. Engineers shall not solicit or accept an engineering contract from a governmental body on which a principal, officer or employee of their organization serves as a member.

h. When, as a result of their studies, Engineers believe a project will not be successful, they shall so advise their employer or client.

i. Engineers shall treat information coming to them in the course of their assignments as confidential, and shall not use such information as a means of making personal profit if such action is adverse to the interests of their clients, their employers or the public.

   (1) They will not disclose confidential information concerning the business affairs or technical processes of any present or former employer or client or bidder under evaluation without his consent.

   (2) They shall not reveal confidential information or findings of any commission or board of which they are members.

   (3) When they use designs supplied to them by clients, those designs shall not be duplicated by the Engineers for others without express permission.

   (4) While in the employ of others, Engineers will not enter promotional efforts or negotiations for work or make arrangements for other employment as principals or to practice in connection with specific projects for which they have gained particular and specialized knowledge without the consent of all interested parties.

j. The Engineer shall act with fairness and justice to all parties when administering a construction (or other) contract.

k. Before undertaking work for others in which the Engineer may make improvements, plans, designs, inventions, or other records which may justify copyrights, or patents, they shall enter into a positive agreement regarding ownership.

l. Engineers shall admit and accept their own errors when proven wrong and refrain from distorting or altering the facts to justify their decisions.

m. Engineers shall not accept professional employment outside of their regular work or interest without the knowledge of their employers.

n. Engineers shall not attempt to attract an employee from another employer by false and misleading representations.

o. Engineers shall not review the work of other Engineers except with the knowledge of such Engineers, or unless the assignments/or contractual agreements for the work have been terminated.

   (1) Engineers in governmental, industrial or educational employment are entitled to review and evaluate the work of other engineers when so required by their duties.

   (2) Engineers in sales or industrial employment are entitled to make engineering comparison or their products with the products of other suppliers.

   (3) Engineers in sales employment shall not offer nor give engineering consultation or designs or advice other than specifically applying to equipment, materials or systems being sold or offered for sale by them.

5. Engineers shall build their professional reputation on the merit of their services and shall not compete unfairly with others.

   a. Engineers shall not pay nor offer to pay, either directly or indirectly, any commission, political contribution, or a gift, or other consideration in order to secure work, exclusive of securing salaried positions through employment agencies.

   b. Engineers should negotiate contracts for professional services fairly and only on the basis of demonstrated competence and qualifications for the type of professional services required.

   c. Engineers should negotiate a method and rate of compensation commensurate with the agreed upon scope of services. A meeting of the minds of the parties to the contract is essential to the mutual confidence. The public interest requires that he cost of engineering services be fair and reasonable, but not the controlling consideration in selection of individuals or firms to provide these services.

      (1) These principles shall be applied by Engineers in obtaining the services of other professionals.

   d. Engineers shall not attempt to supplant other Engineers in a particular employment after becoming aware that definite steps have been taken toward the others' employment or after they have been employed.

      (1) They shall not solicit employment from clients who already have Engineers under contract for the same work.

      (2) They shall not accept employment from clients who already have Engineers for the same work not yet completed or not yet paid for unless the performance or payment requirements in the contract are being litigated or the contracted Engineers' services have been terminated in writing by either party.

      (3) In the case of termination of litigation, the prospective Engineers before accepting the assignment shall advise the Engineers being terminated or involved in litigation.

   e. Engineers shall not request, propose nor accept professional commissions on a contingent basis under circumstances under which their professional judgments may be compromised, or when a contingency provision is used as a device for promoting or securing a professional commission.

f. Engineers shall not falsify nor permit misrepresentation of their, or their associates', academic or professional qualifications. They shall not misrepresent nor exaggerate their degree of responsibility in or for the subject matter of prior assignments. Brochures or other presentations incident to the solicitation of employment shall not misrepresent pertinent facts concerning employers, employees, associates, joint ventures, or their past accomplishments with the intent and purpose of enhancing their qualifications and work.

g. Engineers may advertise professional services only as a means of identification and limited to the following:

(1) Professional cards and listings in recognized and dignified publications, provided they are consistent in size and are in a section of the publication regularly devoted to such professional cards and listings. The information displayed must be restricted to the firm name, address, telephone number, appropriate symbol, names of the principal participants and the fields of practice in which the firm is qualified.

(2) Signs on equipment, offices and at the site of the projects for which they render services, limited to the name of the firm, address, telephone number and type of service, as appropriate.

(3) Brochures, business cards, letterheads and other factual representations of experience, facilities, personnel and capacity to render service, providing the same are not misleading relative to the extent of participation in the project cited and are not indiscriminately distributed.

(4) Listings in the classified section of telephone directories, limited to the name, address, telephone number, and specialties in which the firm is qualified without resorting to special or bold type.

h. Engineers may use display advertising in recognized dignified business and professional publications, providing it is factual and relates only to engineering, is free of ostentation, contains no laudatory expressions or implications, is not misleading with respect to the Engineers' extent of participation in the services or projects described.

i. Engineers may prepare articles for the lay or technical press which are factual, dignified and free from ostentations or laudatory implications. Such articles shall not imply other than their direct participation in the work described unless credit is given to others for their share of the work.

j. Engineers may extend permission for their name to be used in commercial advertisements, such as may be published by manufactures, contractors, material suppliers, etc., only by means of a modest dignified notation acknowledging their participation and the scope thereof in the project or product described. Such permission shall not include public endorsement of proprietary products.

k. Engineers may advertise for recruitment of personnel in appropriate publications or by special distribution. The information presented must be displayed in a dignified manner, restricted to the firm name, address, telephone number, appropriate symbol, names of principal participants, the fields of practice in which the firm is qualified and factual descriptions of the positions available, qualifications required and benefits available.

l. Engineers shall not enter competitions for designs for the purpose of obtaining commissions for specific projects, unless provision is made for reasonable compensation for all designs submitted.

m. Engineers shall not maliciously or falsely, directly or indirectly, injure the professional reputation, prospects, practice or employment of another engineer, nor shall they indiscriminately criticize another's work.

n. Engineers shall not undertake nor agree to perform any engineering service on a free basis, except for professional services which are advisory in nature for civic, charitable, religious or non-profit organizations. When serving as members of such organizations, engineers are entitled to utilize their personal engineering knowledge in service of these organizations.

o. Engineers shall not use equipment, supplies, laboratory nor office facilities of their employers to carry on outside private practice without consent.

p. In case of tax-free or tax-aided facilities, engineers should not use student services at less than rates of other employees of comparable competence, including fringe benefits.

6. Engineers shall act in such a manner as to uphold and enhance the honor, integrity and dignity of the profession.

a. Engineers shall not knowingly associate with nor permit the use of their names nor firm names in business ventures by any person or firm which they know, or have reason to believe, are engaging in business or professional practices of a fraudulent or dishonest nature.

b. Engineers shall not use association with non-engineers, corporations, nor partnerships as 'cloaks' for unethical acts.

7. Engineers shall continue their professional development throughout their careers, and shall provide opportunities for the professional development of those engineers under their supervision.

a. Engineers shall encourage their engineering employees to further their education.

b. Engineers shall encourage their engineering employees to become registered at the earliest possible date.

c. Engineers shall encourage their engineering employees to attend and present papers at professional and technical society meetings.

d. Engineers should support the professional and technical societies of their disciplines.

e. Engineers shall give proper credit for engineering work to those to whom credit is due, and recognize the proprietary interests of others. Whenever possible, they shall name the person or persons who may be responsible for designs, inventions, writings, or other accomplishments.

f. Engineers shall endeavor to extend the public knowledge of engineering, and shall not participate in the dissemination of untrue, unfair or exaggerated statements regarding engineering.

g. Engineers shall uphold the principle of appropriate and adequate compensation for those engaged in engineering work.

h. Engineers should assign professional engineering employees with complete information on working conditions and their proposed status of employment, and after employment shall keep them informed of any changes.

Accreditation Board for Engineering and Technology, Inc.
111 Market Place, Suite 1050 --- Baltimore, MD 212302-4012

THE CHALLENGER ACCIDENT

Engineers design and build many different products each year that are components of large and complex sociotechnical systems. Usually these products and the systems are conservatively designed with adequate safety factors, carefully tested, and perform well in service for extended periods of time. They provide a much needed service without endangering either individual or public safety. However, from time to time mistakes are made in the initial design. These mistakes are usually detected in prototype testing, and eliminated prior to releasing the product to the market place. In very rare circumstances, a mistake or mistakes are overlooked for a variety of reasons, and the system is released and placed in service with an unacceptably high probability for failure resulting in a catastrophic accident. The space shuttle system clearly falls into this category. If you are a space supporter, you may argue that the risks are worth the benefits. But if this is the case, unprepared and untrained high school teachers should not be invited to ride along to enhance the image of the space program. It is one thing to order a career astronaut into peril, but totally a different proposition to invite an uninformed civilian to participate in a very dangerous project.

We will be discussing, in considerable detail, the probability of failure (or an accident) in another chapter. Here you need to understand that there is always some risk of failure when designing high performance systems. We must accept a trade-off between risk, safety and performance. In most complex sociotechnical systems (air transportation for example) there is a small but finite risk for failure with a subsequent loss of life and property. This risk must be known to the public. It is the responsibility of the engineering community, industry and the government to alert the potential customers to the dangers involved. Knowing the risk, you can decide whether or not you want to use the system. Some people worry more than others and place a very high value on safety. They require a probability of failure of very nearly zero. Do you know someone who will not travel by flying? (John Madden, the popular football announcer for the Fox network, travels from game to game each week on a special bus). Other folks love the thrill of a risk, and they are willing to accept a much higher probability of an accident. They think hang gliding is great.

With this background on risk, safety and performance, let's begin our discussion of the Challenger accident [9]. The Challenger was one of the original four orbiters that was built by National Air and Space Administration (NASA) to serve the space shuttle system. The rocket fuel supply (liquid hydrogen) on the Challenger 51-L mission exploded 73 seconds after launch on Tuesday January 28, 1986. The crew of six and a civilian passenger were killed, and the space shuttle was lost.

We have selected the Challenger accident as a case study for this chapter because it illustrates several ethical issues in the engineering and management of large, complex and inherently dangerous systems: Some issues that will be raised are:

1. The design of the shuttle and the selection of the contractors involved a lot of political considerations [9].
2. The lack of communications between people and organizations was a significant factor in the accident.
3. The interface between the upper level administrators (business managers) and the engineers was an important element in the decision to launch on that disastrous morning.
4. Public attention and opinion, not safety, affected the decision making process.

5. The risk potential was not known by the public, and not appreciated by key administrators directly involved in the launch decision.

6. Christa McAuliffe, a high school teacher and mother of two children, was killed in the accident. Why?

7. Dr. Judith Resnik, a career astronaut and an electrical engineer, was also killed. She is very special to us because she earned her Ph. D. from the University of Maryland at College Park.

BACKGROUND INFORMATION ABOUT THE SHUTTLE

To set the stage for the accident, we have to go back to the early 1970s. NASA had been very successful with the Apollo missions (in spite of Apollo 13), and looked forward to larger and more aggressive space endeavors. They proposed an integrated space system that would include a space station, space shuttle, space tug, and manned bases on both Mars and the moon. Sounds great until you look at the price tag. The public did a quick look and wanted no part of it. A poll indicated that the public believed that Apollo had been too costly. The politicians, always driven by the polls, took note and reduced NASA's budget. NASA recognized the need for a new, cost effective project to follow Apollo that the public (and politicians) would buy. Responding to these political pressures NASA proposed the space shuttle which would serve the military, the scientific community, and the rapidly growing commercial business of placing satellites in orbit. The space shuttle system, illustrated in Fig. 2, was marketed as a relatively routine space transport system. Even the name shuttle connects with the airline shuttle services that fly every hour on the hour from one large city to another.

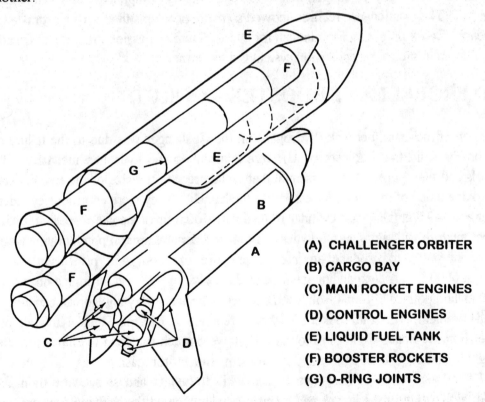

(A) CHALLENGER ORBITER
(B) CARGO BAY
(C) MAIN ROCKET ENGINES
(D) CONTROL ENGINES
(E) LIQUID FUEL TANK
(F) BOOSTER ROCKETS
(G) O-RING JOINTS

Fig. 2 Illustration showing the main components on the NASA space shuttle.

To make the space shuttle system a commercial success, NASA proposed a fleet of four orbiters [10] which would eventually fly on a weekly basis (NASA initially set a goal of 160 hours for the turn around time for an orbiter.) They planned on nearly 600 flights in the period from 1980 to 1991. The early estimates of the cost of a launch was $ 28 million with a payload delivery cost of $100 to $270 per pound. After some operational experience, the cost estimates proved to be much too optimistic. Launches actually cost of the order of $ 280 million, and the cost to place a pound of payload in orbit on the shuttle was in excess of $5200 [11]. Early experience with the space shuttle system indicated that NASA could not hold their schedules, and their costs were running more than a factor of 10 higher than the original estimates. Not a very good report card, and NASA was struggling to improve it's image as 1985 ended.

Prior to the launch of the Challenger on January 28, 1986, 24 shuttle flights had been made. These flights were all successful in that they returned to earth with all crew members safe. They also showed the operational capabilities of the shuttle system in placing commercial satellites into orbit, repairing satellites in space, and salvaging malfunctioning satellites. However, these flights also showed that the shuttle system had many serious problems. The three main liquid rocket engines were too fragile with many critical components. Some of the tiles in the heat shield needed repair and/or replacement after every flight. The computers and the inertial navigation systems experienced some failures. The brakes and landing gear were stressed to the limit when the orbiter (an 80 ton dead stick glider) landed at speeds ranging from 195 to 240 MPH. And finally the seals in the solid fuel booster rockets showed distress (sometimes extensive) in 12 of the 24 previous launches.

The record of the shuttle during the 1981-1985 period showed a consecutive series of successful launches, but with many prolonged delays to repair failing components in a very large, highly stressed system. The maintenance records showed so many problems that NASA estimated it took three man-years of work preparing for a launch for every minute of mission flight time. Clearly the word shuttle to describe such a transportation system is a misnomer.

THE SOLID PROPELLANT BOOSTER ROCKETS

The explosion of the main fuel tank (hydrogen) on the Challenger was due to the failure of the O-ring seals on the solid fuel boosters which were adjacent to the hydrogen fuel tank. To understand the seals and their purpose, it is essential that we appreciate the size of the two booster rockets used to provide much of the thrust necessary for the launch.. They were enormous cylinders 12 feet in diameter and 149 feet tall. Each cylinder is filled with 500 tons of a solid propellant, which consists of a rubber mixture with aluminum powder and the oxidizer ammonium perchlorate. When ignited, the propellant burns to produce an internal pressure of about 450 psi (lbs/in^2) at a temperature of about 6000 °F. The expanding gasses exit the rocket motor case through a nozzle, to produce about 2.6 million pounds of thrust from each booster. These are very big Roman candles.

Big was the problem. The solid rocket boosters were made by Morton Thiokol in Utah, but launched from the Kennedy Space Center in Florida. They were much too long to ship across the country as a single cylinder. To circumvent this problem, the motor casing was fabricated in segments each 27 feet long. The segments were filled with propellant and shipped by train and assembled in a special facility near the launch pad. This procedure solved the shipping problem, but created another. The rocket casing is a pressure vessel that must contain the very hot gasses at a pressure of about 450 psi. The joints, where the cylindrical segments of the motor case were fitted together, needed to be sealed so that these hot gasses would not leak, and cause damage to adjacent components of the launch vehicle.

The seal was made with a pair of rubber O-rings fitted over one finger of a clevis type of joint as shown in Fig. 3. The O-rings, 0.280 inch in diameter, are compressed between the two surfaces effecting a seal that prevents the pressurized fluid from leaking past the joint. A putty like compound serves to keep the hot gases some distance from the O-rings. The pins lock the segments together (axial constraint only), but do not clamp the clevis fingers about the center finger.

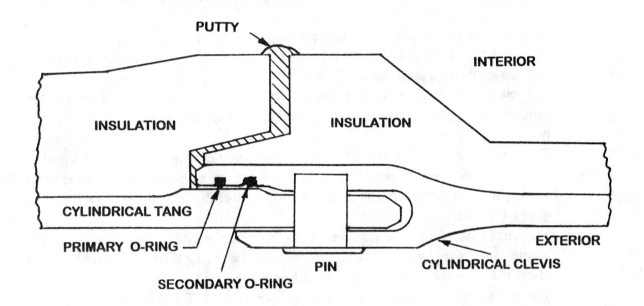

Fig. 3 Design of the O-ring seal used at the circumferential joints of the solid rocket motor cases.

The sealing of the segmented cylinders with the O-rings was not a new concept. A single O-ring had been employed previously on the Titan III rocket effectively sealing the circumferential joints on a smaller diameter motor case. Morton Thiokol, the contractor building the boosters, introduced the second O-ring to provide a margin of safety in the event the first O-ring failed. It all sounded good, and the initial test firings of the booster in Utah were apparently successful.

Unfortunately one test does not validate the safety of even a simple system. To insure high reliability of a component many tests are required. Moreover, this test firing had no bearing on problems associated with the reuse of the solid rocket boosters (up to 20 times) after splash down and recovery from the Atlantic, and shipment back and forth between Florida and Utah.

FAILURE OF THE O-RING SEAL

Photographs taken at launch indicated that the O-ring seals failed almost immediately after ignition of the booster. The hot gasses cut through the O-rings and the wall of the booster. The hole spread and a flaming jet of escaping gasses cut through the wall of the adjacent tank containing the liquid hydrogen, and the game was effectively over. While lots of things happened in the last five seconds before the big explosion, all bad, the penetration of the large, adjacent tank which contained liquid oxygen and liquid hydrogen spelled the disastrous end of the mission.

It was a terrible day, and we as a nation were in shock as we mourned the loss of the crew and the school teacher/mother. It was sometime later, when the facts were brought to the public about the space shuttle program, that we learned the shuttle should not have been permitted to fly that day. The very high probability of failure of the O-ring seals was predictable. Moreover, key engineers at Morton Thiokol tried without success to prevent the launch.

The story containing all of the facts about the failure of the O-ring seal is too long to be covered here, but you are encouraged to read reference [9 - 14] where very complete and readable treatments are given. We describe here the essential elements leading to the catastrophic failure.

1. The circumferential joint changed shape when the motor case was pressurized, and the gap which the O-rings filled increased markedly in size. A schematic illustration of the new gap geometry is presented in Fig. 4.
2. The new gap opening was so large that the back-up O-ring probably could not seal the joint. When the booster case was pressurized the seal depended on a single O-ring, not two.
3. The hot exhaust gasses had eroded the O-rings on 12 of the previous 24 launches indicating some leakage about half of the time, and on a few occasions considerable erosion of the rings. There was also clear evidence that the putty failed to keep the hot gasses from attacking the rubber O-rings.
4. The joints move during the early launch sequence as the orbiter engines and then the booster engines are ignited. This motion requires the O-ring seals to be flexible and to reseat and reseal continuously during these movements.
5. The temperature the night before the launch dropped to 22 °F and had only increased to about 28 °F at the time of the launch.
6. The O-ring seals were not certified to operate below a temperature of 53 °F by Morton Thiokol, the company responsible for the solid rocket boosters.
7. Tests conducted by Morton Thiokol in 1985, six months before the accident, indicated that the O-ring seal was not effective at low temperatures. (At 50 °F the O-rings would not expand and follow the movement in the joint during the period of operation of the booster). In fact at a normal temperature of 75 °F, it required 2.4 seconds for the O-rings to reseat and effect the seal when the gap was opened.

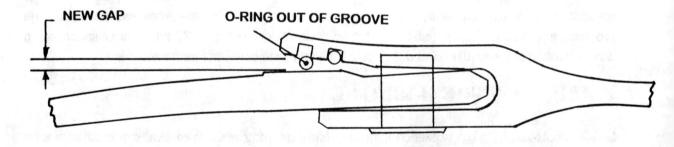

Fig. 4 Rotation of the joint due to cylinder pressurization opens the gap in the seal region.

The seven facts listed above show very clear evidence of two problems. First, the design of the seal in the circumferential joint was marginal and should have been fixed much earlier in the program. You do not look at a piece of burnt and eroded O-ring and walk away from the problem.

Second, the launch should have been postponed due to cold weather for several different reasons. The very low temperatures were well below the limits to which the system had been certified. The rubber in the O-rings becomes very stiff at these low temperatures and cannot respond quickly enough to accommodate joint movement. The O-rings were almost guaranteed to leak. Also there was considerable ice on the rocket motors and the fuel tank. Pieces of this ice might damaged the rocket motors and the tiles on the heat shield of the orbiter when it fell off during the violent vibrations which occur just after ignition but before lift-off..

IGNORING THE PROBLEM

Problems usually do not go away when we ignore them. They persist and sooner or later failure will occur. The seals had failed on 12 of the previous 24 flights. The failures were not catastrophic because the leaks were small and at locations which did not endanger the adjacent components. The boosters on the shuttle only operate for a few minutes before they are cut loose to fall into the Atlantic for recovery at a later time. So the small leaks for short periods of time were tolerated by those concerned with the launch schedule. The trouble with the leak on the Challenger was that it was big, and the jet of hot gasses issuing from the leak were pointed directly at the liquid hydrogen fuel tank. The jet of hot gasses acted just like a cutting torch and the huge fuel tank was penetrated within seconds.

The O-ring seals were inadequate and the joint needed to be redesigned and modified. In fact during the development of the boosters in 1977 - 79, several memos were written by engineers at the Marshall Space Flight Center indicating their concern over the adequacy of the joint. They recommended that the joint be redesigned because "the adequacy of the clevis joint was completely unacceptable". The suppliers of the O-rings indicated "that the O-ring was being required to perform beyond its intended design and that a different type of seal should be considered".

Apparently there was a major break down in communication and none of these memos and reports from the Marshall Space Flight Center were forwarded to Morton Thiokol, the designers and builders of the boosters. A problem, properly detected by engineering at the NASA Center in charge of monitoring the technical aspects of the development of the boosters, was not pursued to its logical conclusion. Instead of insisting on a redesign of the joint, Marshall Space Flight Center approved the boosters in September of 1980. The pressures of cost overruns and repeated schedule slippage sometimes make managers (and engineers) accept flawed designs. It is very short sighted, and not in the best interest of safe design to do so.

OK. NASA accepted a booster rocket with inadequate seals, but the bird flew. On the first launch, with Columbia, there was no reported damage to the seals. The O-ring seals failed on the second flight. One of the O-rings was burnt with about 20 % of the thickness of the ring vaporized. The ring failure did not affect the mission, but it is clear that the putty was not protecting the rings from the hot gasses. Marshall Space Flight Center reacted to the field experience by reclassifying the joint to a "Criticality 1 category". This was the official recognition on the part of NASA that a seal failure could result in "loss of mission, vehicle and crew due to metal erosion, burn through and probable case burst resulting in fire and deflagration." The failure of the O-rings by erosion continued with high frequency (50 % of the missions), but NASA decided to accept the situation and did not recommend remedial action.

It appears that NASA decided to live with the seal problem at least until a new lighter booster motor could be designed with improved joints. These new boosters were to be ready about six months after the Challenger exploded. A very real lesson in too little too late.

RECOGNIZING THE INFLUENCE OF TEMPERATURE

From the initial development phase of the boosters, the O-ring seals had been recognized as a problem. A problem that NASA classified as very serious, but one that they must have erroneously believed did not elevate the risk to unacceptable limits. In January of 1985 a year before the accident, the Challenger was launched on a cold day at a temperature of 51 °F. An examination of the joints after the recovery of the boosters showed that a number of O-rings were severely damaged and evidence of extensive blow by was observed. The O-ring seal problems were reviewed by the contractor, Morton Thiokol, and the monitor, Marshall Space Flight Center. During this review Morton Thiokol noted that low temperatures were a contributing factor to the inability of the O-rings to seal properly. The condition was considered undesirable, but was acceptable. The O-ring problems persisted, and NASA eventually placed a launch constraint on the space shuttle.

The launch constraint would prohibit launches until the problem had been resolved or reviewed in detail prior to each launch. Unfortunately NASA routinely provided launch waivers for each subsequent flight to avoid delays in an unrealistic launching schedule. The problem of O-ring erosion was common, one or more joints on 50 % of the launches, and it had not caused any serious difficulty to date. It was expected and accepted.

FLY EVEN IF IT IS COLD: A MANAGEMENT DECISION

The very cold weather on the night before the fatal launch was no surprise. The weatherman was on the mark in predicting the temperature and the ice which formed on the structures that night. The engineers at Morton Thiokol, in Utah, were very concerned about the effects of the very low temperatures on the ability of the O-rings to function properly. Teleconferences took place between the Thiokol Moron engineers and the NASA program managers. The engineers argued to postpone the launch until the temperature increased to at least 53 °F.

The NASA program managers were very unhappy about the engineers concern. They did not want the extended delay required to wait for warmer weather. Senior managers (Senior Vice Presidents) from Morton Thiokol sensed the displeasure from the customer, and took over the decision process. Senior management decided to keep the customer happy and reversed the decision of the Vice President of Engineering not to launch at these very low temperatures. Morton Thiokol signed off on the launch in spite of the fact that the temperature at launch time was expected to be 25 °F below the lowest temperature (53 °F) which engineering believed the seals would be effective.

The senior management at Morton Thiokol caved in to client pressure. The engineers were unanimous in their opposition to launch. Senior management asked the engineers to prove that the O-ring seals would fail. Of course, they could not state with 100 % certainty that the rings would fail. Failure analysis is in terms of probabilities, and the risk had elevated as the temperature decreased. But the engineers could not prove conclusively that the seals would fail in a manner that would detrimentally affect the mission. Senior managers at Morton Thiokol and program managers at NASA concluded erroneously that the risks were low enough to proceed with the launch.

APPROVALS AT THE TOP

NASA has a four level approval procedure to control the launch of the shuttle. This sounds good, but for a multilevel approval process to be of any value information must flow freely from the bottom of the organization to the top of the chain of command. The technical discussions concerning the ability of the O-rings to function at the very low temperatures took place between the principals at level 4, Morton Thiokol, and level 3 the program managers at Marshall Space Flight Center. While the discussion was extended, with clear polarization between engineering and management, not a word of these concerns was conveyed to the top two levels of management at NASA.

Approvals for the launch were given by the level 2 administrator for the National STS Program in Houston, TX, and the level 1 administrator at NASA's Headquarters in Washington D.C. These approvals were essentially automatic because the administrators in charge had been isolated. They were not privy to the management decision to fly with super cold O-ring seals. They were not informed of the engineering recommendation to postpone the launch, and to wait until the temperature increased to at least 53 °F.

The lesson here is clear --- approvals by the very high level managers are worthwhile only if they are informed decisions. If complete information is not presented to the super bosses to evaluate, then their approval is meaningless. These administrators are not knowledgeable, and as such they add no value to a go no go decision process.

SUMMARY

We have briefly described the degradation of ethical behavior during the past three generations, and given some of the reasons for these changes. While there are many controversial issues which can be classified as "dilemma" ethics, there are several "basic" characteristics where the difference between right and wrong is clearly defined. The list of basic virtues include:

- Prudence
- Justice
- Fortitude
- Temperance
- Faith
- Hope
- Charity
- Humility
- Respect

We have laws written to prohibit, by punishment, a well defined set of vices. These laws control behavior, but only extremely repulsive actions on the part of individuals or corporations.

Individual behavior is important, but as we learned in the Challenger explosion, it is not sufficient. Corporations (business) must operate with the highest ethical standards. The chief executive officer (CEO) establishes the standards and the lower level managers act to establish the corporate character.

Honor codes exist on some college campuses to guide ethical behavior (lying, cheating and stealing), and they are respected by the students and the faculty. Character development and honor codes go together in producing a meaningful educational experience. We list here some of the more important attributes in character development:

1. Excellence
2. Respect
3. Integrity
4. Decisiveness
5. Responsible
6. Temperate
7. Fortitude
8. Understanding

We treated professional ethics by introducing the Code of Ethics of Engineers that is sponsored by the Accreditation Board for Engineering and Technology (ABET). The code is short with four fundamental principles and seven basic canons. The canons are explained in an extensive set of guidelines which are reproduced verbatim in this chapter.

Finally, we have discussed the fatal accident of the Challenger space shuttle which exploded during the launch in January 1986. Many ethical issues dealing with individual and corporate behavior are apparent as we review the events leading to the fatal accident. We trust that this example will aid you in developing both your individual and professional character.

REFERENCES

1. The Day America Told the Truth
2. Sommers, C. H., "Teaching the Virtues," *Public Interest*, No. 111, Spring 1993, pp. 3-13.
3. Woodward, K. L. "What is Virtue," *Newsweek*, June 13, 1994.
4. George, R. P., Making Men Moral: Civil Liberties and Public Morality, Oxford University Press, NY, 1994.
5. Baker, D. F. "Ethical Issues and Decision Making in Business," Vital Speeches of the Day, 1993
6. Anon, United States Air Force Character Development Manual, USAF Academy, C0, December 1994.
7. Abramson, R., "A Matter of Honor," *Los Angeles Times*, April 3, 1994.
8. Lickona, T. A. Educating for Character, Bantam Book, New York, 1991.
9. Jensen, C., No Down Link: A Dramatic Narrative about the Challenger Accident, Farrar, Straus and Gitoux, New York, 1996.
10. *Time*, February 10, 1986
11. Lewis, R. S., The Voyages of Columbia: The First True Space Ship, Columbia University Press, New York, 1984.

12. *The New York Times* April 23, 1986.
13. Feynman, R. P. <u>What Do You Care What Other People Think?</u> Bantam, New York, 1988.
14. Report to the President by the Presidential Commission on the Space Shuttle Challenger Accident, Ayer Co., Salem, 1986.

EXERCISES

1. Our political leaders are often attacked by the media for lax ethical behavior. Please write a short paper describing three recent lapses of ethical behavior on the part of the leadership in either the State or Federal government. Did these officials break the law?. Is it important that they did or did not break the law?

2. List what you consider poor behavior in a college class room.

3. Have you ever cheated on an important exam in high school? What about a college exam? Do you know of someone who cheated? What was your attitude when you observed this cheating?

4. Is there an honor code at the University of Maryland? Have you read it? Do you abide by the rules?

5. Write a paragraph or two explaining your position regarding:
   - Prudence,
   - Justice,
   - Fortitude,
   - Temperance.

6. Write a paragraph or two describing your feelings about:
   - Faith
   - Hope
   - Charity

7. If you could write a law that would go on the books tomorrow and be strictly enforced, what behavior would it prohibit.

8. Why does the CEO of a corporation set the standard for ethical behavior? What are some of the issues that arise on a daily basis that develop corporate character? Is it possible for a corporation with tens of thousands of employees to develop a character like an individual?

9. Write a 400 word essay describing your character. Be sure to cover the weaknesses as well as the strengths of your character.

10. Examine the code of ethics for engineers given in Fig. 1 and describe you opinion of it. Can you live with the expectations if you become a professional engineer? Do you believe it is adequate, or should it be revised to reflect a more modern viewpoint of what is right and wrong?

11. Suppose that you were an engineer working for Morton Thiokol on the evening of January 27, 1986 involved in the discussion of the O-rings. What would you have done:
    - Early in the teleconference.
    - At the critical stage of the teleconference.
    - After management had taken over the decision process.

There is no right or wrong answer to these questions. We want to place you in a professional dilemma. Some day you may be placed in a similar situation where there is a trade-off between safety and corporate business interests. Think about your response in advance and be prepared to deal with a dilemma.